Smart Antennas for Next Generation Wireless Systems

EURASIP Journal on
Wireless Communications and Networking

Smart Antennas for Next Generation Wireless Systems

Guest Editors: Angeliki Alexiou,
Monica Navarro, and Robert W. Heath Jr.

ISBN 978-977-454-019-6

Contents

Hindawi Publishing Corporation
EURASIP Journal on Wireless Communications and Networking
Volume 2007, Article ID 20427, 2 pages
doi:10.1155/2007/20427

Editorial

Smart Antennas for Next Generation Wireless Systems

Angeliki Alexiou,[1] Monica Navarro,[2] and Robert W. Heath Jr.[3]

[1] Bell Laboratories, Lucent Technologies, Suffolk CO10 1LN, UK
[2] Centre Tecnològic de Telecomunicacions de Catalunya (CTTC), 08860 Barcelona, Spain
[3] Department of Electrical and Computer Engineering, The University of Texas at Austin, TX 78712, USA

Received 31 December 2007; Accepted 31 December 2007

The adoption of multiple antenna techniques in future wireless systems is expected to have a significant impact on the efficient use of the spectrum, the minimisation of the cost of establishing new wireless networks, the enhancement of the quality of service, and the realisation of reconfigurable, robust, and transparent operation across multitechnology wireless networks. Although a considerable amount of research effort has been dedicated to the investigation of MIMO systems performance, results, conclusions, and ideas on the critical implementation aspects of smart antennas in future wireless systems remain fragmental.

The objective of this special issue is to address these critical aspects and present the most recent developments in the areas of antenna array design, implementation, measurements, and MIMO channel modelling, robust signal processing for multiple antenna systems and interference-aware system level optimisation.

In the area of *antenna array design*, the paper by T. Cooper et al. presents a tower-top smart antenna calibration scheme, designed for high-reliability tower-top operation and based upon an array of coupled reference elements which sense the array's output. The theoretical limits of the accuracy of this calibration approach are assessed and the expected performance is evaluated by means of initial prototyping of a precision coupler circuit for a 2 × 2 array.

The design of uniform rectangular arrays for MIMO communication systems with strong line-of-sight components is studied in the paper by F. Bøhagen et al., based on an orthogonality requirement inspired by the mutual information. Because the line-of-sight channel is more sensitive to antenna geometry and orientation, a new geometrical model is proposed in this paper that includes the transmit and receive orientation through a reference coordinate system, along with a spherical wave propagation model to provide more accurate propagation predictions for the line-of-sight channel. It is shown that the separation distance between antennas can be optimized in several cases and that these configurations are robust in Ricean channels with different K-factors.

The paper by M. Mowlér et al. considers estimation of the mutual coupling matrix for an adaptive antenna array from calibration measurements. This paper explicitly incorporates the lack of information about the phase centre and the element factor of the array by including them in an iterative joint optimization. In particular, the element factor of the array is approximated through a basis expansion—the coefficients of the expansion are estimated by the algorithm. Analysis, simulations, and experimental results demonstrate the efficacy of the proposed estimator.

Diversity characterisation of two-antenna systems for UMTS terminals by means of measurements performed concurrently with the help of a reverberation chamber and a Wheeler Cap setup is addressed in the paper by A. Diallo et al. It is shown that even if the envelope correlation coefficients of these systems are very low, having antennas with high isolation improves the total efficiency by increasing the effective diversity gain.

In the paper by S. Savazzi et al., the authors address the design problem of linear antenna array optimization to enhance the overall throughput of an interference-limited system. They focus on the design of linear antenna arrays with nonuniform spacing between antenna array elements, which explicitly takes into account the cellular layout and the propagation model, and show the potential gains with respect to the conventional half wavelength systems. For such purposes two optimisation criteria are considered: one based on the minimization of the average interference power at the output of a conventional beamformer, for which a closed-form solution is derived and from which the justification for un-equal spacings is inferred; a second targets the maximization of the ergodic capacity and resorts to numerical results.

Addressing the *robust MIMO signal processing* aspect, the paper by P. Theofilakos and A. Kanatas considers the use of subarrays as a way to improve performance in MIMO communication systems. In this approach, each radio frequency chain is coupled to the antenna through a beamforming vector on a subarray of antenna elements. Bounds on capacity for Rayleigh fading channels highlight the benefits of subarray formation while low-complexity algorithms for grouping antennas into subarrays illustrate how to realize this concept in practical systems.

The paper by I. Cosovic and G. Auer presents an analytical framework for the assessment of the pilot grid for MIMO-OFDM in terms of overhead and power allocation. The optimum pilot grid is identified based on the criterion of the capacity for OFDM operating in time-variant frequency selective channels. A semianalytical procedure is also proposed to maximize the capacity with respect to the considered estimator for realizable and suboptimal estimation schemes.

The paper by A. Del Coso and C. Ibars analyses a point-to-point multiple relay Gaussian channel that uses a decode-and-forward relaying strategy under a half duplex constraint. It derives the instantaneous achievable rate under perfect CSI assumption and obtains the relay selection algorithm and power allocation strategy within the two consecutive time slots that maximises the achievable rate. Furthermore the study provides upper and lower bounds on the ergodic achievable rate, and derives the asymptotic behaviour in terms of the number of relays, showing that for any random distribution of relays the ergodic capacity of the multiple relay channel under AWGN grows asymptotically with the logarithm of the Lambert function of the total number of relays.

In the paper by A. Ikhlef and D. Le Guennec, the problem of signal detection in MIMO systems is addressed focusing on blind reduced complexity schemes. In particular, the authors propose the use of blind source separation techniques, which avoid the use of training sequences, for blind recovery of QAM and PSK signals in MIMO channels. The proposed low-complexity algorithm is a simplified version of the CMA algorithm that operates over a single signal dimension, that is, either on the real or imaginary part and of which convergence is also proved in the paper.

In the area of *system level optimisation*, the paper by C. Sun et al. considers the application of switched directional beams at the transmitter and receiver of a MIMO communication link. The beams provide a capacity gain by focusing on different dominant wave clusters in the environment; switching between beams gives additional diversity benefits. Electronically steerable parasitic array radiators are suggested as a means to implement the beamforming in the RF. Performance is particularly enhanced at low SNRs compared to a conventional MIMO system that requires an RF chain for each antenna.

In the paper by N. Jalden et al., the inter- and intrasite correlation properties of shadow fading and power-weighted angular spread at both the mobile station and the base station are studied, for different interbase station distances, uti-lizing narrow band multisite MIMO measurements in the 1800 MHz band.

A. M. Kuzminskiy and H. R. Karimi show in their paper the potential increase in throughput when multiantenna interference cancellation techniques are considered to complement the multiple access control protocol. The work evaluates the gains that multiantenna interference cancellation schemes provide in the context of WLAN systems which implement the CSMA/CA MAC protocol.

Transmit diversity techniques and the resulting gains at the cell border in a cellular MC-CDMA environment using smart base stations are addressed in the paper by S. Plass et al. Cellular cyclic delay diversity (C-CDD) and cellular Alamouti technique (CAT) are proposed, that improves the performance at the cell borders by enhancing macrodiversity and reducing the overall intercell interference.

The paper by B. Bougard et al. investigates the transceiver energy efficiency of multiantenna broadband transmission schemes and evaluates such transceiver power consumption for an adaptive system. In particular, the paper evaluates the tradeoff between the net throughput at the MAC layer versus the average power consumption, that an adaptive system switching between a space-division multiplexing, space-time coding or single antenna transmission achieves. Authors provide a model that aims to capture channel state information in a compact way, and from which a simple policy-based adaptation scheme can be implemented.

In the last paper, W. Sheng and S. D. Blostein formulate the problem of admission control for a CDMA beamforming system as a cross-layer design problem. In the proposed framework, the parameters of a truncated automatic retransmission algorithm and a packet level admission control policy are jointly optimized to maximize throughput subject to quality-of-service requirements. Numerical examples show that throughput can be increased substantially in the low packet error rate regime.

The theme of this special session was inspired by the joint research collaboration in the area of smart antennas within the ACE project, a Network of Excellence under the FP6 European Commission's Information Society Technologies Initiative. The objective of this issue is to share some insight and encourage more research on the critical implementation aspects for the adoption of smart antennas in future wireless systems.

We would like to thank the authors, the reviewers and Hindawi staff for their efforts in the preparation of this special issue.

Angeliki Alexiou
Monica Navarro
Robert W. Heath Jr.

Hindawi Publishing Corporation
EURASIP Journal on Wireless Communications and Networking
Volume 2007, Article ID 41941, 12 pages
doi:10.1155/2007/41941

Research Article

Tower-Top Antenna Array Calibration Scheme for Next Generation Networks

Justine McCormack, Tim Cooper, and Ronan Farrell

Centre for Telecommunications Value-Chain Research, Institute of Microelectronics and Wireless Systems, National University of Ireland, Kildare, Ireland

Received 1 November 2006; Accepted 31 July 2007

Recommended by A. Alexiou

Recently, there has been increased interest in moving the RF electronics in basestations from the bottom of the tower to the top, yielding improved power efficiencies and reductions in infrastructural costs. Tower-top systems have faced resistance in the past due to such issues as increased weight, size, and poor potential reliability. However, modern advances in reducing the size and complexity of RF subsystems have made the tower-top model more viable. Tower-top relocation, however, faces many significant engineering challenges. Two such challenges are the calibration of the tower-top array and ensuring adequate reliability. We present a tower-top smart antenna calibration scheme designed for high-reliability tower-top operation. Our calibration scheme is based upon an array of coupled reference elements which sense the array's output. We outline the theoretical limits of the accuracy of this calibration, using simple feedback-based calibration algorithms, and present their predicted performance based on initial prototyping of a precision coupler circuit for a 2×2 array. As the basis for future study a more sophisticated algorithm for array calibration is also presented whose performance improves with array size.

1. INTRODUCTION

Antennas arrays have been commercially deployed in recent years in a range of applications such as mobile telephony, in order to provide directivity of coverage and increase system capacity. To achieve this, the gain and phase relationship between the elements of the antenna array must be known. Imbalances in these relationships can arise from thermal effects, antenna mutual coupling, component aging, and finite manufacturing tolerance [1]. To overcome these issues, calibration is required [2, 3]. Traditionally, calibration would have been undertaken at the manufacturer, address static effects arising from the manufacturing tolerances. However, imbalances due to dynamic effects require continual or dynamic calibration.

Array calibration of cellular systems has been the subject of much interest over the last decade (e.g., [4–6]), and although many calibration processes already exist, the issue of array calibration has, until now, been studied in a "tower-bottom" smart antenna context (e.g., tsunami(II) [2]). Industry acceptance of smart antennas has been slow, principally due to their expense, complexity, and stringent relia-

bility requirements. Therefore, alternative technologies have been used to increase network performance, such as cell splitting and tower-bottom hardware upgrades [7, 8].

To address the key impediments to industry acceptance of complexity and expense, we have been studying the feasibility of a self-contained, self-calibrating "tower-top" base transceiver station (BTS). This system sees the RF and mixed signal components of the base station relocated next to the antennas. This provides potential capital and operational savings from the perspective of the network operator due to the elimination of the feeder cables and machined duplexer filter. Furthermore, the self-contained calibration electronics simplify the issue of phasing the tower-top array from the perspective of the network provider.

Recent base station architectures have seen some departure from the conventional tower-bottom BTS and tower-top antenna model. First, amongst these was the deployment of tower-top duplexer low-noise amplifiers (TT-LNA), demonstrating a tacit willingness on the part of the network operator to relocate equipment to the tower-top if performance gains proved adequate and sufficient reliability could be achieved [9]. This willingness can be seen with the

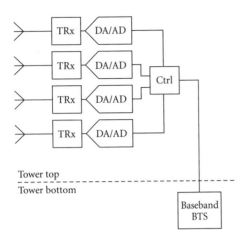

FIGURE 1: The hardware division between tower top and bottom for the tower-top BTS.

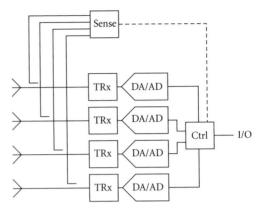

FIGURE 2: A simplified block schematic diagram of a typical array calibration system.

exploration of novel basestation architectures, with examples such as reduced RF feeder structures utilising novel switching methodologies [10, 11], and the development of basestation hotelling with remote RF heads [12]. Such approaches aim to reduce capital infrastructure costs, and also site rental or acquisition costs [13].

In this paper, we present our progress toward a reliable, self-contained, low-cost calibration system for a tower-top cellular BTS. The paper initially presents a novel scheme for the calibration of an arbitray-sized rectilinear array using a structure of interlaced reference elements. This is followed in Section 3 by a theoretical analysis of this scheme and predicted performance. Section 4 presents a description of a prototype implementation with a comparison between experimental and predicted performance. Section 5 presents some alternative calibration approaches utilising the same physical structure.

2. RECTILINEAR ARRAY CALIBRATION

2.1. Array calibration

To yield a cost-effective solution for the cellular BTS market, we have been studying the tower-top transceiver configuration shown in Figure 1. This configuration has numerous advantages over the tower-bottom system but, most notably, considerably lower hardware cost than a conventional tower-bottom BTS may be achieved [14].

We define two varieties of array calibration. The first, radiative calibration, employs free space as the calibration path between antennas. The second, where calibration is performed by means of a wired or transmission line path and any radiation from the array in the process of calibration is ancillary, is refered to as "nonradiative" calibration. The setup of Figure 2 is typically of a nonradiative calibration process [2]. This process is based upon a closed feedback loop between the radiative elements of the array and a sensor. This sensor provides error information on the array output and generates an error signal. This error signal is fed back to correctively weight the array element's input (transmit cal-

ibration) or output (receive calibration). It is important to observe that this method of calibration does not correct for errors induced by antenna mutual coupling. Note that in our calibration scheme, a twofold approach will be taken to compensate for mutual coupling. The first is to minimise mutual coupling by screening neighbouring antennas—and perhaps using electromagnetic (EM) bandgap materials to reduce surface wave propagation to distant antennas in large arrays. The second is the use of EM modelling-based mitigation such as that demonstrated by Dandekar et al. [6]. Further discussion of mutual coupling compensation is beyond the scope of this paper.

While wideband calibration is of increasing interest, it remains difficult to implement. On the other hand, narrowband calibration schemes are more likely to be practically implemented [1]. The calibration approach presented here is directed towards narrowband calibration. However, the methodology supports wideband calibration through sampling at different frequencies.

2.2. Calibration of a 2×2 array

Our calibration process employs the same nonradiative calibration principle as shown in Figure 2. The basic building block, however, upon which our calibration system is based is shown in Figure 3. This features four radiative array transceiver elements, each of which is coupled by transmission line to a central, nonradiative reference element.

In the case of transmit calibration (although by reciprocity receive calibration is also possible), the transmit signal is sent as a digital baseband signal to the tower-top and is split (individually addressed) to each transmitter for SISO (MIMO) operation. This functionality is subsumed into the control (Ctrl) unit of Figure 3.

Remaining with our transmit calibration example, the reference element sequentially receives the signals in turn from the feed point of each of the radiative array elements. This enables the measurement of their phase and amplitude relative to some reference signal. This information on the

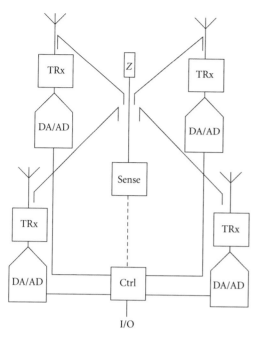

FIGURE 3: A central, nonradiative reference sensor element coupled to four radiative array transceiver elements.

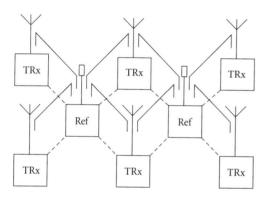

FIGURE 4: A pair of reference elements, used to calibrate a 2×3 array.

relative phase and amplitude imbalance between the feed points of each of the transceivers is used to create an error signal. This error signal is fed back and used to weight the input signal to the transceiver element—effecting calibration. Repeating this procedure for the two remaining elements calibrates our simple 2 × 2 array. This baseband feedback system is to be implemented in the digital domain, at the tower-top. The functionality of this system and the attendant computing power, energy, and cost requirements of this system are currently under investigation.

2.3. Calibration of an n × n array

By repeating this basic 2 × 2 pattern with a central reference element, it becomes possible to calibrate larger arrays [15]. Figure 4 shows the extension of this basic calibration principle to a 2 × 3 array.

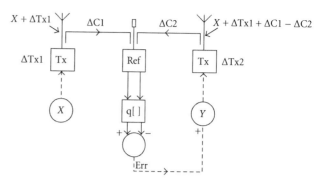

FIGURE 5: Propagation of error between calibrating elements.

To calibrate a large, $n \times n$, antenna array, it is easy to see how this tessellation of array transceivers and reference elements could be extended arbitrarily to make any rectilinear array geometry.

From the perspective of a conventional array, this has the effect of interleaving a second array of reference sensor elements between the lines of radiative transceiver elements, herein referred to as "interlinear" reference elements, to perform calibration. Each reference is coupled to four adjacent radiative antenna elements via the six-port transmission line structure as before. Importantly, because there are reference elements shared by multiple radiative transceiver elements, a sequence must be imposed on the calibration process. Thus, each transceiver must be calibrated relative to those already characterised.

Cursorily, this increase in hardware at the tower-top due to our interlinear reference elements has the deleterious effect of increasing the cost, weight, and power inefficiency of the radio system. The reference element hardware overhead, however, produces three important benefits in a tower-top system: (i) many shared reference elements will enhance the reliability of the calibration scheme—a critical parameter for a tower-top array; (ii) the array design is inherently scalable to large, arbitrary shape, planar array geometries; (iii) as we will show later in this paper, whilst these reference nodes are functional, the multiple calibration paths between them may potentially be used to improve the calibration accuracy of the array. For now, however, we consider basic calibration based on a closed loop feedback mechanism.

3. RECTILINEAR CALIBRATION—THEORY OF OPERATION

3.1. Basic calibration

Figure 5 shows a portion of an $n \times n$ array where two of the radiative elements of our array are coupled to a central reference transceiver. As detailed in Section 2.2, the calibration begins by comparing the output of transceiver 1 with transceiver 2, via the coupled interlinear reference element. Assuming phase only calibration of a SISO system, at a single frequency and with perfect impedance matching, each of the arbitrary phase errors incured on the signals, that are sent through the calibration system, may be considered additive constants (Δi, where i is the system element in question). Where there is no variation between the coupled paths and the accuracy of the phase measurement process is arbitrarily high, then, as can be seen in Figure 5, the calibration process is essentially perfect.

However, due to finite measurement accuracy and coupler balance, errors propagate through the calibration scheme. Initial sensitivity analysis [16] showed that when the resolution of the measurement accuracy, q[ ], is greater than or equal to 14 bits (such as that attainable using modern DDS, e.g., AD9954 [17] for phase control), the dominant source of error is the coupler imbalance.

From Figure 5 it is clear that an error, equal in magnitude to the pair of coupler imbalances that the calibration signal encounters, is passed on to the feed point of each calibrated transceiver. If this second transceiver is then used in subsequent calibration operations, this error is passed on. Clearly, this cumulative calibration error is proportional to the number of the calibration couplers in a given calibration path. For simple calibration algorithms such as that shown in Figure 5, the array geometry and calibration path limit the accuracy with which the array may be calibrated.

3.2. Theoretical calibration accuracy

3.2.1. Linear array

Figure 6(a) shows the hypothetical calibration path taken in phasing a linear array of antennas. Each square represents a radiative array element. Each number denotes the number of coupled calibration paths accrued in the calibration of that element, relative to the first element numbered 0 (here the centremost). If we choose to model the phase and amplitude imbalance of the coupler (σ_{c_k}) as identically distributed Gaussian, independent random variables, then the accuracy of calibration for the linear array of N elements relative to the centre element, σ_{a_k}, will be given by the following:

even N:

$$\sigma_{a_k}^2 = \frac{2\sigma_{c_k}^2}{N-1} \sum_{i=1}^{N/2} 2i, \tag{1}$$

odd N:

$$\sigma_{a_k}^2 = \frac{2\sigma_{c_k}^2}{N-1} \left(\left[\sum_{i=1}^{N/2} 2i \right] + 1 \right), \tag{2}$$

where the subscript $k = A$ or ϕ for amplitude or phase error. With this calibration topology, linear arrays are the hardest to accurately phase as they encounter the highest cumulative error. This can be mitigated in part (as shown here) by starting the calibration at the centre of the array.

3.2.2. Square array

Based on this observation, a superior array geometry for this calibration scheme is a square. Two example square arrays calibration methods are shown in Figures 6(b) and 6(c). The former initiates calibration relative to the top-left hand

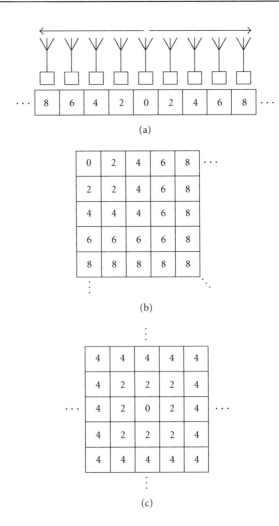

FIGURE 6: Calibration paths through (a) the linear array. Also the square array starting from (b) the top left and (c) the centre of the array.

transceiver element. The calibration path then propagates down through to the rest of the array taking the shortest path possible. Based upon the preceding analysis, the predicted calibration accuracy due to coupler imbalance of an $n \times n$ array is given by

$$\sigma_{a_k}^2 = \frac{2\sigma_{c_k}^2}{N-1} \sum_{i=1}^{n} (2i-1)(i-1) \tag{3}$$

with coupler error variance $\sigma_{c_k}^2$, centred around a mean equal to the value of the first element.

Figure 6(c) shows the optimal calibration path for a square array, starting at the centre and then radiating to the periphery of the array by the shortest path possible. The closed form expressions for predicting the overall calibration accuracy of the array relative to element 0 are most conveniently expressed for the odd and even n, where $n^2 = N$:

even n:

$$\sigma_{a_k}^2 = \frac{2\sigma_{c_k}^2}{N-1} \left(\left[\sum_{i=1}^{n/2-1} (8i)(2i) \right] \right) + \frac{2n-1}{N-1} n\sigma_{c_k}^2, \tag{4}$$

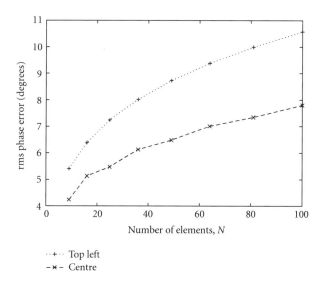

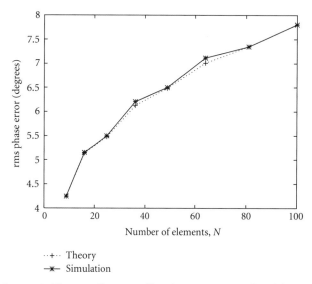

TABLE 1

Component (i)	μ_{i_A}	σ_{i_A}	μ_{i_ϕ}	σ_{i_ϕ}
Tx S$_{21}$	50 dB	3 dB	10°	20°
Ref S$_{21}$	60 dB	3 dB	85°	20°
Cal S$_{21}$	−40 dB	0.1 dB	95°	3°

···+··· Top left
-*- Centre

FIGURE 7: Comparison of the theoretical phase accuracy predicted by the closed form expressions for the square array calibration schemes, with $\sigma_{c_\phi} = 3°$.

···+··· Theory
-*- Simulation

FIGURE 9: The overall array calibration accuracy predicted by (4) and the calibration simulation for $\sigma_{c_\phi} = 3°$.

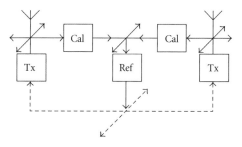

FIGURE 8: Block schematic diagram of the array calibration simulation used to test the accuracy of the theoretical predictions.

odd n:

$$\sigma_{a_k}^2 = \frac{2\sigma_{c_k}^2}{N-1} \sum_{i=1}^{n/2-1/2} (8i)(2i). \tag{5}$$

A graph of the relative performance of each of these two calibration paths as a function of array size (for square arrays only) is shown in Figure 7. This shows, as predicted, that the phasing error increases with array size. The effect of this error accumulation is reduced when the number of coupler errors accrued in that calibration is lower—that is, when the calibration path is shorter. Hence, the performance of the centre calibrated array is superior and does not degrade as severely as the top-left calibrated array for large array sizes.

As array sizes increase, the calibration path lengths will inherently increase. This will mean that the outer elements will tend to have a greater error compared to those near the reference element. While this will have impact on the array performance, for example, in beamforming, it is difficult to quantify. However, in a large array the impact of a small number of elements with relatively large errors is reduced.

3.3. Simulation

3.3.1. Calibration simulation system

To determine the accuracy of our theoretical predictions on array calibration, a simulation comprising the system shown in Figure 8 was implemented. This simulation was based on the S-parameters of each block of the system, again assuming perfect impedance matching and infinite measurement resolution. Attributed to each block of this schematic was a mean performance (μ_{i_k}) and a normally distributed rms error (σ_{i_k}), which are shown in Table 1.

3.3.2. Results

For each of the square array sizes, the results of 10 000 simulations were complied to obtain a statistically significant sample of results. For brevity and clarity, only the phase results for the centre-referenced calibration are shown, although comparable accuracy was also attained for both the amplitude output and the "top-left" algorithm. Figure 9 shows the phase accuracy of the centre-referenced calibration algorithm. Here we can see good agreement between theory and simulation. The reason for the fluctuation in both the theoretical and simulated values is because of the difference between the even and odd n predictions for the array accuracy. This difference arises because even n arrays do not have a centre element, thus the periphery of the array farthest from the nominated centre element incurs slightly higher error.

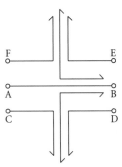

FIGURE 10: Schematic representation of the six-port, precision directional coupler.

3.3.3. Practical calibration accuracy

These calibration schemes are only useful if they can calibrate the array to within the limits useful for adaptive beamforming. The principle criterion on which this usefulness is based is on meeting the specifications of 1 dB peak amplitude error and 5° rms phase error [16]. The preceding analysis has shown that, in the absence of measurement error,

$$\lim_{\sigma_c \to 0} \sigma_a \longrightarrow 0, \qquad (6)$$

where σ_a is the rms error of the overall array calibration error. Because of this, limiting the dominant source of phase and amplitude imbalance, that of the array feed-point coupler structure, will directly improve the accuracy of the array calibration.

4. THE CALIBRATION COUPLER

4.1. 2 × 2 array calibration coupler

The phase and amplitude balance of the six-port coupler structure at the feed point of every transceiver and reference element in Figure 4 is crucial to the performance of our calibration scheme. This six-port coupler structure is shown schematically in Figure 10. In the case of the reference element, the output (port B) is terminated in a matched load (antenna) and the input connected to the reference element hardware (port A). Ports C–F of the coupler feed adjacent transceiver or reference elements. Similarly, for the radiative transceiver element, port B is connected to the antenna element and port A the transceiver RF hardware. For the individual coupler shown in Figure 10 using conventional low-cost, stripline, board fabrication techniques, phase balance of 0.2 dB and 0.9° is possible [18]. By interconnecting five of these couplers, then the basic 2 × 2 array plus single reference sensor element building block of our scheme is formed. It is this pair of precision six-port directional couplers whose combined error will form the individual calibration paths between transceiver and reference element.

A schematic representation of the 2 × 2 array coupler is shown in Figure 11. This forms the feed-point coupler structure of Figure 4, with the central coupler (port 1) connected to the reference element and the load (port 2). Each peripheral couplers is connected to a radiative transceiver element

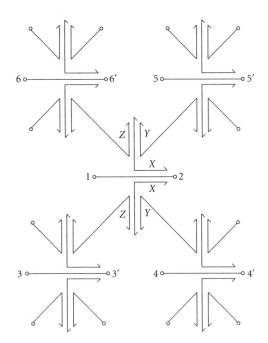

FIGURE 11: Five precision couplers configured for 2 × 2 array calibration.

(ports 3–6). By tiling identical couplers at half integer wavelength spacing, our objective was to produce a coupler network with very high phase and amplitude balance.

4.2. Theoretical coupler performance

The simulation results for our coupler design, using ADS momentum, are shown in Figure 12 [19]. Insertion loss at the design frequency of 2.46 GHz is predicted as 0.7 dB. The intertransceiver isolation is high—a minimum of 70.4 dB between transceivers. In the design of the coupler structure, a tradeoff exists between insertion loss and transceiver isolation. By reducing the coupling factor between the antenna feeder transmission line and the coupled calibration path (marked X on Figure 11), higher efficiency may be attained. However, weaker calibration coupling than −40 dBm is undesirable from the perspective of calibration reference element efficiency and measurement reliability. This necessitates stronger coupling between the calibration couplers—this stronger coupling in the second coupler stage (marked Y or Z on Figure 11) will reduce transceiver isolation. It is for this reason that −20 dB couplers are employed in all instances (X, Y, and Z).

The ADS simulation predicts that the calibration path will exhibit a coupling factor of −44.4 dB, slightly higher than desired.

The phase and amplitude balance predicted by the simulation is shown in Figures 13 and 14. This is lower than reported for a single coupler. This is because the individual coupler exhibits a natural bias toward high phase balance between the symmetrical pairs of coupled lines—ports D,E and C,F of Figure 10. In placing the couplers as shown in Figure 11, the error in the coupled path sees the sum of an

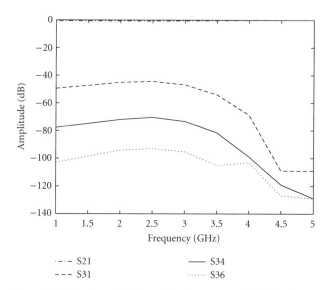

FIGURE 12: The theoretically predicted response of the ideal 2×2 coupler.

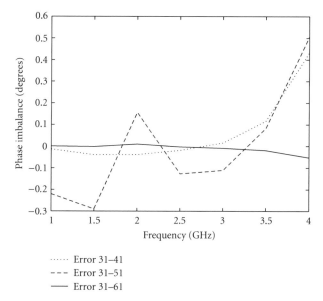

FIGURE 13: The predicted phase imbalance of an ideal 2×2 coupler.

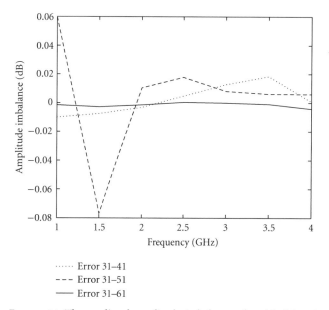

FIGURE 14: The predicted amplitude imbalance of an ideal 2×2 coupler.

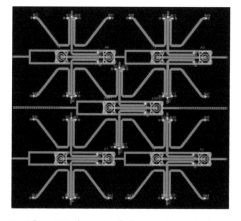

FIGURE 15: The PCB layout of the centre stripline controlled impedance conductor layer.

A,D (X,Z) type error and an A,C (X,Y) type error. This has the overall effect of reducing error. Were there to be a diagonal bias toward the distribution of error, then the error would accumulate.

Also visible in these results is a greater phase and amplitude balance between the symmetrically identical coupler pairs. For example, the phase and amplitude imbalance between ports 3 and 6 is very high. This leads to efforts to increase symmetry in the design, particularly the grounding via screens.

4.3. Measured coupler performance

Our design for Figure 11 was manufactured on a low-cost FR-4 substrate using a stripline design produced in Eagle [20]—see Figure 15. Additional grounding strips, connected by blind vias to the top and bottom ground layers, are visible which provide isolation between the individual couplers. A photograph of the finished 2×2 coupler manufactured by ECS circuits [21] is shown in Figure 16. Each of the coupler arms is terminated in low-quality surface mount 47 Ω resistors.

The 2×2 coupler was then tested using an R&S ZVB20 vector network analyser [22]. The results of this measurement with an input power of 0 dBm and 100 kHz of resolution bandwidth are shown in Figure 17. The coupler insertion loss is marginally higher than the theoretical prediction at 1.2 dB. This will affect the noise performance of the receiver and the transmit efficiency and hence must be budgeted for in our tower-top transceiver design. The

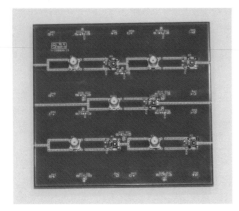

FIGURE 16: A photograph of the transceiver side of the calibration coupler board. The opposite side connects to the antenna array and acts as the ground plane.

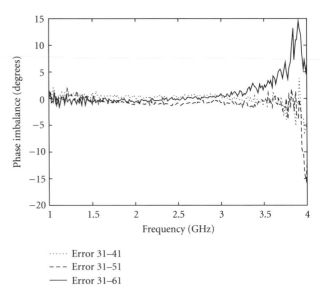

FIGURE 18: The measured phase imbalance of the 2 × 2 coupler.

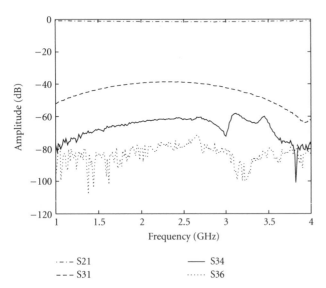

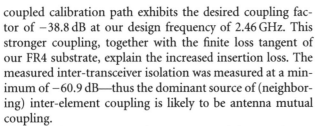

FIGURE 17: The measured performance of the prototype 2 × 2 coupler.

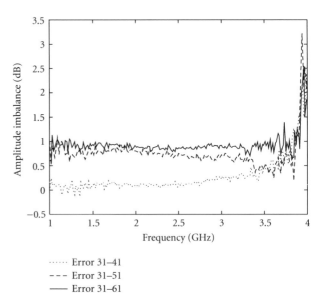

FIGURE 19: The measured amplitude imbalance of the 2 × 2 coupler.

coupled calibration path exhibits the desired coupling factor of −38.8 dB at our design frequency of 2.46 GHz. This stronger coupling, together with the finite loss tangent of our FR4 substrate, explain the increased insertion loss. The measured inter-transceiver isolation was measured at a minimum of −60.9 dB—thus the dominant source of (neighboring) inter-element coupling is likely to be antenna mutual coupling.

The other important characteristics of the coupler, its phase and amplitude balance, are shown in Figures 18 and 19 respectively. Phase balance is significantly poorer than indicated by the theoretical value. The maximum phase error recorded at our design frequency of 2.46 GHz for this coupler is 0.938°—almost an order of magnitude worse than the predicted imbalance shown in Figure 13.

The amplitude balance results, Figure 19, are similarly inferior to the ADS predictions (contrast with Figure 14). The greatest amplitude imbalance is between S31 and S61 of 0.78 dB—compared with 0.18 dB in simulation. However, clearly visible in the amplitude response, and hidden in the phase error response, is the grouping of error characteristics between the paths S31-S41 and S51-S61.

Because the coupler error did not cancel as predicted by the ADS simulation, but is closer in performance to the series connection of a pair of individual couplers, future simulation of the calibration coupler should include Monte Carlo analysis based upon fabrication tolerance to improve the accuracy of phase and amplitude balance predictions.

Clearly a single coupler board cannot be used to characterise all couplers. To improve the statistical relevance of our

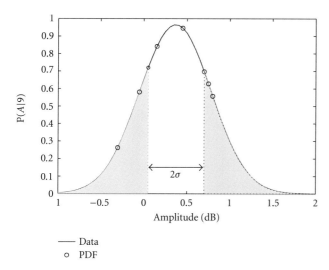

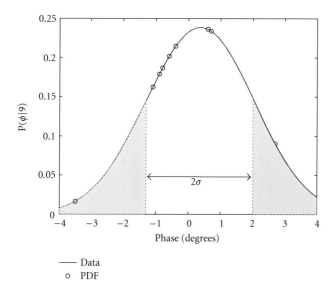

FIGURE 20: The measured coupler amplitude imbalance fitted a Gaussian probability density function, $\sigma_A = 0.4131$ dB, $\mu_A = 0.366$ dB.

FIGURE 21: The measured coupler phase imbalance fitted to a Gaussian probability density function $\sigma_\phi = 1.672°$, $\mu_\phi = 0.371°$.

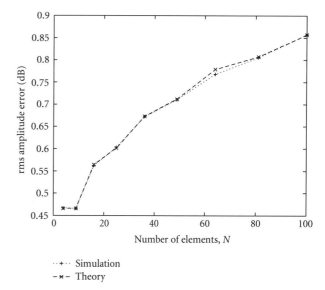

FIGURE 22: The theoretical prediction of overall array amplitude calibration accuracy based upon the use of the coupler hardware of Section 4.1.

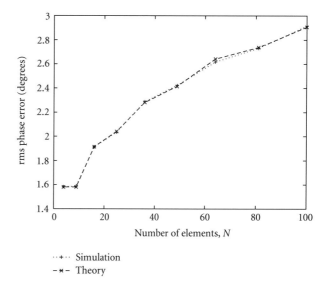

FIGURE 23: The theoretical prediction of overall array phase calibration accuracy based upon the use of the coupler hardware of Section 4.1.

results, three 2 × 2 coupler boards were manufactured and the phase and amplitude balance of each of them recorded at our design frequency of 2.46 GHz. These results are plotted against the Gaussian distribution to which the results were fitted for the amplitude and phase (Figures 20 and 21 correspondingly). Whilst not formed from a statistically significant sample (only nine points were available for each distribution), these results are perhaps representative of the calibration path imbalance in a small array. The mean and standard deviation of the coupler amplitude imbalance distribution are $\mu_{c_A} = 0.366$ dB and $\sigma_{c_A} = 0.4131$ dB. This error is somewhat higher than predicted by our theoretical study. Work toward improved amplitude balance is ongoing. The phase balance, with an rms error of 1.672°, is of the order anticipated given the performance of the individual coupler.

With this additional insight into the statistical distribution of error for a single coupled calibration path, we may make inferences about the overall array calibration accuracy possible with such a system.

4.4. Predicted array calibration performance

To investigate the utility, or otherwise, of our practical array calibration system, the coupler statistics derived from our hardware measurements were fed into both the centre-referenced calibration algorithm simulation and the theoretical prediction of Section 3. The results of this simulation are shown in Figures 22 and 23.

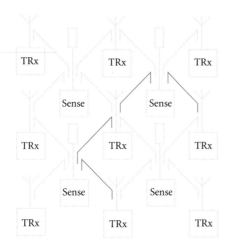

FIGURE 24: The redundant coupled calibration paths which may be useful in enhancing the quality of calibration.

The results from these figures show that the approach yields a highly accurate calibration, with rms phase errors for a typical 16-element array of less than 2° and a gain imbalance of less than 0.55 dB. As arrays increase in size, the errors do increase. For phase calibration, the increase is small even for very large arrays. Gain calibration is more sensitive to size and a 96-element array would have a 0.85 dB rms error. Ongoing work is focused upon improving the gain calibration performance for larger arrays. The following section is presenting some initial results for alternative calibration schemes which utilise the additional information from the redundant calibration paths.

5. FUTURE WORK

5.1. Redundant coupler paths

In each of the calibration algorithms discussed thus far, only a fraction of the available coupled calibration paths is employed. Figure 24 shows the coupled paths which are redundant in the "top-left" calibration scheme of Figure 6(b). The focus of future work will be to exploit the extra information which can be obtained from these redundant coupler paths.

5.2. Iterative technique

5.2.1. Operation

Given that we cannot measure the array output without incurring error due to the imbalance of each coupler, we have devised a heuristic method for enhancing the antenna array calibration accuracy. This method is designed to exploit the additional, unused coupler paths and information about the general distribution and component tolerance of the errors within the calibration system, to improve calibration accuracy. One candidate technique is based loosely on the iterative algorithmic processes outlined in [23]. Our method is a heuristic, threshold-based algorithm and attempts to infer the actual error in each component of the calibration system—allowing them to be compensated for.

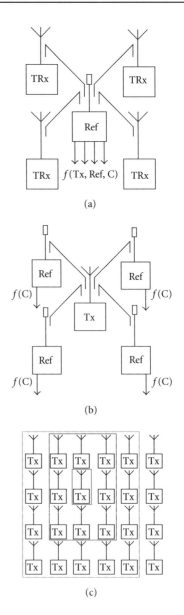

FIGURE 25: The two main processes of our heuristic method: (a) reference characterisation and (b) transmitter characterisation. (c) The error dependency spreads from the neighbouring elements with each iteration of the heuristic process.

Figure 25 illustrates the two main processes of our iterative heuristic algorithm. The first stage, Figure 25(a), is the measurement of each of the transmitters by the reference elements connected to them. The output of these measurements, for each reference, then have the mean performance of each neighbouring measured blocks subtracted. This results in four error measurements (per reference element) that are a function of the proximate coupler, reference and transmitter errors. Any error measurements which are greater than one standard deviation from the mean transmitter and coupler output are discarded. The remaining error measurements, without the outliers, are averaged and are used to estimate the reference element error.

The second phase, Figure 25(b), repeats the process described above, this time for each transmitter. Here the functionally equivalent step of measuring each transmitter by the four neighbouring references is performed. Again, the mean performance of each block in the signal path is calculated and subtracted. However, during this phase the reference error is treated as a *known* quantity—using the inferred value from the previous measurement. Based on this assumption, the resultant error signal is a function of the coupler error and the common transceiver element alone.

By extrapolating the transmitter error, using the same process as for the reference element, the coupler errors may be calculated and compensated for by weighting the transceiver input. This process is repeated. In each subsequent iteration, the dependency of the weighting error signal is dependent upon successive concentric array elements as illustrated in Figure 25(c).

The iterative process continues for much greater than n iterations, until either subsequent corrective weightings are within a predefined accuracy, or until a time limit is reached.

Cognisant of the negative effect that the peripheral elements of the array will have on the outcome of this calibration scheme, these results are discarded. For the results presented here, this corresponds to the connection of an additional ring of peripheral reference elements to the array. Future work will focus on the combining algorithmic and conventional calibration techniques to negate the need for this additional hardware.

5.2.2. Provisional results

To test the performance of this calibration procedure, the results are of 1000 simulations of a 10×10 array, each performed for 100 calibration iterations, was simulated using the system settings of Section 4.4. The centre calibration scheme gave an overall rms array calibration accuracy (σ_a) of 0.857 dB and 2.91°. The iterative calibration procedure gives a resultant phase accuracy of 1.32° and amplitude accuracy of 0.7148 dB. Figure 26 shows how the amplitude accuracy of the iterative calibration varies with each successive iteration. The horizontal line indicates the performance of the centre-referenced calibration. A characteristic of the algorithm is its periodic convergence. This trait, shared by simulated annealing algorithms, prevents convergence to (false) local minima early in the calibration process. This, unfortunately, also limits the ultimate accuracy of the array calibration. For instance, the phase accuracy of this array (Figure 27) degrades by 0.1° to 1.32° from its minimum value, reached on the 37th iteration. Future work will focus on tuning the algorithm's performance, perhaps to attenuate this oscillation in later iterations with a temperature parameter (T) and associated reduction function $f(T)$. Hybrid algorithms—targeting different calibration techniques at different sections of the array—are also currently under investigation.

6. CONCLUSION

In this paper, we have presented a new scheme for tower-top array calibration, using a series of nonradiative, interlinear

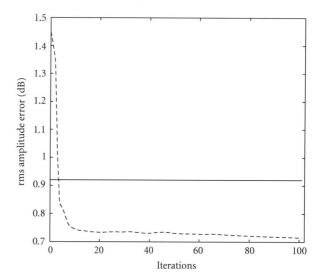

FIGURE 26: Resultant array amplitude feed-point calibration accuracy (σ_{a_A}) for a single $N = 100$ array, plotted versus the number of calibration iterations.

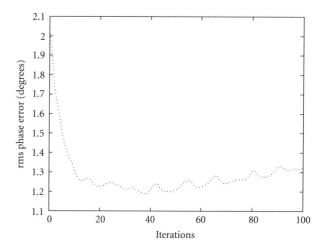

FIGURE 27: Resultant array phasing feed-point calibration accuracy (σ_{a_ϕ}) for a single $N = 100$ array, plotted versus the number of calibration iterations.

reference elements to sense the output of the array. The accuracy of this calibration scheme is a function of the array size, the calibration path taken in calibrating the array, and the coupler performance. Where the measurement accuracy is unlimited, then the accuracy of this calibration is dependent upon the number of couplers in a given calibration path.

The basic building block of this calibration scheme is the 2×2 array calibration coupler. We have shown that using low-cost fabrication techniques and low-quality FR-4 substrate, a broadband coupler network with rms phase balance of 1.1175° and amplitude balance of 0.3295 dB is realisable.

Based upon this coupler hardware, we have shown that phase calibration accurate enough for cellular smart antenna applications is possible. Although amplitude accuracy is still outside our initial target, work is ongoing on improving the precision coupler network and on the development of calibration algorithms to further reduce this requirement.

Finally, we presented examples of one such algorithm—whose performance, unlike that of the conventional feedback algorithms, improves with array size. Moreover, this calibration algorithm, which is based upon exploiting randomness within the array, outperforms conventional calibration for large arrays. Future work will focus on use of simulated annealing and hybrid calibration algorithms to increase calibration accuracy.

ACKNOWLEDGMENT

The authors would like to thank Science Foundation Ireland for their generous funding of this project through the Centre for Telecommunications Value-Chain Research (CTVR).

REFERENCES

[1] N. Tyler, B. Allen, and H. Aghvami, "Adaptive antennas: the calibration problem," *IEEE Communications Magazine*, vol. 42, no. 12, pp. 114–122, 2004.

[2] C. M. Simmonds and M. A. Beach, "Downlink calibration requirements for the TSUNAMI (II) adaptive antenna testbed," in *Proceedings of the 9th IEEE International Symposium on Personal, Indoor and Mobile Radio Communications (PIMRC '98)*, vol. 3, pp. 1260–1264, Boston, Mass, USA, September 1998.

[3] K. Sakaguchi, K. Kuroda, J.-I. Takada, and K. Araki, "Comprehensive calibration for MIMO system," in *Proceedings of the 5th International Symposium on Wireless Personal Multimedia Communications (WPMC 3'02)*, vol. 2, pp. 440–443, Honolulu, Hawaii, USA, October 2002.

[4] C. M. S. See, "Sensor array calibration in the presence of mutual coupling and unknown sensor gains and phases," *Electronics Letters*, vol. 30, no. 5, pp. 373–374, 1994.

[5] R. Sorace, "Phased array calibration," *IEEE Transactions on Antennas and Propagation*, vol. 49, no. 4, pp. 517–525, 2001.

[6] K. R. Dandekar, L. Hao, and X. Guanghan, "Smart antenna array calibration procedure including amplitude and phase mismatch and mutual coupling effects," in *Proceedings of the IEEE International Conference on Personal Wireless Communications (ICPWC '00)*, pp. 293–297, Hyderabad, India, December 2000.

[7] T. Kaiser, "When will smart antennas be ready for the market? Part I," *IEEE Signal Processing Magazine*, vol. 22, no. 2, pp. 87–92, 2005.

[8] F. Rayal, "Why have smart antennas not yet gained traction with wireless network operators?" *IEEE Antennas and Propagation Magazine*, vol. 47, no. 6, pp. 124–126, 2005.

[9] G. Brown, "3G base station design and wireless network economics," *Unstrung Insider*, vol. 5, no. 10, pp. 1–30, 2006.

[10] J. D. Fredrick, Y. Wang, and T. Itoh, "A smart antenna receiver array using a single RF channel and digital beamforming," *IEEE Transactions on Microwave Theory and Techniques*, vol. 50, no. 12, pp. 3052–3058, 2002.

[11] S. Ishii, A. Hoshikuki, and R. Kohno, "Space hopping scheme under short range Rician multipath fading environment," in *Proceedings of the 52nd Vehicular Technology Conference (VTC '00)*, vol. 1, pp. 99–104, Boston, Mass, USA, September 2000.

[12] A. J. Cooper, "'Fibre/radio' for the provision of cordless/mobile telephony services in the access network," *Electronics Letters*, vol. 26, no. 24, pp. 2054–2056, 1990.

[13] G. Brown, "Open basestation bonanza," *Unstrung Insider*, vol. 4, no. 7, pp. 1–20, 2005.

[14] T. Cooper and R. Farrell, "Value-chain engineering of a towertop cellular base station system," in *Proceedings of the IEEE 65th Vehicular Technology Conference (VTC '07)*, pp. 3184–3188, Dublin, Ireland, April 2007.

[15] T. S. Cooper, R. Farrell, and G. Baldwin, "Array Calibration," Patent Pending S2006/0482.

[16] T. Cooper, J. McCormack, R. Farrell, and G. Baldwin, "Toward scalable, automated tower-top phased array calibration," in *Proceedings of the IEEE 65th Vehicular Technology Conference (VTC '07)*, pp. 362–366, Dublin, Ireland, April 2007.

[17] Analog Devices Datasheet, "400 MSPS 14-Bit DAC 1.8V CMOS Direct Digital Synthesizer," January 2003.

[18] T. S. Cooper, G. Baldwin, and R. Farrell, "Six-port precision directional coupler," *Electronics Letters*, vol. 42, no. 21, pp. 1232–1234, 2006.

[19] Agilent EEsof, Palo Alto, Calif, USA. Advanced Design System, Momentum.

[20] CadSoft Computer, 801 South Federal Hwy., Suite 201, Delray Beach, FL 33483-5185. Eagle.

[21] ECS Circuits, Unit 2, Western Business Park, Oak Close, Dublin 12, Ireland.

[22] Rhode & Schwartz Vertiriebs-GmbH, Muehldorfstrasse 15, 81671 Muenchen, Germany.

[23] J. Hromkovic, *Algorithmics for Hard Problems*, Springer, Berlin, Germany, 2nd edition, 2004.

Hindawi Publishing Corporation
EURASIP Journal on Wireless Communications and Networking
Volume 2007, Article ID 45084, 10 pages
doi:10.1155/2007/45084

Research Article

Optimal Design of Uniform Rectangular Antenna Arrays for Strong Line-of-Sight MIMO Channels

Frode Bøhagen,[1] Pål Orten,[2] and Geir Øien[3]

[1] Telenor Research and Innovation, Snarøyveien 30, 1331 Fornebu, Norway
[2] Department of Informatics, UniK, University of Oslo (UiO) and Thrane & Thrane, 0316 Oslo, Norway
[3] Department of Electronics and Telecommunications, Norwegian University of Science and Technology (NTNU),
  7491 Trandheim, Norway

Received 26 October 2006; Accepted 1 August 2007

Recommended by Robert W. Heath

We investigate the optimal design of *uniform rectangular arrays* (URAs) employed in *multiple-input multiple-output* communications, where a strong *line-of-sight* (LOS) component is present. A general geometrical model is introduced to model the LOS component, which allows for any orientation of the transmit and receive arrays, and incorporates the uniform linear array as a special case of the URA. A spherical wave propagation model is used. Based on this model, we derive the optimal array design equations with respect to mutual information, resulting in orthogonal LOS subchannels. The equations reveal that it is the distance between the antennas projected onto the plane perpendicular to the transmission direction that is of importance with respect to design. Further, we investigate the influence of nonoptimal design, and derive analytical expressions for the singular values of the LOS matrix as a function of the quality of the array design. To evaluate a more realistic channel, the LOS channel matrix is employed in a Ricean channel model. Performance results show that even with some deviation from the optimal design, we get better performance than in the case of uncorrelated Rayleigh subchannels.

1. INTRODUCTION

Multiple-input multiple-output (MIMO) technology is a promising tool for enabling spectrally efficient future wireless applications. A lot of research effort has been put into the MIMO field since the pioneering work of Foschini and Gans [1] and Telatar [2], and the technology is already hitting the market [3, 4]. Most of the work on wireless MIMO systems seek to utilize the decorrelation between the subchannels introduced by the multipath propagation in the wireless environment [5]. Introducing a strong *line-of-sight* (LOS) component for such systems is positive in the sense that it boosts the *signal-to-noise ratio* (SNR). However, it will also have a negative impact on MIMO performance as it increases the correlation between the subchannels [6].

In [7], the possibility of enhancing performance by proper antenna array design for MIMO channels with a strong LOS component was investigated, and it was shown that the performance can actually be made superior for pure LOS subchannels compared to fully decorrelated Rayleigh subchannels with equal SNR. The authors of the present

paper have previously studied the optimal design of *uniform linear arrays* (ULAs) with respect to *mutual information* (MI) [8, 9], and have given a simple equation for the optimal design. Furthermore, some work on the design of *uniform rectangular arrays* (URAs) for MIMO systems is presented in [10], where the optimal design for the special case of two broadside URAs is found, and the optimal throughput performance was identified to be identical to the optimal Hadamard bound. The design is based on taking the spherical nature of the electromagnetic wave propagation into account, which makes it possible to achieve a high rank LOS channel matrix [11]. Examples of real world measurements that support this theoretical work can be found in [12, 13].

In this paper, we extend our work from [8], and use the same general procedure to investigate URA design. We introduce a new general geometrical model that can describe any orientation of the transmit (Tx), receive (Rx) URAs, and also incorporate ULAs as a special case. Again, it should be noted that a spherical wave propagation model is employed, in contrast to the more commonly applied approximate plane-wave model. This model is used to derive new equations for the optimal design of the URAs with respect to MI. The results are more general than those presented in an earlier work, and the cases of two ULAs [8] and two broadside URAs [10] can be identified as two special cases. The proposed principle is best suited for fixed systems, for example, fixed wireless access and radio relay systems, because the optimal design is dependent on the Tx-Rx distance and on the orientation of the two URAs. Furthermore, we include an analysis of the influence of nonoptimal design, and analytical expressions for the singular values of the LOS matrix are derived as a function of the quality of the array design. The results are useful for system designers both when designing new systems, as well as when evaluating the performance of existing systems.

The rest of the paper is organized as follows. Section 2 describes the system model used. In Section 3, we present the geometrical model from which the general results are derived. The derivation of the optimal design equations is given in Section 4, while the eigenvalues of the LOS channel matrix are discussed in Section 5. Performance results are shown in Section 6, while conclusions are drawn in Section 7.

2. SYSTEM MODEL

The wireless MIMO transmission system employs N Tx antennas and M Rx antennas when transmitting information over the channel. Assuming *slowly varying* and *frequency-flat* fading channels, we model the MIMO transmission in complex baseband as [5]

$$\mathbf{r} = \sqrt{\eta} \cdot \mathbf{Hs} + \mathbf{n}, \tag{1}$$

where $\mathbf{r} \in \mathbb{C}^{M \times 1}$ is the received signal vector, $\mathbf{s} \in \mathbb{C}^{N \times 1}$ is the transmitted signal vector, $\mathbf{H} \in \mathbb{C}^{M \times N}$ is the normalized channel matrix linking the Tx antennas with the Rx antennas, η is the common power attenuation over the channel, and $\mathbf{n} \in \mathbb{C}^{M \times 1}$ is the additive white Gaussian noise (AWGN) vector. $\mathbf{n}$ contains i.i.d. circularly symmetric complex Gaussian elements with zero mean and variance σ_n^2, that is, $\mathbf{n} \sim \mathcal{CN}(\mathbf{0}_{M \times 1}, \sigma_n^2 \cdot \mathbf{I}_M)$,[1] where $\mathbf{I}_M$ is the $M \times M$ identity matrix.

As mentioned above, $\mathbf{H}$ is the normalized channel matrix, which implies that each element in $\mathbf{H}$ has unit average power; consequently, the average SNR is independent of $\mathbf{H}$. Furthermore, it is assumed that the total transmit power is P, and all the subchannels experience the same path loss as accounted for in η, resulting in the total average received SNR at one Rx antenna being $\bar{\gamma} = \eta P / \sigma_n^2$. We apply $\mathbf{s} \sim \mathcal{CN}(\mathbf{0}_{N \times 1}, (P/N) \cdot \mathbf{I}_N)$, which means that the MI of a MIMO transmission described by (1) becomes [2][2]

$$\mathcal{I} = \sum_{p=1}^{U} \log_2\left(1 + \frac{\bar{\gamma}}{N}\mu_p\right) \quad \text{bps/Hz}, \tag{2}$$

where $U = \min(M, N)$ and μ_p is the pth eigenvalue of $\mathbf{W}$ defined as

$$\mathbf{W} = \begin{cases} \mathbf{HH}^H, & M \le N, \\ \mathbf{H}^H\mathbf{H}, & M > N, \end{cases} \tag{3}$$

where $(\cdot)^H$ is the Hermitian transpose operator.[3]

One way to model the channel matrix is as a sum of two components: a LOS component: and an *non-LOS* (NLOS) component. The ratio between the power of the two components gives the Ricean K-factor [15, page 52]. We express the normalized channel matrix in terms of K as

$$\mathbf{H} = \sqrt{\frac{K}{1+K}} \cdot \mathbf{H}_{\text{LOS}} + \sqrt{\frac{1}{1+K}} \cdot \mathbf{H}_{\text{NLOS}}, \tag{4}$$

where $\mathbf{H}_{\text{LOS}}$ and $\mathbf{H}_{\text{NLOS}}$ are the channel matrices containing the LOS and NLOS channel responses, respectively. In this paper, $\mathbf{H}_{\text{NLOS}}$ is modeled as an *uncorrelated* Rayleigh matrix, that is, $\text{vec}(\mathbf{H}_{\text{NLOS}}) \sim \mathcal{CN}(\mathbf{0}_{MN \times 1}, \mathbf{I}_{MN})$, where $\text{vec}(\cdot)$ is the matrix vectorization (stacking the columns on top of each other). In the next section, the entries of $\mathbf{H}_{\text{LOS}}$ will be described in detail, while in the consecutive sections, the connection between the URA design and the properties of $\mathbf{H}_{\text{LOS}}$ will be addressed. The influence of the stochastic channel component $\mathbf{H}_{\text{NLOS}}$ on performance is investigated in the results section.

3. THE LOS CHANNEL: GEOMETRICAL MODEL

When investigating $\mathbf{H}_{\text{LOS}}$ in this section, we only consider the direct components between the Tx and Rx. The optimal design, to be presented in Section 4, is based on the fact that the LOS components from each of the Tx antennas arrive at the Rx array with a spherical wavefront. Consequently, the common approximate plane wave model, where the Tx and Rx arrays are assumed to be points in space, is not applicable [11]; thus an important part of the contribution of this paper is to characterize the received LOS components.

The principle used to model $\mathbf{H}_{\text{LOS}}$ is *ray-tracing* [7]. Ray-tracing is based on finding the path lengths from each of the Tx antennas to each of the Rx antennas, and employing these path lengths to find the corresponding received phases. We will see later how these path lengths characterize $\mathbf{H}_{\text{LOS}}$, and thus its rank and the MI.

To make the derivation in Section 4 more general, we do not distinguish between the Tx and the Rx, but rather the side with the most antennas and the side with the fewest antennas (the detailed motivation behind this decision is given in the first paragraph of Section 4). We introduce the notation $V = \max(M, N)$, consequently we refer to the side with V antennas as the Vx, and the side with U antennas as the Ux.

We restrict the antenna elements, both at the Ux and at the Vx, to be placed in plane URAs. Thus the antennas are

---

[1] $\mathcal{CN}(\mathbf{x}, \mathbf{Y})$ denotes a complex symmetric Gaussian distributed random vector, with mean vector $\mathbf{x}$ and covariance matrix $\mathbf{Y}$.

[2] Applying equal power Gaussian distributed inputs in the MIMO system is capacity achieving in the case of a Rayleigh channel, but not necessarily in the Ricean channel case studied here [14]; consequently, we use the term MI instead of capacity.

[3] μ_p also corresponds to the pth singular value of $\mathbf{H}$ squared.

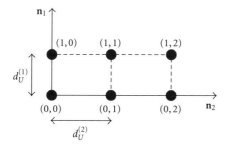

FIGURE 1: An example of a Ux URA with $U = 6$ antennas ($U_1 = 2$ and $U_2 = 3$).

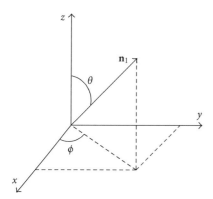

FIGURE 2: Geometrical illustration of the first principal direction of the URA.

placed on lines going in two orthogonal *principal directions*, forming a lattice structure. The two principal directions are characterized with the vectors $\mathbf{n}_1$ and $\mathbf{n}_2$, while the uniform separation in each direction is denoted by $d^{(1)}$ and $d^{(2)}$. The numbers of antennas at the Ux in the first and second principal directions are denoted by U_1 and U_2, respectively, and we have $U = U_1 \cdot U_2$. The position of an antenna in the lattice is characterized by its index in the first and second principal direction, that is, (u_1, u_2), where $u_1 \in \{0, \ldots, U_1 - 1\}$ and $u_2 \in \{0, \ldots, U_2 - 1\}$. As an example, we have illustrated a Ux array with $U_1 = 2$ and $U_2 = 3$ in Figure 1. The same definitions are used at the Vx side for V_1, V_2, v_1, and v_2.

The path length between Ux antenna (u_1, u_2) and Vx antenna (v_1, v_2) is denoted by $l_{(v_1, v_2)(u_1, u_2)}$ (see Figure 4). Since the elements of $\mathbf{H}_{\text{LOS}}$ are assumed normalized as mentioned earlier, the only parameters of interest are the received phases. The elements of $\mathbf{H}_{\text{LOS}}$ then become

$$(\mathbf{H}_{\text{LOS}})_{m,n} = e^{(j2\pi/\lambda)l_{(v_1, v_2)(u_1, u_2)}}, \qquad (5)$$

where $(\cdot)_{m,n}$ denotes the element in row m and column n, and λ is the wavelength. The mapping between m, n, and (v_1, v_2), (u_1, u_2) depends on the dimension of the MIMO system, for example, in the case $M > N$, we get $m = v_1 \cdot V_2 + v_2 + 1$ and $n = u_1 \cdot U_2 + u_2 + 1$. The rest of this section is dedicated to finding an expression for the different path lengths. The procedure employed is based on pure geometrical considerations.

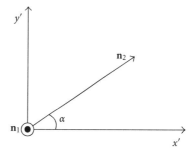

FIGURE 3: Geometrical illustration of the second principal direction of the URA.

We start by describing the geometry of a single URA; afterwards, two such URAs are utilized to describe the communication link. We define the local origo to be at the lower corner of the URA, and the first principal direction as shown in Figure 2, where we have employed spherical coordinates to describe the direction with the angles $\theta \in [0, \pi/2]$ and $\phi \in [0, 2\pi]$. The unit vector for the first principal direction $\mathbf{n}_1$, with respect to the Cartesian coordinate system in Figure 2, is given by [16, page 252]

$$\mathbf{n}_1 = \sin\theta\cos\phi\,\mathbf{n}_x + \sin\theta\sin\phi\,\mathbf{n}_y + \cos\theta\,\mathbf{n}_z, \qquad (6)$$

where $\mathbf{n}_x$, $\mathbf{n}_y$, and $\mathbf{n}_z$ denote the unit vectors in their respective directions.

The second principal direction has to be orthogonal to the first; thus we know that $\mathbf{n}_2$ is in the plane, which is orthogonal to $\mathbf{n}_1$. The two axes in this orthogonal plane are referred to as x' and y'. The plane is illustrated in Figure 3, where $\mathbf{n}_1$ is coming perpendicularly out of the plane, and we have introduced the third angle α to describe the angle between the x'-axis and the second principal direction. To fix this plane described by the x'- and y'-axis to the Cartesian coordinate system in Figure 2, we choose the x'-axis to be orthogonal to the z-axis, that is, placing the x'-axis in the xy-plane. The x' unit vector then becomes

$$\mathbf{n}_{x'} = \frac{1}{|\mathbf{n}_1 \times \mathbf{n}_z|}\mathbf{n}_1 \times \mathbf{n}_z = \sin\phi\mathbf{n}_x - \cos\phi\mathbf{n}_y. \qquad (7)$$

Since origo is defined to be at the lower corner of the URA, we require $\alpha \in [\pi, 2\pi]$. Further, we get the y' unit vector

$$\begin{aligned}\mathbf{n}_{y'} &= \frac{1}{|\mathbf{n}_1 \times \mathbf{n}_{x'}|}\mathbf{n}_1 \times \mathbf{n}_{x'} \\ &= \cos\theta\cos\phi\mathbf{n}_x + \cos\theta\sin\phi\mathbf{n}_y - \sin\theta\mathbf{n}_z.\end{aligned} \qquad (8)$$

Note that when $\theta = 0$ and $\phi = \pi/2$, then $\mathbf{n}_{x'} = \mathbf{n}_x$ and $\mathbf{n}_{y'} = \mathbf{n}_y$. Based on this description, we observe from Figure 3 that the second principal direction has the unit vector

$$\mathbf{n}_2 = \cos\alpha\mathbf{n}_{x'} + \sin\alpha\mathbf{n}_{y'}. \qquad (9)$$

These unit vectors, $\mathbf{n}_1$ and $\mathbf{n}_2$, can now be employed to describe the position of any antenna in the URA. The position difference, relative to the local origo in Figure 2, between

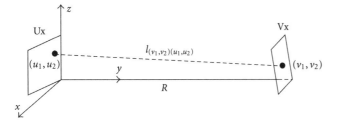

FIGURE 4: The transmission system investigated.

two neighboring antennas placed in the first principal direction is

$$
\begin{aligned}
\mathbf{k}^{(1)} &= d^{(1)}\mathbf{n}_1 \\
&= d^{(1)}\big(\sin\theta\cos\phi\,\mathbf{n}_x + \sin\theta\sin\phi\,\mathbf{n}_y + \cos\theta\mathbf{n}_z\big),
\end{aligned}
\tag{10}
$$

where $d^{(1)}$ is the distance between two neighboring antennas in the first principal direction. The corresponding position difference in the second principal direction is

$$
\begin{aligned}
\mathbf{k}^{(2)} = d^{(2)}\mathbf{n}_2 = d^{(2)}\big(&(\cos\alpha\sin\phi + \sin\alpha\cos\theta\cos\phi)\mathbf{n}_x \\
&+ (\sin\alpha\cos\theta\sin\phi - \cos\alpha\cos\phi)\mathbf{n}_y \\
&- \sin\alpha\sin\theta\mathbf{n}_z\big),
\end{aligned}
\tag{11}
$$

where $d^{(2)}$ is the distance between the antennas in the second principal direction. $d^{(1)}$ and $d^{(2)}$ can of course take different values, both at the Ux and at the Vx; thus we get two pairs of such distances.

We now employ two URAs as just described to model the communication link. When defining the reference coordinate system for the communication link, we choose the lower corner of the Ux URA to be the global origo, and the y-axis is taken to be in the direction from the lower corner of the Ux URA to the lower corner of the Vx URA. To determine the z- and x-axes, we choose the first principal direction of the Ux URA to be in the yz-plane, that is, $\phi_U = \pi/2$. The system is illustrated in Figure 4, where R is the distance between the lower corner of the two URAs. To find the path lengths that we are searching for, we define a vector from the global origo to Ux antenna (u_1, u_2) as

$$
\mathbf{a}_U^{(u_1,u_2)} = u_1 \cdot \mathbf{k}_U^{(1)} + u_2 \cdot \mathbf{k}_U^{(2)},
\tag{12}
$$

and a vector from the global origo to Vx antenna (v_1, v_2) as

$$
\mathbf{a}_V^{(v_1,v_2)} = R \cdot \mathbf{n}_y + v_1 \cdot \mathbf{k}_V^{(1)} + v_2 \cdot \mathbf{k}_V^{(2)}.
\tag{13}
$$

All geometrical parameters in $\mathbf{k}^{(1)}$ and $\mathbf{k}^{(2)}$ $(\theta, \phi, \alpha, d^{(1)}, d^{(2)})$ in these two expressions have a subscript U or V to distinguish between the two sides in the communication link. We can now find the distance between Ux antenna (u_1, u_2) and

Vx antenna (v_1, v_2) by taking the Euclidean norm of the vector difference:

$$
l_{(v_1,v_2)(u_1,u_2)} = \left\| \mathbf{a}_V^{(v_1,v_2)} - \mathbf{a}_U^{(u_1,u_2)} \right\|
\tag{14}
$$

$$
= \big(l_x^2 + (R + l_y)^2 + l_z^2\big)^{1/2}
\tag{15}
$$

$$
\approx R + l_y + \frac{l_x^2 + l_z^2}{2R}.
\tag{16}
$$

Here, l_x, l_y, and l_z represent the distances between the two antennas in these directions when disregarding the distance between the URAs R. In the transition from (15) to (16), we perform a Maclaurin series expansion to the first order of the square root expression, that is, $\sqrt{1+a} \approx 1 + a/2$, which is accurate when $a \ll 1$. We also removed the $2 \cdot l_y$ term in the denominator. Both these approximations are good as long as $R \gg l_x, l_y, l_z$.

It is important to note that the geometrical model just described is general, and allows any orientation of the two URAs used in the communication link. Another interesting observation is that the geometrical model incorporates the case of ULAs, for example, by employing $U_2 = 1$, the Ux array becomes a ULA. This will be exploited in the analysis in the next section. A last but very important observation is that we have taken the spherical nature of the electromagnetic wave propagation into account, by applying the actual distance between the Tx and Rx antennas when considering the received phase. Consequently, we have not put any restrictions on the rank of $\mathbf{H}_{\mathrm{LOS}}$, that is, $\mathrm{rank}(\mathbf{H}_{\mathrm{LOS}}) \in \{1, 2, \ldots, U\}$ [11].

4. OPTIMAL URA/ULA DESIGN

In this section, we derive equations for the optimal URA/ULA design with respect to MI when transmitting over a pure LOS MIMO channel. From (2), we know that the important channel parameter with respect to MI is the $\{\mu_p\}$. Further, in [17, page 295], it is shown that the maximal MI is achieved when the $\{\mu_p\}$ are all equal. This situation occurs when all the vectors $\mathbf{h}_{(u_1,u_2)}$ (i.e., columns (rows) of $\mathbf{H}_{\mathrm{LOS}}$ when $M > N$ ($M \le N$)), containing the channel response between one Ux antenna (u_1, u_2) and all the Vx antennas, that is,

$$
\begin{aligned}
&\mathbf{h}_{(u_1,u_2)} \\
&= \big[e^{(j2\pi/\lambda)l_{(0,0)(u_1,u_2)}}, e^{(j2\pi/\lambda)l_{(0,1)(u_1,u_2)}}, \ldots, e^{(j2\pi/\lambda)l_{((V_1-1),(V_2-1))(u_1,u_2)}}\big]^T,
\end{aligned}
\tag{17}
$$

are orthogonal to each other, resulting in $\mu_p = V$, for $p \in \{1, \ldots, U\}$. Here, $(\cdot)^T$ is the vector transpose operator. This requirement is actually the motivation behind the choice to distinguish between Ux and Vx instead of Tx and Rx. By basing the analysis on Ux and Vx, we get one general solution, instead of getting one solution valid for $M > N$ and another for $M \le N$.

When the orthogonality requirement is fulfilled, all the U subchannels are orthogonal to each other. When doing spatial multiplexing on these U orthogonal subchannels, the optimal detection scheme actually becomes the matched filter, that is, $\mathbf{H}_{\mathrm{LOS}}^H$. The matched filter results in no interference between the subchannels due to the orthogonality, and at the same time maximizes the SNR on each of the subchannels (maximum ratio combining).

A consequence of the orthogonality requirement is that the inner product between any combination of two different such vectors should be equal to zero. This can be expressed as $\mathbf{h}_{(u_{1_b}, u_{2_b})}^H \mathbf{h}_{(u_{1_a}, u_{2_a})} = 0$, where the subscripts a and b are employed to distinguish between the two different Ux antennas. The orthogonality requirement can then be written as

$$\sum_{v_1=0}^{V_1-1} \sum_{v_2=0}^{V_2-1} e^{j2\pi/\lambda (l_{(v_1,v_2)(u_{1_a}, u_{2_a})} - l_{(v_1,v_2)(u_{1_b}, u_{2_b})})} = 0. \quad (18)$$

By factorizing the path length difference in the parentheses in this expression with respect to v_1 and v_2, it can be written in the equivalent form

$$\sum_{v_1=0}^{V_1-1} e^{j2\pi(\hat{\beta}_{11}+\hat{\beta}_{12})v_1} \cdot \sum_{v_2=0}^{V_2-1} e^{j2\pi(\hat{\beta}_{21}+\hat{\beta}_{22})v_2} = 0, \quad (19)$$

where $\hat{\beta}_{ij} = \beta_{ij}(u_{j_b} - u_{j_a})$, and the different β_{ij}s are defined as follows:[4]

$$\beta_{11} = \frac{d_V^{(1)} d_U^{(1)} V_1}{\lambda R} \cos\theta_V \cos\theta_U, \quad (20)$$

$$\beta_{12} = \frac{d_V^{(1)} d_U^{(2)} V_1}{\lambda R} [\sin\theta_V \cos\phi_V \cos\alpha_U \\ - \cos\theta_V \sin\alpha_U \sin\theta_U], \quad (21)$$

$$\beta_{21} = -\frac{d_V^{(2)} d_U^{(1)} V_2}{\lambda R} \sin\alpha_V \sin\theta_V \cos\theta_U, \quad (22)$$

$$\beta_{22} = \frac{d_V^{(2)} d_U^{(2)} V_2}{\lambda R} [\cos\alpha_U \cos\alpha_V \sin\phi_V \\ + \cos\alpha_U \sin\alpha_V \cos\theta_V \cos\phi_V \\ + \sin\alpha_V \sin\alpha_U \sin\theta_V \sin\theta_U]. \quad (23)$$

The orthogonality requirement in (19) can be simplified by employing the expression for a geometric sum [16, page 192] and the relation $\sin x = (e^{jx} - e^{-jx})/2j$ [16, page 128] to

$$\underbrace{\frac{\sin[\pi(\hat{\beta}_{11}+\hat{\beta}_{12})]}{\sin[(\pi/V_1)(\hat{\beta}_{11}+\hat{\beta}_{12})]}}_{=\zeta_1} \cdot \underbrace{\frac{\sin[\pi(\hat{\beta}_{21}+\hat{\beta}_{22})]}{\sin[(\pi/V_2)(\hat{\beta}_{21}+\hat{\beta}_{22})]}}_{=\zeta_2} = 0. \quad (24)$$

Orthogonal subchannels, and thus maximum MI, are achieved if (24) is fulfilled for all combinations of (u_{1_a}, u_{2_a}) and (u_{1_b}, u_{2_b}), except when $(u_{1_a}, u_{2_a}) = (u_{1_b}, u_{2_b})$.

_____

[4] This can be verified by employing the approximate path length from (16) in (18).

The results above clearly show how achieving orthogonal subchannels is dependent on the geometrical parameters, that is, the design of the antenna arrays. By investigating (20)–(23) closer, we observe the following inner product relation:

$$\beta_{ij} = \frac{V_i}{\lambda R} \hat{\mathbf{k}}_U^{(j)T} \hat{\mathbf{k}}_V^{(i)} \quad \forall i,j \in \{1,2\}, \quad (25)$$

where $\hat{\mathbf{k}}^{(i)} = k_x^{(i)} \mathbf{n}_x + k_z^{(i)} \mathbf{n}_z$, that is, the vectors defined in (10) and (11) where the y-term is set equal to zero. Since solving (24) is dependent on applying correct values of β_{ij}, we see from (25) that it is the extension of the arrays in the x- and z-direction that are crucial with respect to the design of orthogonal subchannels. Moreover, the optimal design is independent of the array extension in the y-direction (direction of transmission). The relation in (25) will be exploited in the analysis to follow to give an alternative projection view on the results.

Both ζ_1 and ζ_2, which are defined in (24), are $\sin(x)/\sin(x/V_i)$ expressions. For these to be zero, the $\sin(x)$ in the nominator must be zero, while the $\sin(x/V_i)$ in the denominator is non-zero, which among other things leads to requirements on the dimensions of the URAs/ULAs, as will be seen in the next subsections. Furthermore, ζ_1 and ζ_2 are periodic functions, thus (24) has more than one solution. We will focus on the solution corresponding to the smallest arrays, both because we see this as the most interesting case from an implementation point of view, and because it would not be feasible to investigate all possible solutions of (24). From (20)–(23), we see that the array size increases with increasing β_{ij}, therefore, in this paper, we will restrict the analysis to the case where the relevant $|\beta_{ij}| \le 1$, which are found, by investigating (24), to be the smallest values that produce solutions. In the next four subsections, we will systematically go through the possible different combinations of URAs and ULAs in the communications link, and give solutions of (24) if possible.

4.1. ULA at Ux and ULA at Vx

We start with the simplest case, that is, both Ux and Vx employing ULAs. This is equivalent to the scenario we studied in [8]. In this case, we have $U_2 = 1$ giving $\hat{\beta}_{12} = \hat{\beta}_{22} = 0$, and $V_2 = 1$ giving $\zeta_2 = 1$, therefore, we only need to consider $\hat{\beta}_{11}$. Studying (24), we find that the only solution with our array size restriction is $|\beta_{11}| = 1$, that is,

$$d_V^{(1)} d_U^{(1)} = \frac{\lambda R}{V_1 \cos\theta_V \cos\theta_U}, \quad (26)$$

which is identical to the result derived in [8]. The solution is given as a product $d_U d_V$, and in accordance with [8], we refer to this product as the *antenna separation product* (ASP). When the relation in (26) is achieved, we have the optimal design in terms of MI, corresponding to orthogonal LOS subchannels.

Projection view

Motivated by the observation in (25), we reformulate (26) as $(d_V^{(1)} \cos \theta_V) \cdot (d_U^{(1)} \cos \theta_U) = \lambda R / V_1$. Consequently, we observe that the the product of the antenna separations projected along the local z-axis at both sides of the link should be equal to $\lambda R / V_1$. The z-direction is the only direction of relevance due to the fact that it is only the array extension in the xz-plane that is of interest (cf. (25)), and the fact that the first (and only) principal direction at the Ux is in the yz-plane (i.e., $\phi_U = \pi/2$).

4.2. URA at Ux and ULA at Vx

Since Vx is a ULA, we have $V_2 = 1$ giving $\zeta_2 = 1$, thus to get the optimal design, we need $\zeta_1 = 0$. It turns out that with the aforementioned array size restriction ($|\beta_{ij}| \leq 1$), it is not possible to find a solution to this problem, for example, by employing $|\beta_{11}| = |\beta_{12}| = 1$, we observe that $\zeta_1 = 0$ for most combinations of Ux antennas, except when $u_{1_a} + u_{2_a} = u_{1_b} + u_{2_b}$, which gives $\zeta_1 = V_1$. By examining this case a bit closer, we find that the antenna elements in the URA that are correlated, that is, giving $\zeta_1 = V_1$, are the diagonal elements of the URA. Consequently, the optimal design is not possible in this case.

Projection view

By employing the projection view, we can reveal the reason why the diagonal elements become correlated, and thus why a solution is not possible. Actually, it turns out that the diagonal of the URA projected on to the xz-plane is perpendicular to the ULA projected on to the xz-plane when $|\beta_{11}| = |\beta_{12}| = 1$. This can be verified by applying (25) to show the following relation:

$$\underbrace{\left(\hat{\mathbf{k}}_U^{(1)} - \hat{\mathbf{k}}_U^{(2)}\right)^T}_{\text{diagonal of URA}} \cdot \hat{\mathbf{k}}_V^{(1)} = 0. \qquad (27)$$

Moreover, the diagonal of the URA can be viewed as a ULA, and when two ULAs are perpendicular aligned in space, the ASP goes towards infinity (this can be verified by employing $\theta_V \to \pi/2$ in (26)). This indicates that it is not possible to do the optimal design when this perpendicularity is present.

4.3. ULA at Ux and URA at Vx

As mentioned earlier, a ULA at Ux gives $U_2 = 1$, resulting in $(u_{2_b} - u_{2_a}) = 0$, and thus $\hat{\beta}_{12} = \hat{\beta}_{22} = 0$. Investigating the remaining expression in (24), we see that the optimal design is achieved when $|\beta_{11}| = 1$ if $V_1 \geq U$, giving $\zeta_1 = 0$, or $|\beta_{21}| = 1$ if $V_2 \geq U$, giving $\zeta_2 = 0$, that is,

$$d_V^{(1)} d_U^{(1)} = \frac{\lambda R}{V_1 \cos \theta_V \cos \theta_U} \quad \text{if } V_1 \geq U, \text{ or} \qquad (28)$$

$$d_V^{(2)} d_U^{(1)} = \frac{\lambda R}{V_2 \sin \theta_V |\sin \alpha_V| \cos \theta_U} \quad \text{if } V_2 \geq U. \qquad (29)$$

Furthermore, the optimal design is also achieved if both the above ASP equations are fulfilled simultaneously, and either $q/V_1 \notin \mathbb{Z}$ or $q/V_2 \notin \mathbb{Z}$, for all $q < U$. This guarantees either $\zeta_1 = 0$ or $\zeta_2 = 0$ for all combinations of u_{1_a} and u_{1_b}.

Projection view

A similar reformulation as performed in Section 4.1 can be done for this scenario. We see that both ASP equations, (28) and (29), contain the term $\cos \theta_U$, which projects the antenna distance at the Ux side on the z-axis. The other trigonometric functions project the Vx antenna separation on to the z-axis, either based on the first principal direction (28) or based on the second principal direction (29).

4.4. URA at Ux and URA at Vx

In this last case, when both Ux and Vx are URAs, we have $U_1, U_2, V_1, V_2 > 1$. By investigating (20)–(24), we observe that in order to be able to solve (24), at least one β_{ij} must be zero. This indicates that the optimal design in this case is only possible for some array orientations, that is, values of θ, ϕ, and α, giving one $\beta_{ij} = 0$. To solve (24) when one $\beta_{ij} = 0$, we observe the following requirement on the β_{ij} s: $|\beta_{11}| = |\beta_{22}| = 1$ and $V_1 \geq U_1$, $V_2 \geq U_2$ or $|\beta_{12}| = |\beta_{21}| = 1$ and $V_1 \geq U_2$, $V_2 \geq U_1$.

This is best illustrated through an example. For instance, we can look at the case where $\alpha_V = 0$, which results in $\beta_{21} = 0$. From (24), we observe that when $\beta_{21} = 0$ and $|\beta_{22}| = 1$, we always have $\zeta_2 = 0$ if $V_2 \geq U_2$, except when $(u_{2_b} - u_{2_a}) = 0$. Thus to get orthogonality in this case as well, we need $|\beta_{11}| = 1$ and $V_1 \geq U_1$. Therefore, the optimal design for this example becomes

$$d_V^{(1)} d_U^{(1)} = \frac{\lambda R}{V_1 \cos \theta_V \cos \theta_U}, \quad V_1 \geq U_1, \qquad (30)$$

$$d_V^{(2)} d_U^{(2)} = \frac{\lambda R}{V_2 |\cos \alpha_U \sin \phi_V|}, \quad V_2 \geq U_2. \qquad (31)$$

The special case of two broadside URAs is revealed by further setting $\alpha_U = 0$, $\theta_U = 0$, $\theta_V = 0$, and $\phi_V = \pi/2$ in (30) and (31). The optimal ASPs are then given by

$$d_V^{(1)} d_U^{(1)} = \frac{\lambda R}{V_1}, \quad V_1 \geq U_1; \qquad d_V^{(2)} d_U^{(2)} = \frac{\lambda R}{V_2}, \quad V_2 \geq U_2. \qquad (32)$$

This corresponds exactly to the result given in [10], which shows the generality of the equations derived in this work and how they contain previous work as special cases.

Projection view

We now look at the example where $\alpha_V = 0$ with a projection view. We observe that in (30), both antenna separations in the first principal directions are projected along the z-axis at Ux and Vx, and the product of these two distances should be equal to $\lambda R / V_1$. In (31), the antenna separations along the second principal direction are projected on the x-axis at Ux and Vx, and the product should be equal to $\lambda R/V_2$. These results clearly show that it is the extension of the arrays in the plane perpendicular to the transmission direction that is crucial. Moreover, the correct extension in the xz-plane is dependent on the wavelength, transmission distance, and dimension of the Vx.

4.5. Practical considerations

We observe that the optimal design equations from previous subsections are all on the same form, that is, $d_V d_U = \lambda R/V_i X$, where X is given by the orientation of the arrays. A first comment is that utilizing the design equations to achieve high performance MIMO links is best suited for fixed systems (such as wireless LANs with LOS conditions,[5] broadband wireless access, radio relay systems, etc.) since the optimal design is dependent on both the orientation and the Tx-Rx distance. Another important aspect is the size of the arrays. To keep the array size reasonable,[6] the product λR should not be too large, that is, the scheme is best suited for high frequency and/or short range communications. Note that these properties agree well with systems that have a fairly high probability of having a strong LOS channel component present. The orientation also affects the array size, for example, if $X \rightarrow 0$, the optimal antenna separation goes towards infinity. As discussed in the previous sections, it is the array extension in the xz-plane that is important with respect to performance, consequently, placing the arrays in this plane minimizes the size required.

Furthermore, we observe that in most cases, even if one array is fully specified, the optimal design is still possible. For instance, from (30) and (31), we see that if $d^{(1)}$ and $d^{(2)}$ are given for one URA, we can still do the optimal design by choosing appropriate values for $d^{(1)}$ and $d^{(2)}$ for the other URA. This is an important property for centralized systems utilizing *base stations* (BSs), which allows for the optimal design for the different communication links by adapting the subscriber units' arrays to the BS array.

5. EIGENVALUES OF W

As in the previous two sections, we focus on the pure LOS channel matrix in this analysis. From Section 4, we know that in the case of optimal array design, we get $\mu_p = V$ for all p, that is, all the eigenvalues of $\mathbf{W}$ are equal to V. An interesting question now is: What happens to the μ_ps if the design deviates from the optimal as given in Section 4? In our analysis of nonoptimal design, we make use of $\{\beta_{ij}\}$. From above, we know that the optimal design, requiring the smallest antenna arrays, was found by setting the relevant $|\beta_{ij}|$ equal to zero or unity, depending on the transmission scenario. Since $\{\beta_{ij}\}$ are functions of the geometrical parameters, studying

the deviation from the optimal design is equivalent to studying the behavior of $\{\mu_p\}_{p=1}^U$, when the relevant β_{ij}s deviate from the optimal ones. First, we give a simplified expression for the eigenvalues of $\mathbf{W}$ as functions of β_{ij}. Then, we look at an interesting special case where we give explicit analytical expressions for $\{\mu_p\}_{p=1}^U$ and describe a method for characterizing nonoptimal designs.

We employ the path length found in (16) in the $\mathbf{H}_{\text{LOS}}$ model. As in [8], we utilize the fact that the eigenvalues of the previously defined Hermitian matrix $\mathbf{W}$ are the same as for a real symmetric matrix $\widehat{\mathbf{W}}$ defined by $\mathbf{W} = \mathbf{B}^H \widehat{\mathbf{W}} \mathbf{B}$, where $\mathbf{B}$ is a unitary matrix.[7] For the URA case studied in this paper, it is straightforward to show that the elements of $\widehat{\mathbf{W}}$ are (cf. (24))

$$(\widehat{\mathbf{W}})_{k,l} = \frac{\sin\left[\pi(\hat{\beta}_{11} + \hat{\beta}_{12})\right]}{\sin\left[(\pi/V_1)(\hat{\beta}_{11} + \hat{\beta}_{12})\right]} \cdot \frac{\sin\left[\pi(\hat{\beta}_{21} + \hat{\beta}_{22})\right]}{\sin\left[(\pi/V_2)(\hat{\beta}_{21} + \hat{\beta}_{22})\right]}, \tag{33}$$

where $k = u_{1_a} U_2 + u_{2_a} + 1$ and $l = u_{1_b} U_2 + u_{2_b} + 1$. We can now find the eigenvalues $\{\mu_p\}_{p=1}^U$ of $\mathbf{W}$ by solving $\det(\widehat{\mathbf{W}} - \mathbf{I}_U \mu) = 0$, where $\det(\cdot)$ is the matrix determinant operator. Analytical expressions for the eigenvalues can be calculated for all combinations of ULA and URA communication links by using a similar procedure to that in Section 4. The eigenvalue expressions become, however, more and more involved for increasing values of U.

5.1. Example: $\beta_{11} = \beta_{22} = 0$ or $\beta_{12} = \beta_{21} = 0$

As an example, we look at the special case that occurs when $\beta_{11} = \beta_{22} = 0$ or $\beta_{12} = \beta_{21} = 0$. This is true for some geometrical parameter combinations, when employing URAs both at Ux and Vx, for example, the case of two broadside URAs. In this situation, we see that the matrix $\widehat{\mathbf{W}}$ from (33) can be written as a Kronecker product of two square matrices [10], that is,

$$\widehat{\mathbf{W}} = \widehat{\mathbf{W}}_1 \otimes \widehat{\mathbf{W}}_2, \tag{34}$$

where

$$(\widehat{\mathbf{W}}_i)_{k,l} = \frac{\sin\left(\pi\beta_{ij}(k - l)\right)}{\sin\left(\pi(\beta_{ij}/V_i)(k - l)\right)}. \tag{35}$$

Here, $k, l \in \{1, 2, \ldots, U_j\}$ and the subscript $j \in \{1, 2\}$ is dependent on β_{ij}. If $\widehat{\mathbf{W}}_1$ has the eigenvalues $\{\mu_{p_1}^{(1)}\}_{p_1=1}^{U_j}$, and $\widehat{\mathbf{W}}_2$ has the eigenvalues $\{\mu_{p_2}^{(2)}\}_{p_2=1}^{U_j}$, we know from matrix theory that the matrix $\widehat{\mathbf{W}}$ has the eigenvalues

$$\mu_p = \mu_{p_1}^{(1)} \cdot \mu_{p_2}^{(2)}, \quad \forall p_1, p_2. \tag{36}$$

---

[5] This is of course not the case for all wireless LANs.

[6] What is considered as reasonable, of course, depends on the application, and may, for example, vary for WLAN, broadband wireless access, and radio relay systems.

---

[7] This implies that $\det(\mathbf{W} - \lambda \mathbf{I}) = 0 \Rightarrow \det(\mathbf{B}^H \widehat{\mathbf{W}} \mathbf{B} - \lambda \mathbf{B}^H \mathbf{I} \mathbf{B}) = 0 \Rightarrow \det(\widehat{\mathbf{W}} - \lambda \mathbf{I}) = 0$.

Expressions for $\mu_{p_i}^{(i)}$ were given in [8] for $U_j = 2$ and $U_j = 3$. For example, for $U_j = 2$ we get the eigenvalues

$$\mu_1^{(i)} = V_i + \frac{\sin(\beta_{ij}\pi)}{\sin(\beta_{ij}(\pi/V_i))}, \qquad \mu_2^{(i)} = V_i - \frac{\sin(\beta_{ij}\pi)}{\sin(\beta_{ij}(\pi/V_i))}.$$
(37)

In this case, we only have two nonzero β_{ij}s, which we from now on, denote β_1 and β_2, that fully characterize the URA design. The optimal design is obtained when both $|\beta_i|$ are equal to unity, while the actual antenna separation is too small (large) when $|\beta_i| > 1$ ($|\beta_i| < 1$). This will be applied in the results section to analyze the design.

There can be several reasons for $|\beta_i|$ to deviate from unity (0 dB). For example, the optimal ASP may be too large for practical systems so that a compromise is needed, or the geometrical parameters may be difficult to determine with sufficient accuracy. A third reason for nonoptimal array design may be the wavelength dependence. A communication system always occupies a nonzero bandwidth, while the antenna distance can only be optimal for one single frequency. As an example, consider the 10.5 GHz-licensed band (10.000–10.680 GHz [18]). If we design a system for the center frequency, the deviation for the lower frequency yields $\lambda_{\text{low}}/\lambda_{\text{design}} = f_{\text{design}}/f_{\text{low}} = 10.340/10.000 = 1.034 = 0.145$ dB. Consequently, this bandwidth dependency only contribute to a 0.145 dB deviation in the $|\beta_i|$ in this case, and in Section 6, we will see that this has almost no impact on the performance of the MIMO system.

6. RESULTS

In this section, we will consider the example of a 4×4 MIMO system with URAs both at Ux and Vx, that is, $U_1 = U_2 = V_1 = V_2 = 2$. Further, we set $\theta_V = \theta_U = 0$, which gives $\beta_{12} = \beta_{21} = 0$; thus we can make use of the results from the last subsection, where the nonoptimal design is characterized by β_1 and β_2.

Analytical expressions for $\{\mu_p\}_{p=1}^4$ in the pure LOS case are found by employing (37) in (36). The square roots of the eigenvalues (i.e., singular values of $\mathbf{H}_{\text{LOS}}$) are plotted as a function of $|\beta_1| = |\beta_2|$ in Figure 5. The lines represent the analytical expressions, while the circles are determined by using a numerical procedure to find the singular values when the *exact* path length from (15) is employed in $\mathbf{H}_{\text{LOS}}$. The parameters used in the exact path length case are as follows: $\phi_V = \pi/2$, $\alpha_U = \pi$, $\alpha_V = \pi$, $R = 500$ m, $d_U^{(1)} = 1$ m, $d_U^{(2)} = 1$ m, $\lambda = 0.03$ m, while $d_V^{(1)}$ and $d_V^{(2)}$ are chosen to get the correct values of $|\beta_1|$ and $|\beta_2|$.

The figure shows that there is a perfect agreement between the analytical singular values based on approximate path lengths from (16), and the singular values found based on exact path lengths from (15). We see how the singular values spread out as the design deviates further and further from the optimal (decreasing $|\beta_i|$), and for small $|\beta_i|$, we get rank($\mathbf{H}_{\text{LOS}}$) = 1, which we refer to as a total design mismatch. In the figure, the solid line in the middle represents two singular values, as they become identical in the present case ($|\beta_1| = |\beta_2|$). This is easily verified by observing the

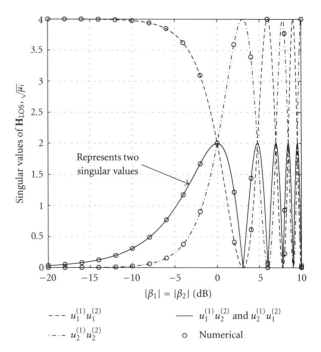

FIGURE 5: The singular values of $\mathbf{H}_{\text{LOS}}$ for the 4×4 MIMO system as a function of $|\beta_1| = |\beta_2|$, both exactly found by a numerical procedure and the analytical from Section 5.

symmetry in the analytical expressions for the eigenvalues. For $|\beta_i| > 1$, we experience some kind of periodic behavior; this is due to the fact that (24) has more than one solution. However, in this paper, we introduced a size requirement on the arrays, thus we concentrate on the solutions where $|\beta| \leq 1$.

When $K \neq \infty$ in (4), the MI from (2) becomes a random variable. We characterize the random MI by the MI *cumulative distribution function* (CDF), which is defined as the probability that the MI falls below a given threshold, that is, $F(\mathcal{I}_{\text{th}}) = \Pr[\mathcal{I} < \mathcal{I}_{\text{th}}]$ [5]. All CDF curves plotted in the next figures are based on 50 000 channel realizations.

We start by illustrating the combined influence of $|\beta_i|$ and the Ricean K-factor. In Figure 6, we show $F(\mathcal{I}_{\text{th}})$ for the optimal design case ($|\beta_1| = |\beta_2| = 0$ dB), and for the total design mismatch ($|\beta_1| = |\beta_2| = -30$ dB).

The figure shows that the design of the URAs becomes more and more important as the K-factor increases. This is because it increases the influence of $\mathbf{H}_{\text{LOS}}$ on $\mathbf{H}$ (cf. (4)). We also observe that the MI increases for the optimal design case when the K-factor increases, while the MI decreases for increasing K-factors in the total design mismatch case. This illustrates the fact that the pure LOS case outperforms the uncorrelated Rayleigh case when we do optimal array design (i.e., orthogonal LOS subchannels).

In Figure 7, we illustrate how $F(\mathcal{I}_{\text{th}})$ changes when we have different combinations of the two $|\beta_i|$. We see how the MI decreases when $|\beta_i|$ decreases. In this case, the Ricean K-factor is 5 dB, and from Figure 6, we know that the MI would be even more sensitive to $|\beta_i|$ for larger K-factors. From the figure, we observe that even with some deviation from the

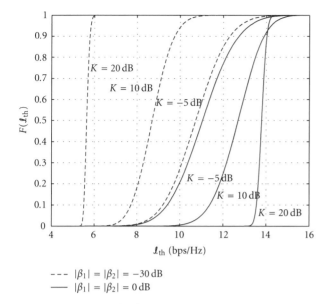

--- $|\beta_1| = |\beta_2| = -30$ dB
—— $|\beta_1| = |\beta_2| = 0$ dB

FIGURE 6: The MI CDF for the 4×4 MIMO system when $\bar{\gamma} = 10$ dB.

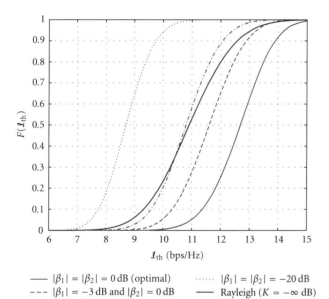

—— $|\beta_1| = |\beta_2| = 0$ dB (optimal)   ····· $|\beta_1| = |\beta_2| = -20$ dB
--- $|\beta_1| = -3$ dB and $|\beta_2| = 0$ dB   —— Rayleigh ($K = -\infty$ dB)
·-·- $|\beta_1| = |\beta_2| = -3$ dB

FIGURE 7: The MI CDF for the 4×4 MIMO system when $\bar{\gamma} = 10$ dB and $K = 5$ dB (except for the Rayleigh channel where $K = -\infty$ dB).

optimal design, we get higher MI compared to the case of uncorrelated Rayleigh subchannels.

7. CONCLUSIONS

Based on the new general geometrical model introduced for the *uniform rectangular array* (URA), which also incorporates the *uniform linear array* (ULA), we have investigated the optimal design for *line-of-sight* (LOS) channels with respect to mutual information for all possible combinations of URA and ULA at transmitter and receiver. The optimal de-

sign based on correct separation between the antennas (d_U and d_V) is possible in several interesting cases. Important parameters with respect to the optimal design are the wavelength, the transmission distance, and the array dimensions in the plane perpendicular to the transmission direction.

Furthermore, we have characterized and investigated the consequence of nonoptimal design, and in the general case, we gave simplified expressions for the pure LOS eigenvalues as a function of the design parameters. In addition, we derived explicit analytical expressions for the eigenvalues for some interesting cases.

ACKNOWLEDGMENTS

This work was funded by Nera with support from the Research Council of Norway (NFR), and partly by the BEATS project financed by the NFR, and the NEWCOM Network of Excellence. Some of this material was presented at the IEEE Signal Processing Advances in Wireless Communications (SPAWC), Cannes, France, July 2006.

REFERENCES

[1] G. J. Foschini and M. J. Gans, "On limits of wireless communications in a fading environment when using multiple antennas," *Wireless Personal Communications*, vol. 6, no. 3, pp. 311–335, 1998.

[2] E. Telatar, "Capacity of multiantenna Gaussian channels," Tech. Memo, AT&T Bell Laboratories, Murray Hill, NJ, USA, June 1995.

[3] T. Kaiser, "When will smart antennas be ready for the market? Part I," *IEEE Signal Processing Magazine*, vol. 22, no. 2, pp. 87–92, 2005.

[4] T. Kaiser, "When will smart antennas be ready for the market? Part II—results," *IEEE Signal Processing Magazine*, vol. 22, no. 6, pp. 174–176, 2005.

[5] D. Gesbert, M. Shafi, D.-S. Shiu, P. J. Smith, and A. Naguib, "From theory to practice: an overview of MIMO space-time coded wireless systems," *IEEE Journal on Selected Areas in Communications*, vol. 21, no. 3, pp. 281–302, 2003.

[6] D. Gesbert, "Multipath: curse or blessing? A system performance analysis of MIMO wireless systems," in *Proceedings of the International Zurich Seminar on Communications (IZS '04)*, pp. 14–17, Zurich, Switzerland, February 2004.

[7] P. F. Driessen and G. J. Foschini, "On the capacity formula for multiple input-multiple output wireless channels: a geometric interpretation," *IEEE Transactions on Communications*, vol. 47, no. 2, pp. 173–176, 1999.

[8] F. Bøhagen, P. Orten, and G. E. Øien, "Design of optimal high-rank line-of-sight MIMO channels," *IEEE Transactions on Wireless Communications*, vol. 6, no. 4, pp. 1420–1425, 2007.

[9] F. Bøhagen, P. Orten, and G. E. Øien, "Construction and capacity analysis of high-rank line-of-sight MIMO channels," in *Proceedings of the IEEE Wireless Communications and Networking Conference (WCNC '05)*, vol. 1, pp. 432–437, New Orleans, La, USA, March 2005.

[10] P. Larsson, "Lattice array receiver and sender for spatially orthonormal MIMO communication," in *Proceedings of the IEEE 61st Vehicular Technology Conference (VTC '05)*, vol. 1, pp. 192–196, Stockholm, Sweden, May 2005.

[11] F. Bøhagen, P. Orten, and G. E. Øien, "On spherical vs. plane wave modeling of line-of-sight MIMO channels," to appear in *IEEE Transactions on Communications.*

[12] H. Xu, M. J. Gans, N. Amitay, and R. A. Valenzuela, "Experimental verification of MTMR system capacity in controlled propagation environment," *Electronics Letters*, vol. 37, no. 15, pp. 936–937, 2001.

[13] J.-S. Jiang and M. A. Ingram, "Spherical-wave model for short-range MIMO," *IEEE Transactions on Communications*, vol. 53, no. 9, pp. 1534–1541, 2005.

[14] D. Hosli and A. Lapidoth, "How good is an isotropic Gaussian input on a MIMO Ricean channel?" in *Proceedings IEEE International Symposium on Information Theory (ISIT '04)*, p. 291, Chicago, Ill, USA, June-July 2004.

[15] G. L. Stüber, *Principles of Mobile Communication*, Kluwer Academic Publishers, Norwell, Mass, USA, 2nd edition, 2001.

[16] L. Råde and B. Westergren, *Mathematics Handbook for Science and Engineering*, Springer, Berlin, Germany, 5th edition, 2004.

[17] D. Tse and P. Viswanath, *Fundamentals of Wireless Communication*, Cambridge University Press, Cambridge, UK, 1st edition, 2005.

[18] IEEE 802.16-2004, "IEEE standard for local and metropolitan area networks part 16: air interface for fixed broadband wireless access systems," October 2004.

Hindawi Publishing Corporation
EURASIP Journal on Wireless Communications and Networking
Volume 2007, Article ID 30684, 9 pages
doi:10.1155/2007/30684

Research Article

Joint Estimation of Mutual Coupling, Element Factor, and Phase Center in Antenna Arrays

Marc Mowlér,[1] Björn Lindmark,[1] Erik G. Larsson,[1, 2] and Björn Ottersten[1]

[1] *ACCESS Linnaeus Center, School of Electrical Engineering, Royal Institute of Technology (KTH), 100 44 Stockholm, Sweden*
[2] *Department of Electrical Engineering (ISY), Linköping University, 58183 Linköping, Sweden*

Received 17 November 2006; Revised 20 June 2007; Accepted 1 August 2007

Recommended by Robert W. Heath Jr.

A novel method is proposed for estimation of the mutual coupling matrix of an antenna array. The method extends previous work by incorporating an unknown phase center and the element factor (antenna radiation pattern) in the model, and treating them as nuisance parameters during the estimation of coupling. To facilitate this, a parametrization of the element factor based on a truncated Fourier series is proposed. The performance of the proposed estimator is illustrated and compared to other methods using data from simulations and measurements, respectively. The Cramér-Rao bound (CRB) for the estimation problem is derived and used to analyze how the required amount of measurement data increases when introducing additional degrees of freedom in the element factor model. We find that the penalty in SNR is 2.5 dB when introducing a model with two degrees of freedom relative to having zero degrees of freedom. Finally, the tradeoff between the number of degrees of freedom and the accuracy of the estimate is studied. A linear array is treated in more detail and the analysis provides a specific design tradeoff.

1. INTRODUCTION

Adaptive antenna arrays in mobile communication systems promise significantly improved performance [1–3]. However, practical limitations in the antenna arrays, for instance, interelement coupling, are not always considered. The array is commonly assumed ideal which means that the radiation patterns for the individual array elements are modelled as isotropic or omnidirectional with a far-field phase corresponding to the geometrical location. Unfortunately, this is not true in practice which leads to reduced performance as reported by [4–6]. One of the major contributors to the nonideal behavior is the mutual coupling between the antenna elements of the array [7, 8] and the result is a reduced performance [9, 10].

To model the mutual coupling, a matrix representation has been proposed whose inverse may be used to compensate the received data in order to extract the true signal [5]. For basic antenna types and array configurations, the coupling matrix can be obtained from electromagnetic calculations. Alternatively, calibration measurement data can be collected and the coupling matrix may be extracted from the data [11–13]. In [14], compensation with a coupling matrix was found superior to using dummy columns in the case of a 4-column dual polarized array. One difficulty that arises when estimating the coupling matrix from measurements is that other parameters such as the element factor and the phase center of the antenna array need to be estimated. These have been reported to influence the coupling matrix estimate [14].

Our work extends previous work [5, 11–14] by treating the radiation pattern and the array phase center as unknowns during the coupling estimation. We obtain a robust and versatile *joint* method for estimation of the coupling matrix, the element factor, and the phase center. The proposed method does not require the user to provide any a priori knowledge of the location of the array center or about the individual antenna elements.

The performance of the proposed joint estimator is compared via simulations of an 8-element antenna array to a previous method developed by the authors [13]. In addition, we illustrate the estimation performance based on actual measurements on an 8-element antenna array. Furthermore, a CRB analysis is presented which can be used as a performance benchmark. Finally, a tradeoff between the complexity of the model and the performance of the estimator is examined for an 8-element antenna array with element factor $\mathbf{E}_{\text{true}} = \cos(2\theta)$.

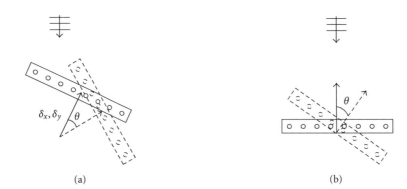

FIGURE 1: Schematic drawing of the antenna array with (a) and without (b) the displacement of the phase center relative to the origin of the coordinate system. Rotating the antenna array (solid) an angle θ gives a relative change (dashed) that depends on the distance to the origin (δ_x, δ_y).

2. DATA MODEL AND PROBLEM FORMULATION

Consider a uniform linear antenna array with M elements having an interelement spacing of d. A narrowband signal, $s(t)$, is emitted by a point source with direction-of-arrival θ relative to the broadside of the receiving array. The data collected by the array with true array response $\tilde{a}(\theta)$ is [5, 14]

$$x = \tilde{\mathbf{a}}(\theta)s(t) + \mathbf{w}, \quad \mathbf{x} \in \mathbb{C}^{M \times 1}. \tag{1}$$

We model $\tilde{a}(\theta)$ as follows:

$$\tilde{\mathbf{a}}(\theta) \simeq \mathbf{C}\mathbf{a}(\theta)e^{jk(\delta_x \cos\theta + \delta_y \sin\theta)} f(\theta), \quad \tilde{\mathbf{a}} \in \mathbb{C}^{M \times 1}, \tag{2}$$

where the following conditions hold.

(i) $\tilde{\mathbf{a}}(\theta)$ is the true radiation pattern in the far-field including mutual coupling, edge effects, and mechanical errors. $\tilde{\mathbf{a}}(\theta)$ is modelled as suggested by (2).

(ii) $\mathbf{C}$ is the $M \times M$ coupling matrix. This matrix is complex-valued and unstructured. The main focus of this paper is to estimate $\mathbf{C}$.

(iii) $\{\delta_x, \delta_y\}$ is the array phase center. Figure 1 illustrates how the displacement of the antenna array is modelled by $\delta_x \hat{x} + \delta_y \hat{y}$ from the origin. The proposed approach also estimates $\{\delta_x, \delta_y\}$. Note that $\{\delta_x, \delta_y\}$ is not included in the models used in [5, 14].

(iv) $f(\theta) \geq 0$ is the element factor (antenna radiation pattern describing the real-valued amplitude of the electric field pattern) in direction θ without mutual coupling and describes the amount of radiation from the individual antenna elements of the array in different directions θ. $f(\theta)$ is real-valued and includes no direction-dependent phase shift [7]. The proposed estimator also estimates $f(\theta)$. Note that $f(\theta)$ was not included in the models used in [5, 14]. All antenna elements are implicitly assumed to have equal radiation patterns when not in an array configuration, that is, $f(\theta)$ is the same for all M elements. When the elements are placed in an array, the radiation patterns of each element are not the same due to the mutual coupling. However, $f(\theta)$ is allowed to be nonisotropic, which is also the case in practice.

(v) $\mathbf{a}(\theta)$ is a Vandermonde vector whose mth element is

$$a_m(\theta) = e^{jkd \sin\theta(m - M/2 - 1/2)}. \tag{3}$$

(vi) d is the distance between the antenna elements.

(vii) k is the wave number.

(viii) w is i.i.d complex Gaussian noise with zero-mean and variance σ^2 per element.

Based on calibration data, when a signal, $s(t) = 1$, is transmitted from N known direction-of-arrivals[1] $\{\theta_1, \dots, \theta_N\}$, we collect one data vector $\mathbf{x}_n$ of measurements for each angle θ_n. The measurements are arranged in a matrix, $\mathbf{X} = [\mathbf{x}_1 \ \cdots \ \mathbf{x}_N]$, as follows:

$$\mathbf{X} = \mathbf{CAD}(\delta_x, \delta_y)\mathbf{E}(f(\theta)) + \mathbf{W},$$
$$\mathbf{A} = [\mathbf{a}(\theta_1) \ \cdots \ \mathbf{a}(\theta_N)],$$
$$\mathbf{E}(f(\theta)) = \text{diag}\{f(\theta_1), \dots, f(\theta_N)\},$$
$$\mathbf{D}(\delta_x, \delta_y) = \text{diag}\{e^{jk(\delta_x \cos\theta_1 + \delta_y \sin\theta_1)}, \dots, e^{jk(\delta_x \cos\theta_N + \delta_y \sin\theta_N)}\}. \tag{4}$$

The matrix $\mathbf{W}$ represents measurement noise, which is assumed to be i.i.d zero-mean complex Gaussian with variance σ^2 per element. In this paper, we propose a method to estimate $\mathbf{C}$ when δ_x, δ_y, and $f(\theta)$ are *unknown*.

One of the novel aspects of the proposed estimator is the inclusion of the element factor, $f(\theta)$, as a jointly unknown parameter during the estimation of the coupling matrix, $\mathbf{C}$. This requires a parametrization that provides a flexible and mathematically appealing representation of an a priori unknown element factor. We have chosen to model $f(\theta)$ as a linear combination of sinusoidal basis functions according to

$$f(\theta) = \sum_{k=1}^{K} \alpha_k \cos(k-1)\theta, \quad |\theta| < \frac{\pi}{2}, \tag{5}$$

---

[1] The orientation of the array is assumed to be perfectly known, while the exact position of the phase center is typically unknown during the antenna calibration.

where K is a known (small) integer, and α are unknown and real-valued constants. Equation (5) is effectively equivalent to a truncated Fourier series where the coefficients are to be estimated. Even though the chosen parametrization may assume negative values, it is introduced to allow the element factor, $\mathbf{E}$, to assume arbitrary shapes that can match the true pattern of the measured antenna array. This can increase the accuracy in the estimate of $\mathbf{C}$ compared to only assuming an omnidirectional element factor that would correspond to $E = I$.

Other alternative parameterizations of $f(\theta)$ exist as well. A piecewise constant function of θ is one example. The proposed parametrization, on the other hand, is particularly attractive since the basis functions are orthogonal and at the same time smooth. Additionally, the unknown coefficients, α_k, enter the model linearly. Based on (5), the element factor can be expressed as

$$
\begin{aligned}
\mathbf{E} &= \sum_{k=1}^{K} \alpha_k \mathbf{Q}_k, \\
\boldsymbol{\alpha} &= \begin{bmatrix} \alpha_1 & \cdots & \alpha_K \end{bmatrix}^T, \\
\mathbf{Q}_k &= \mathrm{diag}\{\cos(k-1)\theta_1, \ldots, \cos(k-1)\theta_N\}.
\end{aligned}
\tag{6}
$$

The coefficients, $\boldsymbol{\alpha}$, will be jointly estimated together with the coupling matrix and phase center by the estimator proposed in this paper.

3. THE PROPOSED ESTIMATOR

We propose to estimate $\mathbf{C}$, $\boldsymbol{\alpha}$, δ_x, and δ_y from $\mathbf{X}$ by using a least-squares criterion[2] on the data model expressed in (4) according to

$$
\min_{\mathbf{C}, \boldsymbol{\alpha}, \delta_x, \delta_y} \left\| \mathbf{X} - \mathbf{C}\mathbf{A}\mathbf{D}(\delta_x, \delta_y)\mathbf{E}(\boldsymbol{\alpha}) \right\|_F^2.
\tag{7}
$$

Under the assumption of Gaussian noise, (7) is the maximum-likelihood estimator. The values of $\mathbf{C}, \boldsymbol{\alpha}, \delta_x, \delta_y$ that minimize the Frobenius norm in (7) are found using an iterative approach. The coupling matrix, the matrix $\mathbf{C}$ is first expressed as if the other parameters were known followed by a minimization over the $\boldsymbol{\alpha}$ parameters while keeping $\mathbf{C}$ and $\{\delta_x, \delta_y\}$ fixed. A second minimization is made over $\{\delta_x, \delta_y\}$ with $\mathbf{C}$ and $\boldsymbol{\alpha}$ treated as constants after which the algorithm loops back to minimize over $\boldsymbol{\alpha}$ again. The steps of the estimator are as follows:

(1) minimize (7) with respect to the coupling matrix, $\mathbf{C}$, while the phase center, $\{\delta_x, \delta_y\}$, and the element factor representation, $\boldsymbol{\alpha}$, are fixed. This is done using the pseudoinverse approach expressed by [4]

$$
\hat{\mathbf{C}} = \mathbf{X}(\mathbf{A}\mathbf{D}\mathbf{E})^H \left[\mathbf{A}\mathbf{D}\mathbf{E}(\mathbf{A}\mathbf{D}\mathbf{E})^H \right]^{-1};
\tag{8}
$$

(2) minimize (7) with respect to $\boldsymbol{\alpha}$ while $\mathbf{C}$, δ_x, and δ_y are fixed. In this step, the value of $\mathbf{C}$ found in the previous step is used as the assumed constant value for the coupling matrix; the minimization over $\boldsymbol{\alpha}$ is then performed using the following manipulations: first, rearrange the measurement matrix, $\mathbf{X}$, and the expression (4) in vectorized form as

$$
\mathbf{x} = \begin{bmatrix} \mathrm{vec}\{\mathrm{Re}\{\mathbf{X}\}\} \\ \mathrm{vec}\{\mathrm{Im}\{\mathbf{X}\}\} \end{bmatrix},
$$

$$
\mathbf{Y} = \begin{bmatrix} \mathrm{vec}\{\mathrm{Re}\{\mathbf{C}\mathbf{A}\mathbf{D}\mathbf{Q}_1\}\} & \cdots & \mathrm{vec}\{\mathrm{Re}\{\mathbf{C}\mathbf{A}\mathbf{D}\mathbf{Q}_K\}\} \\ \mathrm{vec}\{\mathrm{Im}\{\mathbf{C}\mathbf{A}\mathbf{D}\mathbf{Q}_1\}\} & \cdots & \mathrm{vec}\{\mathrm{Im}\{\mathbf{C}\mathbf{A}\mathbf{D}\mathbf{Q}_K\}\} \end{bmatrix};
\tag{9}
$$

the least-squares criterion used is then expressed as

$$
\min_{\boldsymbol{\alpha}} \left\| \mathbf{x} - \mathbf{Y}\boldsymbol{\alpha} \right\|_F^2
\tag{10}
$$

with the solution

$$
\hat{\boldsymbol{\alpha}} = (\mathbf{Y}^T\mathbf{Y})^{-1}\mathbf{Y}^T\mathbf{x},
\tag{11}
$$

which gives the minimizing $\boldsymbol{\alpha}$ parameters;

(3) minimize (7) with respect to δ_x and δ_y while keeping $\mathbf{C}$ and $\boldsymbol{\alpha}$ fixed. Using the $\boldsymbol{\alpha}$ parameters found in the previous step, $\mathbf{C}$, is expressed again according to (8). Assuming the other parameters to be constant, a two-dimensional gradient search is conducted to find the minimizing $\{\delta_x, \delta_y\}$ of (7). Steps 1–3 are iterated until $\|\mathbf{X} - \hat{\mathbf{C}}\mathbf{A}\hat{\mathbf{D}}\hat{\mathbf{E}}\|_F^2$ is within a certain tolerance level.

To provide the algorithm with an initial estimate, in the first iteration, we take $\mathbf{D} = \mathbf{I}$ and $\boldsymbol{\alpha} = [0, 1, 0, \ldots, 0]^T$, which corresponds to the element factor $f(\theta) = \cos\theta$ that was used in [13]. The initialization of the algorithm could of course be done in many different ways and this may affect its performance. The choice considered here can be seen as a refinement of the algorithm previously proposed in [13].

Typically, about 5 iterations of the algorithm are required to reach a local optimum depending on the given tolerance level. Convergence to the global optimum can not be assured; however, the algorithm will converge since the cost function (7) is nonincreasing in each iteration. The topic has been addressed in previous conference papers [13].

4. CRAMÉR-RAO BOUND ANALYSIS

The parameter K, corresponding to the number of terms used in the truncated Fourier series representation of the element factor, is key to the proposed estimator. This value will affect the accuracy of the estimator. Increasing the value will give a better match between the true element factor and the assumed model. At the same time, it will increase the complexity of the model and complicate the estimation. Another aspect is the fact that increasing the value K towards infinity may not be the best way of tuning the algorithm if the true value is much less. This would force the estimator to use a model far more complicated than needed, leading to estimation of additional parameters. Even though the parameters would be close to zero, the performance of the estimator is still affected as indicated by the CRB analysis in this

---

[2] The minimum of this cost function is not unique, see Section 4 for a discussion of this.

section. On the other hand, a smaller number of parameters may give insufficient flexibility to the algorithm and force it into a nonoptimal result.

To compensate for the affected performance associated with a large value of K, the number of measurements N of x_n can be used as well as a higher signal-to-noise ratio (SNR). Let us now study the tradeoff between these three parameters by quantifying how the choice of K relates to N and the SNR ($1/\sigma^2$). The Cramér-Rao bound (CRB) for the estimation problem in (7) is derived and will provide a lower bound on the variance of the unknown parameters [15–17]. All the elements of $\mathbf{C}$ are considered to be unknown complex-valued parameters and therefore separated with respect to real and imaginary parts with C_{ij}^R denoting the real part of the element in the ith row and jth column, and C_{ij}^I denoting the corresponding imaginary part. All the $2M^2 + 2 + K$ real-valued unknown parameters of (4) are collected into the vector

$$\boldsymbol{\xi} = \begin{bmatrix} C_{11}^R \cdots C_{1M}^R \cdots C_{MM}^R \cdots C_{ij}^I \ \delta_x \ \delta_y \ \alpha_1 \cdots \alpha_K \end{bmatrix}^T.$$
$$(12)$$

The CRB for the estimate of $\boldsymbol{\xi}$ is expressed by [15]

$$[\mathrm{CRB}] = [\mathbf{I}_{\mathrm{Fisher}}]^{-1}, \tag{13}$$

where $\mathbf{I}_{\mathrm{Fisher}}$ is the Fisher information matrix given by

$$[\mathbf{I}_{\mathrm{Fisher}}]_{ij} = \frac{2}{\sigma^2} \mathrm{Re} \left\{ \frac{\partial \boldsymbol{\mu}^H}{\partial \xi_i} \frac{\partial \boldsymbol{\mu}}{\partial \xi_j} \right\}, \tag{14}$$

where $\boldsymbol{\mu}$ is the expected value of $\mathbf{x}$ in (9). The problem (7) is unidentifiable as presented. The scaling ambiguity present between $\mathbf{C}$ and $\boldsymbol{\alpha}$ is solved by enforcing a constraint on the problem. Here, we choose to constrain $\|\mathbf{C}\|_F^2 = M$ as opposed to other possibilities such as $\alpha_1 = 1$ or $C_{11}^R = 1$. The latter favors particular elements of $\mathbf{C}$ or $\boldsymbol{\alpha}$ by forcing these to be nonzero. This is undesirable and consequently ruled out. This constraint was also implemented as a normalization in the estimator presented in the previous section, while not mentioned there explicitly.

The CRB under parametric constraints is found by following the results derived in [18]. The constraint is expressed as

$$g(\boldsymbol{\xi}) = \|\mathbf{C}\|_F^2 - M = \mathbf{Tr}\{\mathbf{C}^H\mathbf{C}\} - M = 0. \tag{15}$$

Defining

$$[\mathbf{G}(\boldsymbol{\xi})]_l = \frac{\partial g(\boldsymbol{\xi})}{\partial \xi_l} = \begin{cases} 2\mathrm{Re}\{c_{ij}\} & \text{if } \xi_l = \mathrm{Re}\{c_{ij}\}, \\ 2\mathrm{Im}\{c_{ij}\} & \text{if } \xi_l = \mathrm{Im}\{c_{ij}\}, \\ 0 & \text{otherwise}, \end{cases} \tag{16}$$

the constrained CRB is given by [18]

$$[\mathrm{CRB}] = \mathbf{U}(\mathbf{U}^T\mathbf{I}_{\mathrm{Fisher}}\mathbf{U})^{-1}\mathbf{U}^T, \tag{17}$$

where $\mathbf{U}$ is implicitly defined via $\mathbf{G}(\boldsymbol{\xi})\mathbf{U} = 0$. Note that the CRB is proportional to σ^2, that is, the estimation accuracy is inversely proportional to the SNR.

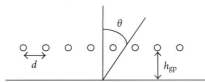

FIGURE 2: Schematic overview of the antenna array consisting of dipoles over an infinite ground plane. The distance from the ground plane is h_{gp} and the spacing between the elements is d. The angle from the normal to the antenna axis is denoted by θ.

5. RESULTS

5.1. Dipole array example

First, let us consider the case of an 8-element dipole array over a ground plane, see Figure 2. In the absence of mutual coupling, each element has the true radiation pattern

$$E_{\mathrm{true}} = \sin(kh_{gp}\cos\theta), \tag{18}$$

where k is the wave number and h_{gp} is the distance to the ground plane. Using known expressions for mutual coupling between dipoles [7], it is straightforward to calculate the embedded element patterns. Figures 3(a) and 3(b) show the radiation pattern for a single element and an 8-element array with mutual coupling for $h_{gp} = 3\lambda/8$. The interelement spacing is $d = \lambda/2$ and the true $\{\delta_x, \delta_y\}$ are $\{0,0\}$.

We now compare our proposed method using a truncated Fourier series expansion of the element factor, see (5), to results obtained by using a cosine-shaped ($E = \cos^n\theta$) element pattern assumption as in [13]. In our proposed method, the algorithm described in Section 3 is used to estimate $\mathbf{C}$, $\{\alpha_1, \alpha_2, \alpha_3\}$, and $\{\delta_x, \delta_y\}$, which corresponds to choosing $K = 3$ in the model for the element factor as expressed in (5). Similarly, the parameter n in the expression $E = \cos^n\theta$ is optimized as part of the procedure when using the method of [13]. In both cases, we make a starting assumption that $\{\delta_x, \delta_y\} = \{0,0\}$, which is also the true value for this case.

With *no* compensation, the uncompensated radiation pattern is represented in Figure 3(b). *With* compensation, where a cosine-shaped element pattern ($E = \cos^n\theta$) is assumed, the resulting radiation pattern is displayed in Figure 3(c). The unknown values in addition to the mutual coupling matrix were estimated to $\hat{n} = 0.5$ and $\{\hat{\delta}_x, \hat{\delta}_y\} = \{0,0\}$, and the performance is significantly improved over the uncompensated case. By allowing the model for the element factor to follow a truncated Fourier series ($K = 3$) according to our method, the performance of the mutual coupling matrix estimate is improved even more resulting in the graph of Figure 3(d). For this case, our method estimated the values for the auxiliary unknown parameters to be $\hat{\boldsymbol{\alpha}} = [-1.1 \ 3.1 \ -1.1]^T$ and $\{\hat{\delta}_x, \hat{\delta}_y\} = \{0,0\}$. The result also agrees well with the true (ideal) radiation pattern when the mutual coupling is neglected as presented in Figure 3(a).

The phase error for the cases of uncompensated and compensated phase diagrams are presented in Figure 4. The top graph represents the uncompensated case, while the

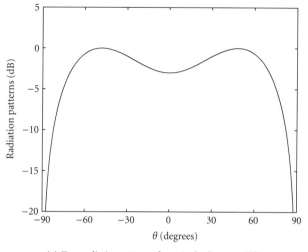

(a) *True* radiation patterns for a single element; $f(\theta)$

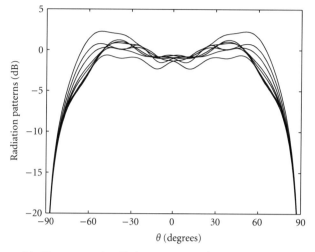

(b) *Uncompensated* radiation patterns with mutual coupling; $\mathbf{x}_1 \cdots \mathbf{x}_M$

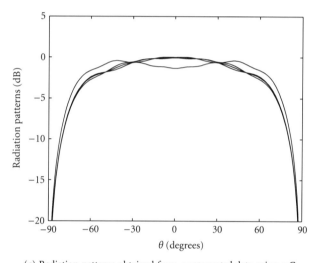

(c) Radiation patterns obtained from *compensated* data using a $\mathbf{C}$ matrix estimated via $\mathbf{E} = \cos^n\theta$ as the parametrization of the element factor [13]; $\mathbf{C}_{\cos}^{-1}\mathbf{x}_1 \cdots \mathbf{C}_{\cos}^{-1}\mathbf{x}_M$

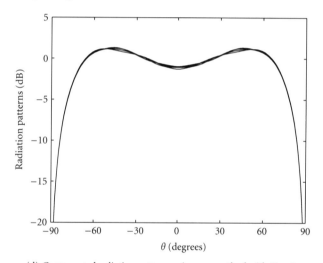

(d) *Compensated* radiation patterns using our method with $K = 3$; $\mathbf{C}^{-1}\mathbf{x}_1 \cdots \mathbf{C}^{-1}\mathbf{x}_M$

FIGURE 3: Compensation for mutual coupling for an array of 8 dipoles placed $h_{gp} = 3\lambda/8$ over a ground plane.

middle and bottom graphs represent compensated cases with a cosine-shaped and the truncated Fourier series parameterizations of the element factors, respectively. This also verifies that our method with $K = 3$ improves the result over constraining the element factor to be cosine-shaped ($E = \cos^n\theta$) as was proposed in previous work [13].

5.2. CRB for dipole array example

The CRB as a function of K (i.e., the order of the parametrization of $f(\theta)$) for an 8-element array of dipoles over ground plane is presented in Figure 5. When generating this figure, we used the theoretical models of [7] for the coupling between antenna elements with inter-element spacing $d = \lambda/2$. The true data used are based on a scenario when $K = 1$, $\mathbf{D} = \mathbf{I}$, and $\mathbf{E} = \mathbf{I}$. The SNR was 0 dB ($\sigma^2 = 1$). The four curves show

(i) $\text{CRB}_C = \text{Tr}\{\{\mathbf{U}_C(\mathbf{U}_C^T\mathbf{I}_C\mathbf{U}_C)^{-1}\mathbf{U}_C^T\}_C\}$, the CRB of $\mathbf{C}$ when both $\mathbf{D}$ and $\mathbf{E}$ are known;

(ii) $\text{CRB}_{CD} = \text{Tr}\{\{\mathbf{U}_{CD}(\mathbf{U}_{CD}^T\mathbf{I}_{CD}\mathbf{U}_{CD})^{-1}\mathbf{U}_{CD}^T\}_C\}$, the CRB of $\mathbf{C}$ when $\mathbf{D}$ is unknown and $\mathbf{E}$ is known;

(iii) $\text{CRB}_{CE} = \text{Tr}\{\{\mathbf{U}_{CE}(\mathbf{U}_{CE}^T\mathbf{I}_{CE}\mathbf{U}_{CE})^{-1}\mathbf{U}_{CE}^T\}_C\}$, the CRB of $\mathbf{C}$ when $\mathbf{D}$ is known and $\mathbf{E}$ is unknown;

(iv) $\text{CRB}_{CDE} = \text{Tr}\{\{\mathbf{U}_{CDE}(\mathbf{U}_{CDE}^T\mathbf{I}_{CDE}\mathbf{U}_{CDE})^{-1}\mathbf{U}_{CDE}^T\}_C\}$, the CRB of $\mathbf{C}$ when $\mathbf{D}$ and $\mathbf{E}$ are unknown.

The results in Figure 5 quantify the increase in achievable estimation performance for the elements of $\mathbf{C}$ when increasing the number of nuisance parameters in the model. In particular, we see that the estimation problem becomes more difficult when more $\boldsymbol{\alpha}$ parameters are introduced in the model. For example, it is more difficult to estimate the coupling matrix $\mathbf{C}$ when the phase center δ_x, δ_y, is unknown. However,

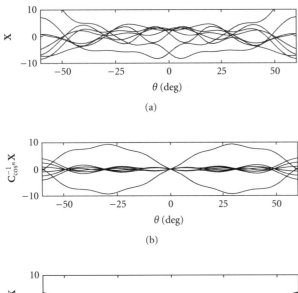

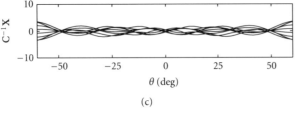

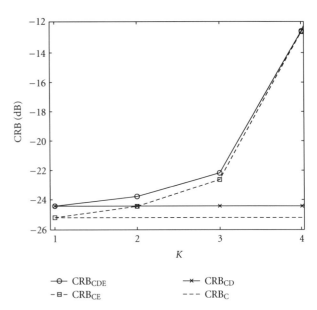

FIGURE 5: The CRB for the elements of $\mathbf{C}$ under different assumptions on whether $\mathbf{D}$, $\mathbf{E}$ are known or not, and for different K. SNR is 0 dB.

FIGURE 4: Uncompensated (a) and compensated (b, c) phase diagrams with $\mathbf{E} = \cos^n\theta$ and $\mathbf{E} = \sum \alpha_k \mathbf{Q}_k$, respectively. The true element factor is $\mathbf{E}_{\text{true}} = \sin(2\pi(3/8)\cos\theta)$, which corresponds to $h_{gp} = 3\lambda/8$.

for $K \geq 3$, the difficulty of identifying $\boldsymbol{\alpha}$ dominates over the problem of estimating the phase center. Since the CRB is proportional to σ^2, the increase in emitter power (i.e., SNR) required to maintain a given performance when the model is increased with more unknowns can be seen in the figure.

In Figure 6, the CRB as a function of K and N is studied. From this figure, we can directly read out how much higher emitter power (or equivalently, lower σ^2) is required to be able to maintain the same estimation performance for $\mathbf{C}$ when K or N vary. For example, if fixing $N = 100$, say, then going from $K = 1$ to $K = 2$ requires 1 dB additional SNR. Going from $K = 2$ to $K = 3$ requires an increase of the SNR level by 1.5 dB. However, going from $K = 3$ to 4 requires 10 dB extra SNR. Thus, $K = 3$ appears to be a reasonable choice. In practice, it is difficult to handle more than $K = 3$ for the element factor in this case.

In Figure 6, we observe that in the limit when the number of angles reaches $N = 15$, the problem becomes unidentifiable. This is so because for $N \leq 15$ the number of unknown parameters in the model exceeds the number of recorded samples. From Figure 6 many other interesting observations can be made. For instance, increasing N from 100 to 200, for a fixed K, is approximately equivalent to increasing the SNR with 3 dB. This holds in general: for $N \gg 1$, doubling the SNR gives the same effect as doubling N.

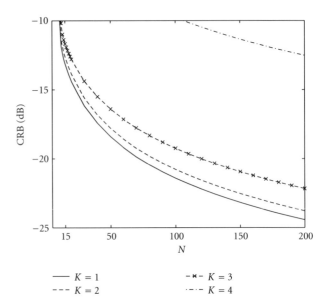

FIGURE 6: The CRB for the elements of $\mathbf{C}$ as a function of K and N when SNR is 0 dB.

5.3. Measured results on a dual polarized array

Next, let us study the performance of the proposed estimator on an actual antenna array. Data from an 8-column antenna array (see Figure 7) were collected at 1900 MHz during calibration measurements with 180 measurement points distributed evenly over the interval $\theta \in \{-90° \cdots 90°\}$. Uncompensated radiation patterns with mutual coupling are presented in Figure 8 for measured data of the array. The estimator presented in this paper was used to estimate the coupling matrix. The estimated coupling matrix was then used

FIGURE 7: Eight-column dual polarized array developed by Powerwave Technologies Inc. Results are presented for this array in Section 5.3. Note that each vertical column of radiators forms an element with pattern $\tilde{a}_m(\theta)$ in the horizontal plane.

to precompensate the data, after which radiation patterns can be obtained.

To modify our estimator (Section 3) to the dual polarized case, we follow [14]. Considering a dual polarized array with $\pm 45°$ polarization and neglecting noise, (4) becomes

$$\begin{bmatrix} \mathbf{x}_{co}^{-45°} & \mathbf{x}_{xp}^{-45°} \\ \mathbf{x}_{xp}^{+45°} & \mathbf{x}_{co}^{+45°} \end{bmatrix} = \begin{bmatrix} \mathbf{C}_{11}\,\mathbf{ADE}_1 & \mathbf{C}_{12}\mathbf{ADE}_2 \\ \mathbf{C}_{21}\,\mathbf{ADE}_2 & \mathbf{C}_{22}\,\mathbf{ADE}_1 \end{bmatrix}, \quad (19)$$

where $\mathbf{x}_{co}^{-45°}$ means measuring the incoming $-45°$ polarized signal with the antenna elements of the same polarization, while $\mathbf{x}_{xp}^{-45°}$ means the data measured at the $-45°$ element when the incoming signal is $+45°$. To estimate the total coupling matrix,

$$\mathbf{C}_{\text{tot}} = \begin{bmatrix} \mathbf{C}_{11} & \mathbf{C}_{12} \\ \mathbf{C}_{21} & \mathbf{C}_{22} \end{bmatrix}, \quad (20)$$

the four blocks of (19) are treated independently according to

$$\mathbf{x}_{co}^{-45°} = \mathbf{C}_{11}\mathbf{ADE}_1, \qquad \mathbf{x}_{xp}^{-45°} = \mathbf{C}_{12}\mathbf{ADE}_2,$$
$$\mathbf{x}_{xp}^{+45°} = \mathbf{C}_{21}\mathbf{ADE}_2, \qquad \mathbf{x}_{co}^{+45°} = \mathbf{C}_{22}\mathbf{ADE}_1. \quad (21)$$

The $\mathbf{D}$ matrices of these four independent equations are equal. The E_1 matrix represents the copolarization blocks of (19), namely, $\mathbf{x}_{co}^{-45°}$ and $\mathbf{x}_{co}^{+45°}$, simultaneously and is modelled according to (5) with a set of α parameters estimated by our method. For the cross-polarization, we assume isotropic element patterns, $\mathbf{E}_2 = \mathbf{I}$. Once the equations in (21) are solved, the total coupling matrix can be expressed using (20) and the radiation pattern of the measured data may be compensated by inverting the coupling matrix according to

$$\begin{bmatrix} \mathbf{x}_{\text{compensated}}^{-45°} \\ \mathbf{x}_{\text{compensated}}^{+45°} \end{bmatrix} = \mathbf{C}_{\text{tot}}^{-1} \begin{bmatrix} \mathbf{x}_{\text{measured}}^{-45°} \\ \mathbf{x}_{\text{measured}}^{+45°} \end{bmatrix}. \quad (22)$$

Figure 8(a) shows the individual radiation patterns of each antenna element as measured during calibration ($\mathbf{x}$). Figure 8(b) shows the radiation patterns after compensation by the coupling matrix ($C_{\text{iso}}^{-1}X$) when isotropic conditions are

assumed by the estimator. This means assuming $\mathbf{E} = \mathbf{I}$, which is equivalent to setting $K = 1$ in our algorithm. The radiation patterns after compensation by the coupling matrix when using the proposed estimator with $K = 3$ are presented in Figure 8(c).

The results using an isotropic assumption on the element factor, Figure 8(b), shows an improvement over the uncompensated data of Figure 8(a). The graphs showing the copolarization (solid) are smoother and more equal to each other, which is what is expected from an array with equal elements when no coupling is present. The cross-polarization (dashed) is suppressed significantly compared to the uncompensated case. The estimated phase center is $\{\hat{\delta}_x, \hat{\delta}_y\} = [0.2\ 0]$.

Further improvement is achieved using our proposed method, Figure 8(c). Using $K = 3$, our method estimates the coefficients in the element factor representation (5) as $\hat{\alpha} = [0.8\ -0.4\ 0.6]$ and the phase center as $\{\hat{\delta}_x, \hat{\delta}_y\} = [0.3\ 0]$. The resulting compensated radiation pattern of Figure 8(c) is even closer to the ideal array response when no coupling is assumed. The copolarization graphs are almost overlapping in the $\pm 60°$ interval showing the radiation patterns of 8 equal elements with cosine-like element factors. The cross-polarization is also improved slightly over the isotropic case.

Phase diagrams representing the average phase error of the coupling matrix before and after compensation with the coupling matrix are presented in Figure 8(d). The phase error after compensation with $K = 3$ (Figure 8(d), bottom), modelling the phase shift and the element factor, is less than without the compensation (Figure 8(d), top). Assuming an isotropic element factor (Figure 8(d), middle) gives a better result than without compensation but not as good as the result of our method. This indicates that the validity of the estimated coupling matrix based on phase considerations increases with the proposed estimator. Furthermore, the results of our method with $K = 3$ show a notable improvement over the results presented in [14].

6. K-VALUE TRADEOFF BASED ON MONTE CARLO SIMULATIONS

We have seen in Section 5.2 that there is a tradeoff between K, N, and the SNR in terms of the CRB. In reality, the error in the estimation is a combination of both model- and noise-induced errors. Let us therefore study the overall performance of our proposed estimator using the root-mean-square error of the mutual coupling matrix:

$$\text{RMS} = \frac{1}{M^2}\|\mathbf{C} - \hat{\mathbf{C}}\|_F = \frac{1}{M^2}\sqrt{\sum_{ij}|C_{ij} - \hat{C}_{ij}|^2}, \quad (23)$$

where M^2 is the number of elements in $\mathbf{C}$. As an example, we use an 8-element linear array with spacing $d = \lambda/2$ and a true element factor given by $\alpha_{\text{true}} = [0\ 0\ 1\ 0\ \dots]^T$. Simulations were conducted with our method for $K = 1 \dots 5$ based on 1000 realizations and an SNR = 30 dB. The result is shown in Figure 9. The CRB for the given SNR level is also

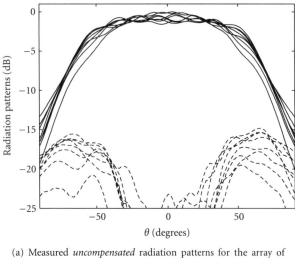

(a) Measured *uncompensated* radiation patterns for the array of Figure 7; $\{\mathbf{x}_{\text{measured}}^{-45°}, \mathbf{x}_{\text{measured}}^{+45°}\}^{T}$

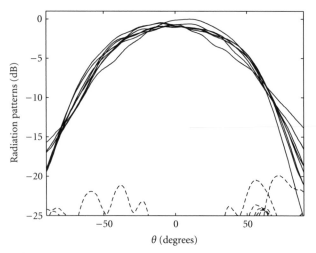

(b) Radiation patterns obtained from *compensated* data using a $\mathbf{C}$ matrix estimated via our algorithm setting $K = 1$ (i.e., forcing $\mathbf{E} \propto \mathbf{I}$); $\mathbf{C}_{\text{iso}}^{-1}\{\mathbf{x}_{\text{measured}}^{-45°}, \mathbf{x}_{\text{measured}}^{+45°}\}^{T}$

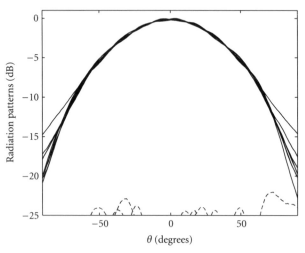

(c) *Compensated* radiation patterns using our method with $K = 3$; $\mathbf{C}^{-1}\{\mathbf{x}_{\text{measured}}^{-45°}, \mathbf{x}_{\text{measured}}^{+45°}\}^{T}$

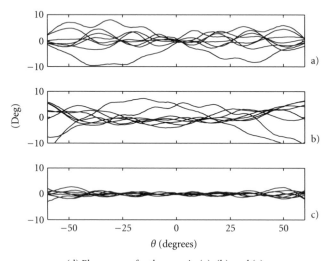

(d) Phase errors for the cases in (a), (b), and (c)

FIGURE 8: Radiation patterns obtained from compensated data with the coupling matrix $\mathbf{C}$ estimated in different ways. The measurements are from the 8-element dual polarized array in Figure 7 with co- (solid) and cross- (dashed) polarization collected during calibration.

presented in the same graph and represents the impact of the noise. This is seen as an increase of the CRB in the region $K > 3$.

Because of the insufficient parametrization of $\mathbf{E}$, the RMS is higher for smaller values of K than the true K. The RMS decreases toward the point where $K = 3$, which is the true value of K. For higher values of K, there is no longer a model error and the noise is the sole contributor to the RMS. This is evident as an increase in RMS when $K > 3$. The same estimation was also made with $\mathbf{E} = \cos^{n}\theta$ [13]. A straight line represents this case in Figure 9 showing the difference in RMS, which is higher compared to using our method with $K = 3$. This shows that the optimum tradeoff for our proposed method, in this case, is $K = 3$. This gives the best

performance when comparing different values of K and the $\mathbf{E} = \cos^{n}\theta$ assumption.

7. CONCLUSIONS

We have introduced a new method for the estimation of the mutual coupling matrix of an antenna array. The main novelty over existing methods was that the array phase center and the element factors were introduced as unknowns in the data model, and treated as nuisance parameters in the estimation of the coupling as well, by being *jointly* estimated together with the coupling matrix.

In a simulated test case, our method outperformed the previously proposed estimator [13] for the case of an

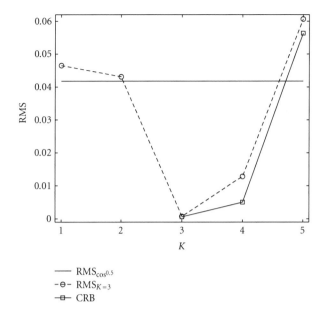

FIGURE 9: CRB and RMS based on Monte Carlo simulations when SNR is 30 dB. The true element factor is $\mathbf{E}_{\text{true}} = \cos 2\theta$, which means $\boldsymbol{\alpha}_{\text{true}} = [0\ 0\ 1\ 0\ 0]$ or $K_{\text{true}} = 3$.

8-element dipole array. The radiation pattern and phase error were significantly improved leading to increased accuracy in any following postprocessing.

Based on a CRB analysis, the SNR penalty associated with introducing a model for the element factor with two degrees of freedom ($K = 3$) was 2.5 dB relative to having zero degrees of freedom. This means that an additional 2.5 dB more power (or a doubling of the number of accumulated samples) must be used to retain the estimation accuracy of the coupling matrix compared to the case when the algorithm assumed omnidirectional elements. To add another degree of freedom (set $K = 4$) costs another 10 dB.

Using measured calibration data from a dual polarized array, we found that the proposed method and the associated estimator could significantly improve the quality of the estimated coupling matrix, and the result of subsequent compensation processing.

Finally, the tradeoff between the complexity of the proposed data model and the accuracy of the estimator was studied via Monte Carlo simulations. For the case of an 8-element linear array, the optimum was found to be at $K = 3$ which coincides with the true value in that case.

ACKNOWLEDGMENTS

This material was presented in part at ICASSP 2007 [19]. Erik G. Larsson is a Royal Swedish Academy of Sciences Research Fellow supported by a grant from the Knut and Alice Wallenberg Foundation.

REFERENCES

[1] D. Tse and P. Viswanath, *Fundamentals of Wireless Communication*, Cambridge University Press, Cambridge, UK, 2005.

[2] A. Swindlehurst and T. Kailath, "A performance analysis of subspace-based methods in the presence of model errors—part I: the MUSIC algorithm," *IEEE Transactions on Signal Processing*, vol. 40, no. 7, pp. 1758–1774, 1992.

[3] M. Jansson, A. Swindlehurst, and B. Ottersten, "Weighted subspace fitting for general array error models," *IEEE Transactions on Signal Processing*, vol. 46, no. 9, pp. 2484–2498, 1998.

[4] B. Friedlander and A. J. Weiss, "Effects of model errors on waveform estimation using the MUSIC algorithm," *IEEE Transactions on Signal Processing*, vol. 42, no. 1, pp. 147–155, 1994.

[5] H. Steyskal and J. S. Herd, "Mutual coupling compensation in small array antennas," *IEEE Transactions on Antennas and Propagation*, vol. 38, no. 12, pp. 1971–1975, 1990.

[6] J. Yang and A. L. Swindlehurst, "The effects of array calibration errors on DF-based signal copy performance," *IEEE Transactions on Signal Processing*, vol. 43, no. 11, pp. 2724–2732, 1995.

[7] C. A. Balanis, *Antenna Theory: Analysis and Design*, John Wiley & Sons, New York, NY, USA, 1997.

[8] T. Svantesson, "The effects of mutual coupling using a linear array of thin dipoles of finite length," in *Proceedings of the 9th IEEE SP Workshop on Statistical Signal and Array Processing (SSAP '98)*, pp. 232–235, Portland, Ore, USA, September 1998.

[9] K. R. Dandekar, H. Ling, and G. Xu, "Effect of mutual coupling on direction finding in smart antenna applications," *Electronics Letters*, vol. 36, no. 22, pp. 1889–1891, 2000.

[10] B. Friedlander and A. Weiss, "Direction finding in the presence of mutual coupling," *IEEE Transactions on Antennas and Propagation*, vol. 39, no. 3, pp. 273–284, 1991.

[11] T. Su, K. Dandekar, and H. Ling, "Simulation of mutual coupling effect in circular arrays for direction-finding applications," *Microwave and Optical Technology Letters*, vol. 26, no. 5, pp. 331–336, 2000.

[12] B. Lindmark, S. Lundgren, J. Sanford, and C. Beckman, "Dual-polarized array for signal-processing applications in wireless communications," *IEEE Transactions on Antennas and Propagation*, vol. 46, no. 6, pp. 758–763, 1998.

[13] M. Mowlér and B. Lindmark, "Estimation of coupling, element factor, and phase center of antenna arrays," in *Proceedings of IEEE Antennas and Propagation Society International Symposium*, vol. 4B, pp. 6–9, Washington, DC, USA, July 2005.

[14] B. Lindmark, "Comparison of mutual coupling compensation to dummy columns in adaptive antenna systems," *IEEE Transactions on Antennas and Propagation*, vol. 53, no. 4, pp. 1332–1336, 2005.

[15] S. M. Kay, *Fundamentals of Statistical Signal Processing: Estimation Theory*, Prentice-Hall, Upper Saddle River, NJ, USA, 1993.

[16] H. L. Van Trees, *Detection, Estimation, and Modulation Theory*, Wiley-Interscience, New York, NY, USA, 2007.

[17] J. Gorman and A. Hero, "Lower bounds for parametric estimation with constraints," *IEEE Transactions on Information Theory*, vol. 36, no. 6, pp. 1285–1301, 1990.

[18] P. Stoica and B. C. Ng, "On the Crameér-Rao bound under parametric constraints," *IEEE Signal Processing Letters*, vol. 5, no. 7, pp. 177–179, 1998.

[19] M. Mowlér, E. G. Larsson, B. Lindmark, and B. Ottersten, "Methods and bounds for antenna array coupling matrix estimation," in *Proceedings of IEEE International Conference on Acoustics, Speech, and Signal Processing (ICASSP '07)*, vol. 2, pp. 881–884, Honolulu, Hawaii, USA, April 2007.

Hindawi Publishing Corporation
EURASIP Journal on Wireless Communications and Networking
Volume 2007, Article ID 37574, 9 pages
doi:10.1155/2007/37574

Research Article

Diversity Characterization of Optimized Two-Antenna Systems for UMTS Handsets

A. Diallo,[1] P. Le Thuc,[1] C. Luxey,[1] R. Staraj,[1] G. Kossiavas,[1] M. Franzén,[2] and P.-S. Kildal[3]

[1] *Laboratoire d'Electronique, Antennes et Télécommunications (LEAT), Université de Nice Sophia-Antipolis,*
  CNRS UMR 6071, 250 rue Albert Einstein, Bât. 4, Les Lucioles 1, 06560 Valbonne, France
[2] *Bluetest AB, Gotaverksgatan 1, 41755 Gothenburg, Sweden*
[3] *Department of Signals and Systems, Chalmers University of Technology, 41296 Gothenburg, Sweden*

Received 16 November 2006; Revised 20 June 2007; Accepted 22 November 2007

Recommended by A. Alexiou

This paper presents the evaluation of the diversity performance of several two-antenna systems for UMTS terminals. All the measurements are done in a reverberation chamber and in a Wheeler cap setup. First, a two-antenna system having poor isolation between its radiators is measured. Then, the performance of this structure is compared with two optimized structures having high isolation and high total efficiency, thanks to the implementation of a neutralization technique between the radiating elements. The key diversity parameters of all these systems are discussed, that is, the total efficiency of the antenna, the envelope correlation coefficient, the diversity gains, the mean effective gain (MEG), and the MEG ratio. The comparison of all these results is especially showing the benefit brought back by the neutralization technique.

1. INTRODUCTION

Nowadays, wireless mobile communications are growing exponentially in several fields of telecommunications. The new generation of mobile phones must be able to transfer large amounts of data and consequently increasing the transfer rate of these data is clearly needed. One solution is to implement a diversity scheme at the terminal side of the communication link. This can be done by multiplying the number of the radiating elements of the handset. In addition, these radiators must be highly isolated to achieve the best diversity performance. Also, the antenna engineers must take into account the radiator's environment of the handset to design suitable multiantenna systems. In practice, the terminal can be considered to operate in a so-called multipath propagation environment: the electromagnetic field will take many simultaneous paths between the transmitter and the receiver. In such a configuration, total efficiency, diversity gain, mean effective gain (MEG), and MEG ratio are the most important parameters for diversity purposes.

Only few papers are actually focusing on the design of a specific technique to address the isolation problem of several planar inverted-F antennas (PIFAs) placed on the same finite-sized ground plane and operating in the same frequency bands. In [1, 2], the authors are evaluating the isolation between identical PIFAs when moving them all along a mobile phone PCB for multiple-input multiple-output (MIMO) applications. The same kind of work is done in [3–6] for different antenna types. The best isolation values are always found when the antennas are spaced by the largest available distance on the PCB, that is, one at the top edge and the other at the bottom. Excellent studies can be found in [7–16], but no specific technique to isolate the elements is described in these papers. One solution is reported in [17], however, for two thin PIFAs for mobile phones operating in different frequency bands (GSM900 and DCS1800). It consists in inserting high-Q-value lumped LC components at the feeding point of one antenna to achieve a blocking filter at the resonant frequency of the other. This solution gives significant results in terms of decoupling but strongly reduces the frequency bandwidth. Another very interesting solution reported in [18, 19] consists in isolating the antennas by a decoupling network, at their feeding ports, this solution suffers from the fact that in small handsets available space is restricted. Finally, a promising solution is described in [20], but in this work the PIFAs are operating around 5 GHz.

Some authors of the current paper have already designed and fabricated several multiantenna structures for mobile

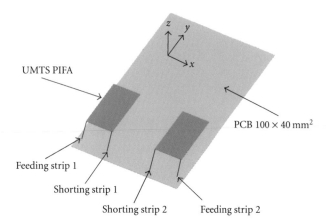

FIGURE 1: 3D view of the initial two-antenna system.

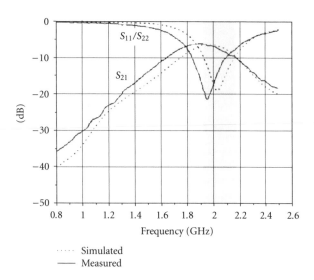

FIGURE 2: Simulated and measured S-parameters of the initial two-antenna system.

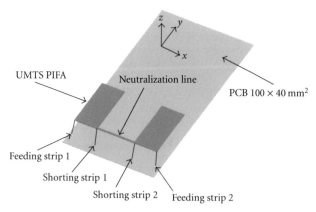

FIGURE 3: 3D view of the two-antenna system with a suspended line between the PIFA shorting strips.

phone applications. In [21], the isolation problem has been addressed for closely spaced PIFAs operating in very close frequency bands with the help of a neutralization technique. Recently, several two-antenna systems operating in the UMTS band (1920–2170 MHz) and especially including neutralization line to achieve high isolation between the feeding ports of their radiating parts have been designed for diversity and MIMO applications [22]. Two prototypes have already been characterized in terms of scattering parameters, total efficiency, and envelope correlation coefficient. The obtained results show that these structures have a strong potential for an efficient implementation of a diversity scheme at the mobile terminal side of a wireless link. However, to completely characterize these prototypes, some particular facilities and the associated expertise are needed [23]. The antenna group of Chalmers Institute of Technology possesses these capabilities through the Bluetest reverberation chamber [24].

This paper is the result of a short-term mission granted by the COST 284. The antenna-design competencies of the LEAT have been combined with the reverberation chamber measurement skills of the antenna group of Chalmers Institute of Technology. Several prototypes have been measured at Chalmers in terms of total efficiency, diversity gain, envelope correlation coefficient, and mean effective gain. Efficiency results are compared with the same measurements obtained through a homemade Wheeler Cap at the LEAT. The envelope correlation coefficient, the MEG, and the MEG ratio calculated from simulated values are also presented and compared [23, 25–27]. We focus on the comparison of the performance of an initial two-antenna system with two different neutralized structures and especially the benefit brought back by the neutralization technique.

2. DESIGNED STRUCTURES AND S-PARAMETER MEASUREMENTS

The multiantenna systems were designed using the electromagnetic software tool IE3D [28]. The initial two-antenna system is presented in Figure 1 (the design procedure was already described in [22]). It consists of two PIFAs symmetrically placed on a 40×100 mm^2 PCB and separated by $0.12\lambda_0$

(18 mm at 2 GHz). They are fed by a metallic strip soldered to an SMA connector and shorted to the PCB by an identical strip. Each PIFA is optimized to cover the UMTS band (1920–2170 MHz) with a return loss goal of −6 dB. The optimized dimensions are of 26.5 mm length and of 8 mm width. A prototype was fabricated using a 0.3-mm-thick nickel silver material (conductivity $\sigma = 4 \times 10^6$ S/m). In Figure 2, we present the simulated and the measured S-parameters of the structure. The absolute value S_{21} reaches a maximum of −5 dB in the middle of the UMTS band.

In the first attempt to improve the isolation between the radiating elements, a suspended line as a neutralization device was inserted between the shorting strips of the two PIFAs (see Figure 3). The optimization of this line was already explained in [21]. Figure 4 shows the S-parameters of this new structure. We can see a good matching and a strong improvement of the isolation in the bandwidth of interest: the measured S_{21} parameter always remains below −15 dB. However, a different isolation can be obtained if we implement the same neutralization technique between the two feeding strips

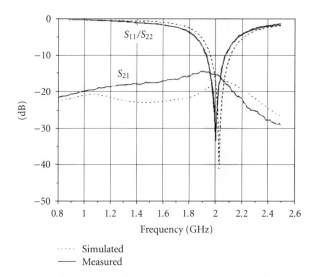

···· Simulated
—— Measured

FIGURE 4: Simulated and measured S-parameters of the two-antenna system with a line between the PIFA shorting strips.

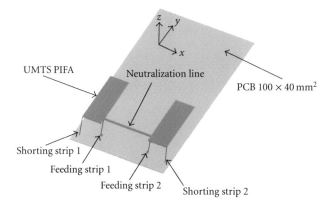

FIGURE 5: 3D view of the two-antenna system with a suspended line between the PIFA feeding strips.

of the PIFAs (see Figure 5). We can observe in Figure 6 that a deep null is now achieved in the middle of the UMTS band. Moreover, the measured S_{21} always remains below -18 dB in the whole UMTS band. All these values seem to be very satisfactory for diversity purposes.

3. COMPARISON OF THE DIVERSITY PERFORMANCE

3.1. Total efficiency

Traditionally, the radiation performance of an antenna is measured outdoors or in an anechoic chamber. In order to obtain the total efficiency, we need to measure the radiation pattern in all directions in space and integrate the received power density to find the total radiated power. This gives the total efficiency when compared to the corresponding radiated power of a known reference antenna. This final result is obtained after a long measurement procedure. This parameter can be measured very much faster and easier in a reverberation chamber. However, it is necessary to measure a reference case (a dipole antenna having an efficiency

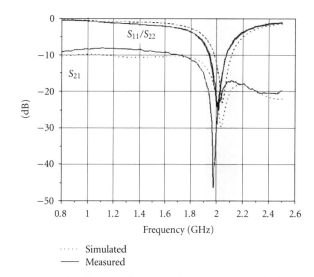

···· Simulated
—— Measured

FIGURE 6: Simulated and measured S-parameters of the two-antenna system with the line between the PIFA feeding strips.

of 96% in our case) and then the antenna system under test (AUT). It is also important that the chamber is loaded in exactly the same way for these both measurements. For the reference case, the transmission between the reference antenna and the excitation antennas is measured in the chamber with the reference antenna in free space that means at least half a wavelength away from any lossy objects and/or the metallic walls of the chamber. As soon as the reference case is completed, we can measure the AUT. From both measurements, we can then compute P_{ref} (1) and P_{AUT} (2):

$$P_{\text{ref}} = \frac{\left| \overline{S_{21,\,\text{ref}}} \right|^2}{\left(1 - \left| \overline{S_{11}} \right|^2 \right)\left(1 - \left| \overline{S_{22,\,\text{ref}}} \right|^2 \right)}, \qquad (1)$$

$$P_{\text{AUT}} = \frac{\left| \overline{S_{21,\,\text{AUT}}} \right|^2}{\left(1 - \left| \overline{S_{11}} \right|^2 \right)\left(1 - \left| \overline{S_{22,\,\text{AUT}}} \right|^2 \right)}, \qquad (2)$$

where $\overline{S_{21}}$ is the averaged transmission power level, $\overline{S_{11}}$ is the free space reflection coefficient of the excitation antenna, and $\overline{S_{22}}$ is the free space reflection coefficient of the reference antenna (or the antenna under test). The $^-$ denotes averaging over n positions of the platform stirrer, polarization stirrer, and mechanical stirrers. The total efficiency can be then calculated from (3)

$$\eta_{\text{tot}} = \left(1 - \left| \overline{S_{22,\,\text{AUT}}} \right|^2 \right)\frac{P_{\text{AUT}}}{P_{\text{ref}}}. \qquad (3)$$

Figure 7 shows the total efficiency in dB of all the antenna systems (without the neutralization line (a), with the line between the feeding strips (b), and with the line between the shorting strips (c)). The simulated curves have been obtained with the help of IE3D which uses the simulated scattering parameters. The experimental curves have been measured in the reverberation chamber and with the help of a homemade Wheeler-Cap setup [16]. With frequency averaging, the standard deviation of the efficiency measurements is

TABLE 1: Comparison of η_{tot} and the MEG of both antennas of the different structures at $f = 2\,\text{GHz}$.

	η_{tot} (dB) antenna1		MEG (dB) antenna1		η_{tot} (dB) antenna2		MEG (dB) antenna2	
	Sim.	RC	Sim.	RC	Sim.	RC	Sim.	RC
Initial	−0.816	−0.75	−3.826	−3.75	−0.81	−1.25	−3.826	−4.25
Line between the feeding strips	−0.10	−0.2	−3.11	−3.2	−0.09	−0.5	−3.108	−3.5
Line between the shorting strips	−0.14	−0.35	−3.152	−3.35	−0.14	−0.65	−3.151	−3.65

TABLE 2: MEG ratio of the antennas for all the prototypes at $f = 2\,\text{GHz}$.

	MEG1/MEG2
Initial	1,12
Line between the feeding strips	1,07
Line between the shorting strips	1,07

given as $+/- 0.5\,\text{dB}$ in the reverberation chamber. The uncertainty of the homemade Wheeler Cap system is assumed to be quite the same. The total efficiency of both antennas from each prototype is presented. It can be seen that they are slightly different in the two measurement cases (dotted lines and solid lines) due to the fact that the fabricated prototypes suffer from small inherent asymmetries. However, only one curve is presented for each simulation case due to perfect symmetries and identical structure on the CAD software. We can observe that all these curves are in a good agreement especially if we compare their maximums. The small frequency shift observed in all the curves with the dotted lines is due to the fact that the antenna was mechanically modified during transportation for measurement, and therefore frequency is detuned. This effect impacts directly the S_{11} and then the frequency location of the maximum of the total efficiency. The improvement brought by the neutralization technique is clearly shown: the maximum total efficiency of the neutralized antennas is around $-0.25\,\text{dB}$, whereas the one of the initial structure is less than $-1\,\text{dB}$.

3.2. Mean effective gain and mean effective gain ratio

In order to characterize the performance of a multichannel antenna in a mobile environment, different parameters as the MEG and the MEG ratio are used. The total efficiency is the average antenna gain in the whole space. Equation (4) shows that it can be calculated from the integration of the radiation pattern cuts

$$\eta_{\text{tot}} = \frac{\int_0^{2\pi} \int_0^{\pi} (G_\theta(\theta,\varphi) + G_\varphi(\theta,\varphi)) \sin\theta d\,\theta d\varphi}{4\pi}, \quad (4)$$

where G_θ and G_φ are the antenna power gain patterns.

The MEG is a statistical measure of the antenna gain in a mobile environment. It is equal to the ratio of the mean received power of the antenna and the total mean incident. It can be expressed by (5) as in [6]:

$$\text{MEG} = \int_0^{2\pi} \int_0^{\pi} \left(\frac{\text{XPR}}{1 + \text{XPR}} G_\theta(\theta,\varphi) P_\theta(\theta,\varphi) \right.$$
$$\left. + \frac{1}{1 + \text{XPR}} G_\varphi(\theta,\varphi) P_\varphi(\theta,\varphi) \right) \sin\theta d\,\theta d\varphi, \quad (5)$$

where P_θ and P_φ are the angular density functions of the incident power, and XPR represents the cross-polarization power gain which is defined in (6):

$$\text{XPR} = \frac{\int_0^{2\pi} \int_0^{\pi} P_\theta(\theta,\varphi) \sin\theta d\,\theta d\varphi}{\int_0^{2\pi} \int_0^{\pi} P_\varphi(\theta,\varphi) \sin\theta d\,\theta d\varphi}. \quad (6)$$

In the case where the antenna is located in a statistically uniform Rayleigh environment (i.e., the case in the reverberation chamber), we have $\text{XPR} = 1$ and $P_\theta = P_\varphi = 1/4\pi$. The MEG is then equal to the total antenna efficiency divided by two or $-3\,\text{dB}$ [27]. Moreover, to achieve good diversity gain, the average received power from each antenna element must be nearly equal: this corresponds to getting the ratio of the MEG between the two antennas close to unity [29]. Table 1 presents η_{tot} and the MEG of both antennas for the three prototypes at $f = 2\,\text{GHz}$. The "Sim." values have been computed using the simulated radiation patterns while the reverberation chamber results "RC" are taken from the previous measurements.

The neutralization line provides an enhancement of the η_{tot} and the MEG as expected from the previous values. The improvement of the MEG is about $0.7\,\text{dB}$ with regard to the initial structure. Table 2 presents the MEG ratio between the two antennas of the different prototypes (computed from the RC MEG) at $2\,\text{GHz}$. It is seen that the antennas have comparable-average-received power because these entire ratios are close to unity. Such a result was somewhat expected due to the symmetric antenna configuration of our prototypes. In fact, the MEG difference only shows here the prototyping errors we made during the fabrication process. Nevertheless, all the results of this section confirm the benefit of using a neutralization technique between the radiators.

3.3. Correlation

For diversity and MIMO applications, the correlation between the signals received by the antennas at the same side of

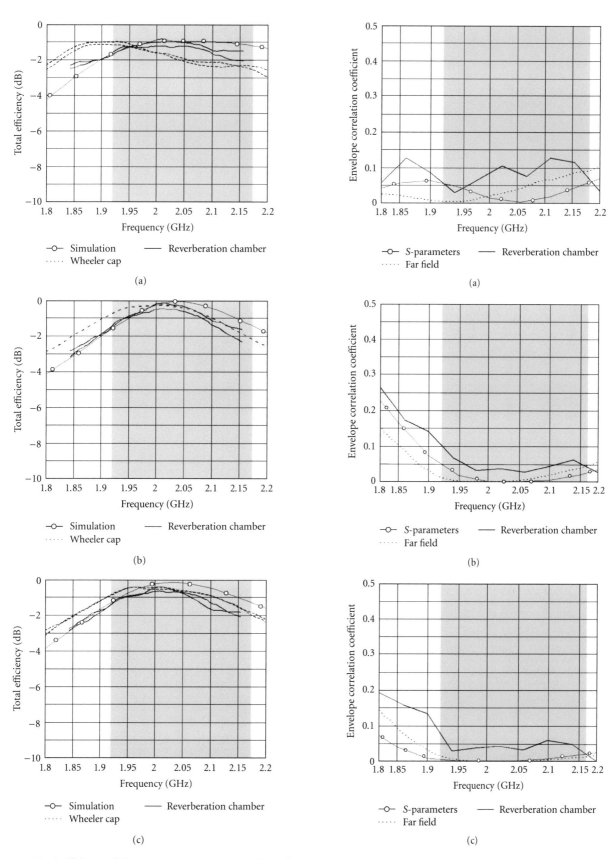

FIGURE 7: Total efficiency of the two-antenna structures: (a) without the neutralization line, (b) with the neutralization line between the feeding strips, and (c) with the neutralization line between the shorting strips.

FIGURE 8: Envelope correlation coefficient versus frequency of the two-antenna systems: (a) without the neutralization line, (b) with the line between the feeding strips, and (c) with the neutralization line between the shorting strips.

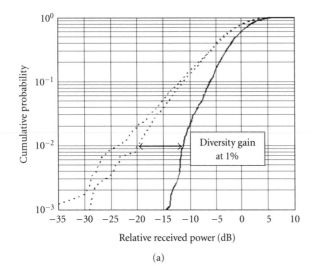

(a)

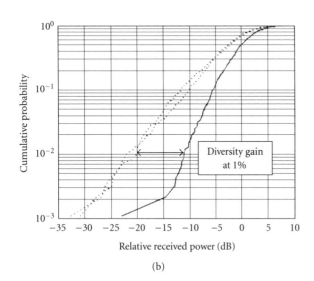

(b)

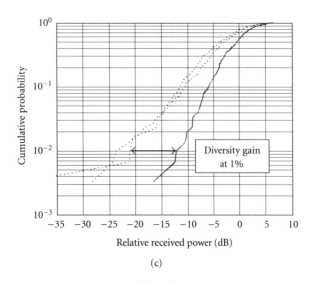

(c)

FIGURE 9: Cumulative probability of the two-antenna systems over a 4 MHz bandwidth at 2 GHz: (a) without the neutralization line, (b) with the neutralization line between the feeding strips, and (c) with the neutralization line between the shorting strips.

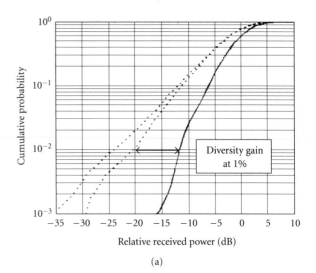

(a)

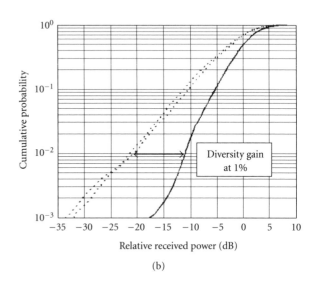

(b)

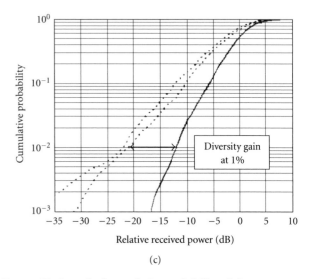

(c)

FIGURE 10: Smoothed cumulative probability of the two-antenna systems over a 4 MHz bandwidth at 2 GHz: (a) without the neutralization line, (b) with the neutralization line between the feeding strips, and (c) with the neutralization line between the shorting strips.

a wireless link is an important figure of merit. Usually, the envelope correlation is presented to evaluate the diversity capabilities of a multiantenna system [30]. This parameter should be preferably computed from 3D-radiation patterns [31, 32], but the process is tedious because sufficient pattern cuts must be taken into account. In the case of a two-antenna system, the envelope correlation ρ_e is given by (7) as in [31, 32]:

$$\rho_e = \frac{\left| \iint_{4\pi} [\vec{F_1}(\theta, \varphi) \bullet \vec{F_2^*}(\theta, \varphi) d\Omega] \right|^2}{\iint_{4\pi} \left| \vec{F_1}(\theta, \varphi) \right|^2 d\Omega \iint_{4\pi} \left| \vec{F_2}(\theta, \varphi) \right|^2 d\Omega}, \qquad (7)$$

where $\vec{F_i} = (\theta, \varphi)$ is the field radiation pattern of the antenna system when the port i is excited, and $\bullet$ denotes the Hermitian product.

However, assuming that the structure will operate in a uniform multipath environment, a convenient and quick alternative consists by using (8) (see [31–33]):

$$\rho_{12} = \frac{\left| S_{11}^* S_{12} + S_{12}^* S_{22} \right|^2}{\left(1 - \left| S_{11} \right|^2 - \left| S_{21} \right|^2 \right)\left(1 - \left| S_{22} \right|^2 - \left| S_{12} \right|^2 \right)}. \qquad (8)$$

It offers a simple procedure compared to the radiation pattern approach, but it should be emphasized that this equation is strictly valid when the three following assumptions are fulfilled:

(i) lossless antenna case that means having antennas with high efficiency and no mutual losses [29, 30];
(ii) antenna system is positioned in a uniform multipath environment which is not strictly the case in real environments, however, the evaluation of some prototypes in different real environments has already shown that there are no major differences in these cases [34];
(iii) load termination of the nonmeasured antenna is 50 Ω. In reality, the radio front-end module does not always achieve this situation, but the 50 Ω evaluation procedure is commonly accepted [35, 36].

All these limitations are clearly showing that in real systems the envelope correlation calculated based on of the help of the S_{ij} parameters is not the exact value, but nevertheless is a good approximation. In addition, it should be noted that antennas with an envelope correlation coefficient less than 0.5 are recognized to provide significant diversity performance [30].

To measure the correlation between the antennas of our systems in the reverberation chamber, each branch is connected to a separate receiver. The two different received signals are recorded, and the envelope correlation can be directly computed. Figure 8 presents the measured envelope correlation coefficients of all the antenna systems. They are compared with those obtained using (7) (computation from the simulated IE3D complex 3D-radiation patterns) and with those obtained using (8) (measured S-parameter values). All these curves are in a moderate agreement, but it can be seen that the envelope correlation coefficients of all the prototypes are always lower than 0.15 on the whole UMTS band: good performance in terms of diversity is thus expected [1]. Here, it is however somewhat difficult to claim

that the neutralization technique provides an improvement of the correlation. It seems rather obvious that with such spaced antennas operating in a uniform multipath environment, low correlation is not very difficult to achieve.

3.4. Apparent diversity gain and actual diversity gain

The concept of diversity means that we make use of two or more antennas to receive a signal and that we are able to combine the replicas of the received signal in a desirable way to improve the communication link performance. One requirement is high isolation between the antennas; otherwise the diversity gain will be low. The apparent diversity gain $G_{\text{div app}}$ relative to antenna1 and the actual diversity gain $G_{\text{div act}}$ are defined in (9)

$$\begin{aligned} G_{\text{div app}} &= \frac{S/N}{S_1/N_1}, \\ G_{\text{div act}} &= \frac{S/N}{S_1/N_1} \eta_{\text{tot1}}, \end{aligned} \qquad (9)$$

where η_{tot1} is the total efficiency of antenna1.

Note that these formulas are valid only if the noise signals N_1 (and N_2 for the second antenna) are independent of the total efficiency. This is the case if the system noise is dominated by those of the receivers or if the antenna noise temperature is the same as the surrounding temperature. The last condition is often close to being satisfied in mobile systems because the antenna is rather omnidirectional and picks up thermal noise mainly from the environment (ground, buildings, trees, human) around the antenna, and less from the low sky temperature.

We can see in Figure 9 the power samples of each two-antenna system (without the neutralization line (a), with the line between the feeding strips (b), and with the line between the shorting strips (c)) averaged over a 20-MHz frequency band at 2 GHz. We can observe that the combined signal curves with the selection combining scheme (solid lines) are steeper than the two curves of the antenna elements taken alone (dotted lines). This is the benefit of combining the two signals received by each antenna of the structure. By just looking at the curves in Figure 9, the uncertainty is undoubtedly very large. This is due to the obvious lack of samples at low-probability levels coming from the measurement procedure.

The apparent diversity gain is determined by the power-level improvement at a certain probability level. In Figures 9(a), 9(b), and 9(c), we have chosen 1% probability. It is then the difference between the strongest antenna element curve and the combined signal curve. The power improvement is 7.6 dB for the system with low isolation, whereas it is 8.8 dB and 9.1 dB for the system with high isolation, respectively, for the line between the shorting strips and the line between the feeding strips. As the total efficiency is not taken into account in the apparent diversity gain, the improvement only comes from the fact that the radiation patterns are slightly different in the case of the two neutralized structures. Especially, an increase of the cross-polarization level occurs in the radiation patterns of the neutralized structures due to the

TABLE 3: Summary of the measured and computed diversity gains of all the antenna systems.

Prototypes	Total efficiency best branch	Apparent diversity gain	Apparent diversity gain, smooth curved	Actual diversity gain	Actual diversity gain, smooth curved
Without any line	−0.75 dB	7.6 dB	8.6 dB	6.3 dB	7.8 dB
Shorting strips link	−0.35 dB	8.8 dB	9.2 dB	8 dB	8.8 dB
Feeding strips link	−0.2 dB	9.1 dB	9.75 dB	8.6 dB	9.5 dB

fact that a strong current is flowing on the line. This increase of the X-pol appears to be beneficial for the diversity gain. When taking into account the total efficiency of the antennas, we can compute the actual diversity gain as 6.3 dB for the initial system, 8 dB and 8.6 dB for the neutralized shorting strips and feeding strips systems, respectively. The data from Figure 9 were also processed with the smooth function of MATLAB [37] in order to evaluate the validity of our measurements. Several "smooth steps" were tried out in this operation and the new curves are presented in Figure 10. It appears that all the apparent diversity gains were formerly underestimated. The new actual diversity gains are now 7.8 dB, 8.8 dB, and 9.5 dB for, respectively, the initial, the neutralized shorting strips and feeding strips systems. A summary of all these values can be found in Table 3.

It seems obvious that the neutralization technique enhances the actual diversity gain. These results are consistent with other publications [38] and even better due to the use of highly efficient antennas here. We should also point out that the apparent diversity gain and the actual diversity gain are not so much different due to the same reason [39].

4. CONCLUSION

In this paper, we have presented different two-antenna systems with poor and high isolations for diversity purposes. The reverberation chamber measurements at the antenna group of Chalmers University of Technology have shown that even if the envelope correlation coefficient of these systems is very low, having antennas with high isolation will improve the total efficiency and the effective diversity gain of the system. The same conclusions have been drawn regarding the MEG values. All these results point out the usefulness of our simple solution to achieve efficient antenna systems at the terminal side of a wireless link for diversity or MIMO applications. Next studies will focus on the effect of the users upon the neutralization technique by positioning the antenna systems next to a phantom head.

ACKNOWLEDGMENT

The authors express their gratitude to the COST284 project for providing the opportunity to make a short-term scientific mission from the LEAT to Chalmers Institute.

REFERENCES

[1] Z. Ying and D. Zhang, "Study of the mutual coupling, correlations and efficiency of two PIFA antennas on a small ground plane," in *Proceedings of IEEE Antennas and Propagation Society International Symposium*, vol. 3, pp. 305–308, Washington, DC, USA, July 2005.

[2] J. Thaysen and K. B. Jakobsen, "MIMO channel capacity versus mutual coupling in multi antenna element system," in *Proceedings of the 26th Annual Meeting & Symposium on Antenna Measurement Techniques Association (AMTA '04)*, Stone Mountain, Ga, USA, October 2004.

[3] M. Karakoikis, C. Soras, G. Tsachtsiris, and V. Makios, "Compact dual-printed inverted-F antenna diversity systems for portable wireless devices," *IEEE Antennas and Wireless Propagation Letters*, vol. 3, no. 1, pp. 9–14, 2004.

[4] K.-L. Wong, C.-H. Chang, B. Chen, and S. Yang, "Three-antenna MIMO system for WLAN operation in a PDA phone," *Microwave and Optical Technology Letters*, vol. 48, no. 7, pp. 1238–1242, 2006.

[5] S. H. Chae, S.-K. Oh, and S.-O. Park, "Analysis of mutual coupling, correlations, and TARC in WiBro MIMO array antenna," *IEEE Antennas and Wireless Propagation Letters*, vol. 6, pp. 122–125, 2007.

[6] H. T. Hui, "Practical dual-helical antenna array for diversity/MIMO receiving antennas on mobile handsets," *IEE Proceedings: Microwaves, Antennas and Propagation*, vol. 152, no. 5, pp. 367–372, 2005.

[7] J. Villanen, P. Suvikunnas, C. Icheln, J. Ollikainen, and P. Vainikainen, "Advances in diversity performance analysis of mobile terminal antennas," in *Proceedings of the International Symposium on Antennas and Propagation (ISAP '04)*, Sendai, Japan, August 2004.

[8] M. Manteghi and Y. Rahmat-Samii, "Novel compact tri-band two-element and four-element MIMO antenna designs," in *Proceedings of the International Symposium on Antennas and Propagations (ISAP '06)*, pp. 4443–4446, Albuquerque, NM, USA, July 2006.

[9] M. Manteghi and Y. Rahmat-Samii, "A novel miniaturized tri-band PIFA for MIMO applications," *Microwave and Optical Technology Letters*, vol. 49, no. 3, pp. 724–731, 2007.

[10] D. Browne, M. Manteghi, M. P. Fits, and Y. Rahmat-Samii, "Experiments with compact antenna arrays for MIMO radio communications," *IEEE Transaction on Antennas and Propagation*, vol. 54, no. 11, part 1, pp. 3239–3250, 2007.

[11] B. Lindmark and L. Garcia-Garcia, "Compact antenna array for MIMO applications at 1800 and 2450 MHz," *Microwave and Optical Technology Letters*, vol. 48, no. 10, pp. 2034–2037, 2006.

[12] R. G. Vaughan and J. B. Andersen, "Antenna diversity in mobile communications," *IEEE Transactions on Vehicular Technology*, vol. 36, no. 4, pp. 149–172, 1987.

[13] B. K. Lau, J. B. Andersen, G. Kristensson, and A. F. Molisch, "Impact of matching network on bandwidth of compact antenna arrays," *IEEE Transactions on Antennas and Propagation*, vol. 54, no. 11, pp. 3225–3238, 2006.

[14] J. B. Andersen and B. K. Lau, "On closely coupled dipoles in a random field," *IEEE Antennas and Wireless Propagation Letter*, vol. 5, no. 1, pp. 73–75, 2006.

[15] J. W. Wallace and M. A. Jensen, "Termination-dependent diversity performance of coupled antennas: network theory analysis," *IEEE Transactions on Antennas and Propagation*, vol. 52, no. 1, pp. 98–105, 2004.

[16] M. A. J. Jensen and J. W. Wallace, "A review of antennas and propagation for MIMO wireless communications," *IEEE Transactions on Antennas and Propagation*, vol. 52, no. 11, pp. 2810–2824, 2004.

[17] J. Thaysen and K. B. Jakobsen, "Mutual coupling reduction using a lumped LC circuit," in *Proceedings of the 13th International Symposium on Antennas (JINA '04)*, pp. 492–494, Nice, France, November 2004.

[18] S. Dossche, S. Blanch, and J. Romeu, "Optimum antenna matching to minimise signal correlation on a two-port antenna diversity system," *IET Electronics Letters*, vol. 40, no. 19, pp. 1164–1165, 2004.

[19] S. Dossche, J. Rodriguez, L. Jofre, S. Blanch, and J. Romeu, "Decoupling of a two-element switched dual band patch antenna for optimum MIMO capacity," in *Proceedings of the International Symposium on Antennas and Propagations (ISAP '06)*, pp. 325–328, Albuquerque, NM, USA, July 2006.

[20] Y. Gao, X. Chen, C. Parini, and Z. Ying, "Study of a dual-element PIFA array for MIMO terminals," in *Proceedings of the International Symposium on Antennas and Propagations (ISAP '06)*, pp. 309–312, Albuquerque, NM, USA, July 2006.

[21] A. Diallo, C. Luxey, P. Le Thuc, R. Staraj, and G. Kossiavas, "Study and reduction of the mutual coupling between two mobile phone PIFAs operating in the DCS1800 and UMTS bands," *IEEE Transactions on Antennas and Propagation*, vol. 54, no. 11, part 1, pp. 3063–3074, 2006.

[22] A. Diallo, C. Luxey, P. Le Thuc, R. Staraj, and G. Kossiavas, "Enhanced diversity antennas for UMTS handsets," in *Proceedings of the European Conference on Antennas and Propagations (EuCAP '06)*, Nice, France, November 2006.

[23] P.-S. Kildal and K. Rosengren, "Correlation and capacity of MIMO systems and mutual coupling, radiation efficiency, and diversity gain of their antennas: simulations and measurements in a reverberation chamber," *IEEE Communications Magazine*, vol. 42, no. 12, pp. 104–112, 2004.

[24] http://www.bluetest.se/.

[25] T. Bolin, A. Derneryd, G. Kristensson, V. Plicanic, and Z. Ying, "Two-antenna receive diversity performance in indoor environment," *Electronics Letters*, vol. 41, no. 22, pp. 1205–1206, 2005.

[26] T. Taga, "Analysis for mean effective gain of mobile antennas in land mobile radio environments," *IEEE Transactions on Vehicular Technology*, vol. 39, no. 2, pp. 117–131, 1990.

[27] K. Kalliola, K. Sulonen, H. Laitinen, O. Kivekäs, J. Krogerus, and P. Vainikainen, "Angular power distribution and mean effective gain of mobile antenna in different propagation environments," *IEEE Transactions on Vehicular Technology*, vol. 51, no. 5, pp. 823–838, 2002.

[28] IE3D, Release 11.15, Zeland software, 2005.

[29] C. C. Chlau, X. Chen, and C. Q. Parinl, "A compact four-element diversity-antenna array for PDA terminals in a mimo system," *Microwave and Optical Technology Letters*, vol. 44, no. 5, pp. 408–412, 2005.

[30] S. C. K. Ko and R. D. Murch, "Compact integrated diversity antenna for wireless communications," *IEEE Transactions on Antennas and Propagation*, vol. 49, no. 6, pp. 954–960, 2001.

[31] I. Salonen and P. Vainikainen, "Estimation of signal correlation in antenna arrays," in *Proceedings of the 12th International Symposium Antennas (JINA '02)*, vol. 2, pp. 383–386, Nice, France, November 2002.

[32] P. Brachat and C. Sabatier, "Réseau d'antennes à 6 Capteurs en Diversité de Polarisation," in *Proceedings of the 13th International Symposium Antennas (JINA '04)*, Nice, France, November 2004.

[33] J. Thaysen and K. B. Jakobsen, "Envelope correlation in (N, N) MIMO antenna array from scattering parameters," *Microwave and Optical Technology Letters*, vol. 48, no. 5, pp. 832–834, 2006.

[34] Z. Ying, V. Plicanic, T. Bolin, G. Kristensson, and A. Derneryd, "Characterization of Multi-Channel Antenna Performance for Mobile Terminal by Using Near Field and Far Field Parameters," COST 273 TD (04)(095), Göteborg, Sweden, June 2004.

[35] A. Derneryd and G. Kristensson, "Signal correlation including antenna coupling," *Electronics Letters*, vol. 40, no. 3, pp. 157–159, 2004.

[36] A. Derneryd and G. Kristensson, "Antenna signal correlation and its relation to the impedance matrix," *Electronics Letters*, vol. 40, no. 7, pp. 401–402, 2004.

[37] http://www.mathworks.fr/.

[38] K. Rosengren and P.-S. Kildal, "Diversity performance of a small terminal antenna for UMTS," in *Proceedings of Nordic Antenna Symposium (Antenn '03)*, Kalmar, Sweden, May 2003.

[39] P.-S. Kildal, K. Rosengren, J. Byun, and J. Lee, "Definition of effective diversity gain and how to measure it in a reverberation chamber," *Microwave and Optical Technology Letters*, vol. 34, no. 1, pp. 56–59, 2002.

Hindawi Publishing Corporation
EURASIP Journal on Wireless Communications and Networking
Volume 2007, Article ID 93421, 9 pages
doi:10.1155/2007/93421

Research Article

Optimal Design of Nonuniform Linear Arrays in Cellular Systems by Out-of-Cell Interference Minimization

S. Savazzi,[1] O. Simeone,[2] and U. Spagnolini[1]

[1] Dipartimento di Elettronica e Informazione, Politecnico di Milano, 20133 Milano, Italy
[2] Center for Wireless Communications and Signal Processing Research (CCSPR), New Jersey Institute of Technology, University Heights, Newark, NJ 07102-1982, USA

Received 13 October 2006; Accepted 11 July 2007

Recommended by Monica Navarro

Optimal design of a linear antenna array with nonuniform interelement spacings is investigated for the uplink of a cellular system. The optimization criterion considered is based on the minimization of the average interference power at the output of a conventional beamformer (matched filter) and it is compared to the maximization of the ergodic capacity (throughput). Out-of-cell interference is modelled as spatially correlated Gaussian noise. The more analytically tractable problem of minimizing the interference power is considered first, and a closed-form expression for this criterion is derived as a function of the antenna spacings. This analysis allows to get insight into the structure of the optimal array for different propagation conditions and cellular layouts. The optimal array deployments obtained according to this criterion are then shown, via numerical optimization, to maximize the ergodic capacity for the scenarios considered here. More importantly, it is verified that substantial performance gain with respect to conventionally designed linear antenna arrays (i.e., uniform $\lambda/2$ interelement spacing) can be harnessed by a nonuniform optimized linear array.

1. INTRODUCTION

Antenna arrays have emerged in the last decade as a powerful technology in order to increase the link or system capacity in wireless systems. Basically, the deployment of multiple antennas at either the transmitter or the receiver side of a wireless link allows the exploitation of two contrasting benefits: diversity and beamforming. Diversity relies on fading uncorrelation among different antenna elements and provides a powerful means to combat the impairments caused by channel fluctuations. In [1] it has been shown that a significant increase in system capacity can be achieved by the use of antenna diversity combined with optimum combining schemes. Independence of fading gains associated to the antennas array can be guaranteed if the scattering environment is rich enough and the antenna elements are sufficiently spaced apart (at least 5–10 λ, where λ denotes the carrier wavelength) [2]. On the other hand, when fading is highly correlated, as for sufficiently small antenna spacings, beamforming techniques can be employed in order to mitigate the spatially correlated noise. Interference rejection through beamforming is conventionally performed by designing a uniform linear array with half wavelength interelement spac-

ings, so as to guarantee that the angle of arrivals can be potentially estimated free of aliasing. Moreover, beamforming is effective in propagation environments where there is a strong line-of-sight component and the system performance is interference-limited [3].

In this paper, we consider the optimization of a linear nonuniform antenna array for the uplink of a cellular system. The study of nonuniform linear arrays dates back to the seventies with the work of Saholos [4] on radiation pattern and directivity. In [5] performance of linear and circular arrays with different topologies, number of elements, and propagation models is studied for the uplink of an interference free system so as to optimize the network coverage. The idea of optimizing nonuniform-spaced antenna arrays to enhance the overall throughput of an interference-limited system was firstly proposed in [6]. Therein, for flat fading channels, it is shown that unequally spaced arrays outperform equally spaced array by 1.5–2 dB. Here, different from [6], a more realistic approach that explicitly takes into account the cellular layout (depending on the reuse factor) and the propagation model (that ranges from line-of-sight to richer scattering according to the ring model [7]) is accounted for.

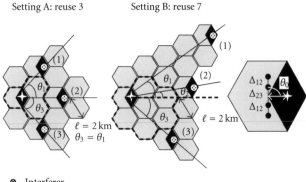

Setting A: reuse 3   Setting B: reuse 7

⊗ Interferer
+ User

FIGURE 1: Two cellular systems with hexagonal cells and trisectorial antennas at the base stations (reuse factor $F = 3$, setting A, on the right and $F = 7$, setting B, on the left). The array is equipped with $N = 4$ antennas. Shaded sectors denote the allowed areas for user and the three interferers belonging to the first ring of interference (dashed lines identify the cell clusters of frequency reuse).

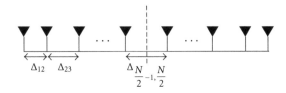

FIGURE 2: Nonuniform symmetric array structure for N even.

For illustration purposes, consider the interference scenarios sketched in Figure 1. Therein, we have two different settings characterized by hexagonal cells and different reuse factors ($F = 3$ for setting A and $F = 7$ for setting B, frequency reuse clusters are denoted by dashed lines). The base station is equipped with a symmetric antenna array[1] containing an even number N of directional antennas ($N = 4$ in the example) to cover an angular sector of 120 deg, other BS antenna array design options are discussed in [8]. Each terminal is provided with one omnidirectional antenna. On the considered radio resource (e.g., time-slot, frequency band, or orthogonal code), it is assumed there is only one active user in the cell, as for TDMA, FDMA, or orthogonal CDMA. The user of interest is located in the respective sector according to the reuse scheme. The contribution of out-of-cell interferers is modelled as spatially correlated Gaussian noise. In Figure 1, the first ring of interference is denoted by shaded cells. The problem we tackle is that of finding the antenna spacings in vector $\boldsymbol{\Delta} = [\Delta_{12} \ \Delta_{23}]^T$ (as shown in the example)

so as to optimize given performance metrics, as detailed below.

Two criteria are considered, namely, the minimization of the average interference power at the output of a conventional beamformer (matched filter) and the maximization of the ergodic capacity (throughput). Since in many applications the position of users and interferers is not known a priori at the time of the antenna deployment or the incell/out-cell terminals are mobile, it is of interest to evaluate the optimal spacings not only for a fixed position of users and interferers but also by averaging the performance metric over the positions of user and interferers within their cells (see Section 2).

Even if the ergodic capacity criterion has to be considered to be the most appropriate for array design in interference-limited scenario, the interference power minimization is analytically tractable and highlights the justification for unequal spacings. Therefore, the problem of minimizing the interference power is considered first and a closed-form expression for this criterion is derived as a function of the antenna spacings (Section 4). This analysis allows to get insight into the structure of the optimal array for different propagation conditions and cellular layouts avoiding an extensive numerical maximization of the ergodic capacity. The optimal array deployments obtained according to the two criteria are shown via numerical optimization to coincide for the considered scenarios (Section 5). More importantly, it is verified that substantial performance gain with respect to conventionally designed linear antenna arrays (i.e., uniform $\lambda/2$ interelement spacing) can be harnessed by an optimized array (up to 2.5 bit/s/Hz for the scenarios in Figure 1).

2. PROBLEM FORMULATION

The signal received by the N antenna array at the base station serving the user of interest can be written as

$$\mathbf{y} = \mathbf{h}_0 x_0 + \sum_{i=1}^{M} \mathbf{h}_i x_i + \mathbf{w}, \tag{1}$$

where $\mathbf{h}_0$ is the $N \times 1$ vector describing the channel gains between the user and the N antennas of the base station, x_0 is the signal transmitted by the user, $\mathbf{h}_i$ and x_i are the corresponding quantities referred to the ith interferer ($i = 1, \ldots, M$), $\mathbf{w}$ is the additive white Gaussian noise with $E[\mathbf{w}\mathbf{w}^H] = \sigma^2 \mathbf{I}$. The channel vectors $\mathbf{h}_0$ and $\{\mathbf{h}_i\}_{i=1}^{M}$ are uncorrelated among each other and assumed to be zero-mean complex Gaussian (Rayleigh fading) with spatial correlation $\mathbf{R}_0 = E[\mathbf{h}_0 \mathbf{h}_0^H]$ and $\{\mathbf{R}_i = E[\mathbf{h}_i \mathbf{h}_i^H]\}_{i=1}^{M}$, respectively. The correlation matrices are obtained according to a widely employed geometrical model that assumes the scatterers as distributed along a ring around the terminal, see Figure 3. This model was thoroughly studied in [2, 7] and a brief review can be found in Section 3. According to this model, the spatial correlation matrices of the fading channel depend on

(1) the set of $N/2$ antenna spacings (N is even) $\boldsymbol{\Delta} = [\Delta_{12} \ \Delta_{23} \ \cdots \ \Delta_{N/2, N/2+1}]^T$, where Δ_{ij} is the distance between the ith and the jth element of the array (the

---

[1] The symmetric array assumption (as in the array structure of Figure 2) has been made mainly for analytical convenience in order to simplify the optimization problem. However, it is expected that for a scenario with a symmetric layout of interference (such as setting A), the assumption of a symmetric array does not imply any loss of optimality, while, on the other hand, for an asymmetric layout (such as setting B), capacity gains could be in principle obtained by deploying an asymmetric array.

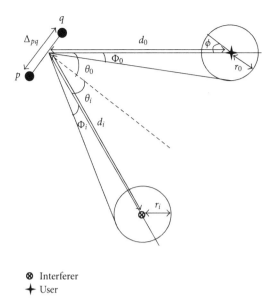

⊗ Interferer
✛ User

Figure 3: Propagation model for user and interferers: the scatterers are distributed on a rings of radii r_i around the terminals.

array is assumed to be symmetric as shown in Figure 2, extension to an odd number of antennas N is straightforward);

(2) the relative positions of user and interferers with respect to the base station of interest (these latter parameters can be conveniently collected into the vector $\boldsymbol{\eta} = [\boldsymbol{\eta}_0^T \ \boldsymbol{\eta}_1^T \ \cdots \ \boldsymbol{\eta}_M^T]^T$, where, as detailed in Figure 3, vector $\boldsymbol{\eta}_0 = [d_0 \ \theta_0]^T$ parametrizes the geometrical location of the in-cell user and vectors $\boldsymbol{\eta}_i = [d_i \ \theta_i]^T$ ($i = 1, \ldots, M$) describe the location of the interferers ($i = 1, \ldots, M$));

(3) the propagation environment is described by the angular spread of the scattered signal received by the base station (ϕ_0 for the user and ϕ_i ($i = 1, \ldots, M$) for the interferers); notice that for ideally $\phi_i \to 0$ all scatterers come from a unique direction so that line-of-sight (LOS) channel can be considered. Shadowing can be possibly modelled as well, see Section 3 for further discussion.

2.1. Interference power minimization

From (1), the instantaneous total interference power at the output of a conventional beamforming (matched filter) is [9]

$$\mathcal{P}(\mathbf{h}_0, \Delta, \boldsymbol{\eta}) = \mathbf{h}_0^H \mathbf{Q} \mathbf{h}_0, \tag{2}$$

where

$$\mathbf{Q} = \mathbf{Q}(\Delta, \boldsymbol{\eta}_1, \ldots, \boldsymbol{\eta}_M) = \sum_{i=1}^{M} \mathbf{R}_i(\Delta, \boldsymbol{\eta}_i) + \sigma^2 \mathbf{I}_N \tag{3}$$

accounts for the spatial correlation matrix of the interferers and for thermal noise with power σ^2. Notice that, for clarity of notation, we explicitly highlighted that the interference correlation matrices depend on the terminals' locations

$\boldsymbol{\eta}$ and the antenna spacings Δ through nonlinear relationships. The first problem we tackle is that of finding the set of optimal spacings $\hat{\Delta}$ that minimizes the average (with respect to fading) interference power, $\mathscr{P}(\Delta, \boldsymbol{\eta}) = E_{\mathbf{h}_0}[\mathcal{P}(\mathbf{h}_0, \Delta, \boldsymbol{\eta})]$, that is,

$$(\text{Problem-1}): \hat{\Delta} = \arg\min_{\Delta} \mathscr{P}(\Delta, \boldsymbol{\eta}) \tag{4}$$

for a fixed given position $\boldsymbol{\eta}$ of user and interferers. Problem 1 is relevant for fixed system with a known layout at the time of antenna deployment. Moreover, its solution will bring insight into the structure of the optimal array, which can be to some extent generalized to a mobile scenario. In fact, in mobile systems or in case the position of users and interferers is not known a priori at the time of the antenna deployment, it is more meaningful to minimize the average interference power for any arbitrary position of in-cell user ($\boldsymbol{\eta}_0$) and out-of-cells interferers ($\boldsymbol{\eta}_1, \boldsymbol{\eta}_2, \ldots, \boldsymbol{\eta}_M$). Denoting the averaging operation with respect to users and interferers positions by $E_{\boldsymbol{\eta}}[\mathscr{P}(\Delta, \boldsymbol{\eta})]$, the second problem (9) can be can be stated as

$$(\text{Problem-2}): \hat{\Delta} = \arg\min_{\Delta} E_{\boldsymbol{\eta}}[\mathscr{P}(\Delta, \boldsymbol{\eta})]. \tag{5}$$

2.2. Ergodic capacity maximization

The instantaneous capacity for the link between the user and the BS reads [2]

$$C(\mathbf{h}_0, \Delta, \boldsymbol{\eta}) = \log_2\left(1 + \mathbf{h}_0^H \mathbf{Q}^{-1} \mathbf{h}_0\right) \text{ [bit/s/Hz]}, \tag{6}$$

and depends on both the antenna spacings Δ and the terminals' locations $\boldsymbol{\eta}$. For fast-varying fading channels (compared to the length of the coded packet) or for delay-insensitive applications, the performance of the system from an information theoretic standpoint is ruled by the ergodic capacity $\mathcal{C}(\Delta, \boldsymbol{\eta})$. The latter is defined as the ensemble average of the instantaneous capacity over the fading distribution,

$$\mathcal{C}(\Delta, \boldsymbol{\eta}) = E_{\mathbf{h}_0}[C(\mathbf{h}_0, \Delta, \boldsymbol{\eta})]. \tag{7}$$

According to the alternative performance criterion herein proposed, the first problem (4) is recast as

$$(\text{Problem 1}): \hat{\Delta} = \arg\max_{\Delta} \mathcal{C}(\Delta, \boldsymbol{\eta}), \tag{8}$$

and therefore requires the maximization of the ergodic capacity for a fixed given position $\boldsymbol{\eta}$ of user and interferers. As before, denoting the averaging operation with respect to users and interferers positions by $E_{\boldsymbol{\eta}}[\mathcal{C}(\Delta, \boldsymbol{\eta})]$, the second problem (5) can be modified accordingly:

$$(\text{Problem 2}): \hat{\Delta} = \arg\max_{\Delta} E_{\boldsymbol{\eta}}[\mathcal{C}(\Delta, \boldsymbol{\eta})]. \tag{9}$$

Different from the interference power minimization approach, in this case, functional dependence of the performance criterion (7) on the antenna spacings Δ is highly nonlinear (see Section 3 for further details) and complicated by the presence of the inverse matrix $\mathbf{Q}^{-1}$ that relies upon Δ and $\boldsymbol{\eta}$. This implies both a large-computational complexity for

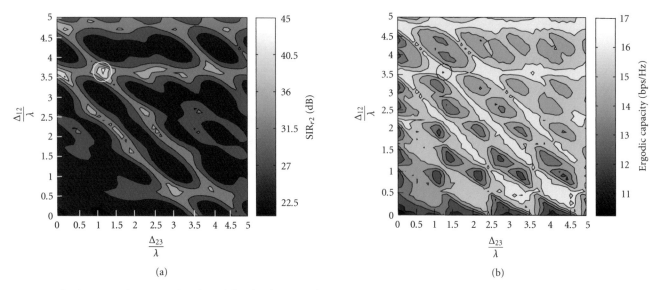

FIGURE 4: Setting-A: rank-2 approximation of the signal-to-interference ratio $\text{SIR}_{r2}(\boldsymbol{\Delta}, \overline{\boldsymbol{\eta}})$ (23) versus Δ_{12}/λ and Δ_{23}/λ (a) compared with ergodic capacity $\mathcal{C}(\boldsymbol{\Delta}, \overline{\boldsymbol{\eta}})$ (b) ($r = 50\,\text{m}$). Circles denote optimal solutions.

the numerical optimization of (8) and (9), and the impossibility to get analytical insight into the properties of the optimal solution. When the number of antenna array is sufficiently small (as in Section 5), optimization can be reasonably dealt with through an extensive search over the optimization domain and without the aid of any sophisticated numerical algorithm. On the contrary, in case of an array with a larger number of antenna elements, more efficient optimization techniques (e.g., simulated annealing) may be employed to reduce the number of spacings to be explored and thus simplify the optimization process. Below we will prove (by numerical simulations) that the limitations of the above optimization (8)-(9) are mitigated by the criteria (4)-(5) still preserving the final result.

3. SPATIAL CORRELATION MODEL

We consider a propagation scenario where each terminal, be it the user or an interferer, is locally surrounded by a large number of scatterers. The signals radiated by different scatterers add independently at the receiving antennas. The scatterers are distributed on a ring of radius r_0 around the terminal (r_i, $i = 1, \ldots, M$ for the interferers) and the resulting angular spread of the received signal at the base station is denoted by $\phi_0 \simeq r_0/d_0$ (or $\phi_i \simeq r_i/d_i$), as in Figure 3. Because of the finite angular spreads $\{\phi_i\}_{i=0}^{M}$, the propagation model appears to be well suited for outdoor channels.

In [7], the spatial correlation matrix of the resulting Rayleigh distributed fading process at the base station is computed by assuming a parametric distribution of the scatterers along the ring, namely, the von Mises distribution (variable $0 \le \vartheta < 2\pi$ runs over the ring, see Figure 3):

$$f(\vartheta) = \frac{1}{2\pi I_0(\kappa)} \exp \left[\kappa \cos(\vartheta) \right]. \tag{10}$$

By varying parameter κ, the distribution of the scatterers ranges from uniform ($f(\vartheta) = 1/(2\pi)$ for $\kappa = 0$) to a Dirac delta around the main direction of the cluster $\vartheta = 0$ (for $\kappa \to \infty$). Therefore, by appropriately adjusting parameter κ and the angular spreads for each user and interferers ϕ_i, a propagation environment with a strong line-of-sight component ($\phi_i \simeq 0$ and/or $\kappa \to \infty$) or richer scattering (larger ϕ_i with κ small enough) can be modelled. The (normalized) spatial correlation matrix has the general expression for both user and interferers (for the (p,q)th element with $p, q = 1, \ldots, N$ and $i = 0, 1, \ldots, M$) [7]:

$$\left[\overline{\mathbf{R}}_i(\theta_i, \Delta) \right]_{pq} = \exp \left[j \frac{2\pi}{\lambda} \Delta_{pq} \sin(\theta_i) \right]$$

$$\cdot \frac{I_0 \left(\sqrt{\kappa^2 - \left((2\pi/\lambda) \Delta_{pq} \phi_i \cos(\theta_i) \right)^2} \right)}{I_0(\kappa)}. \tag{11}$$

It is worth mentioning that spatial channel models based on different geometries such as elliptical or disk models [10, 11] may be considered as well by appropriately modifying the spatial correlation (11). Effects of mutual coupling (not addressed in this paper) between the array elements may be included in our framework too, see [12, 13].

From (11), the spatial correlation matrices $\mathbf{R}_i$ of the user and interferers can be written as

$$\mathbf{R}_i(\boldsymbol{\eta}_i, \Delta) = \rho_i \overline{\mathbf{R}}_i(\theta_i, \Delta) \quad \text{with } \rho_i = \frac{K}{d_i^\alpha}, \tag{12}$$

where K is an appropriate constant that accounts for receiving and transmitting antenna gain and the carrier frequency, and α is the path loss exponent. The contribution of shadowing in (12) will be considered in Section 5.3 as part of an additional log-normal random scaling term.

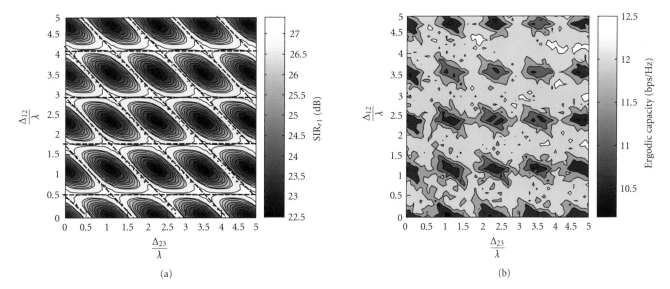

(a)   (b)

FIGURE 5: Setting-A: rank-1 approximation (a) of the signal-to-interference ratio, $\mathrm{SIR}_{r1}(\Delta, \overline{\eta})$, versus Δ_{12}/λ and Δ_{23}/λ. Dashed lines denote the optimality conditions (24) obtained by the rank-1 approximation. As a reference, ergodic capacity is shown (b), for an angular spread approaching zero.

4. REDUCED-RANK APPROXIMATION FOR THE INTERFERENCE POWER

According to a reduced-rank approximation for the spatial correlation matrices of user and interferers $\mathbf{R}_i$ for $i = 0, 1, \ldots, M$, in this section, we derive an analytical closed form expression for the interference power (2) to ease the optimization of the antenna spacings Δ. In Section 4.1, we consider the case where the angular spread for users and interferers ϕ_i is small so that a rank-1 approximation of the spatial correlation matrices can be used. This first case describes line-of-sight channels. Generalization to channel with richer scattering is given in Section 4.2.

4.1. Rank-1 approximation (line-of-sight channels)

If the angular spread is small for both user and interferers[2] (i.e., $\phi_i \ll 1$ for $i = 0, 1, \ldots, M$), the associated spatial correlation matrices $\{\mathbf{R}_i\}_{i=0}^{M}$, can be conveniently approximated by enforcing a rank-1 constraint. For $\phi_i \ll 1$, the following simplification holds in (11): $I_0(\sqrt{\kappa^2 - ((2\pi/\lambda)\Delta_{pq}\phi_i \cos(\theta_i))^2})/I_0(\kappa) \simeq 1$. Therefore, the spatial correlation matrices (12) can be approximated as (we drop the functional dependency for simplicity of notation)

$$\mathbf{R}_i \simeq \rho \cdot \mathbf{v}_i \mathbf{v}_i^H, \qquad (13)$$

where

$$\mathbf{v}_i(\Delta, \mathbf{J}) = \begin{bmatrix} 1 & \exp[-j\omega(\theta_i)\Delta_{12}] & \cdots & \exp[-j\omega(\theta_i)\Delta_{1N}] \end{bmatrix}^T \qquad (14)$$

and $\omega_i(\theta) = 2\pi/\lambda \sin(\theta_i)$.

From (13), the channel vectors for user and interferers can be written as $\mathbf{h}_i = \gamma\sqrt{\rho_i}\mathbf{v}_i$, where $\gamma \sim \mathcal{CN}(0, 1)$. Therefore, within the rank-1 approximation, the interference power reads (the additive noise contribution $\sigma^2 \mathbf{I}_N$ has been dropped since it is immaterial for the optimization problem)

$$\mathcal{P}_1(\Delta, \boldsymbol{\eta}) = \mathbf{v}_0^H \left(\sum_{i=1}^{M} \rho_i \mathbf{v}_i \mathbf{v}_i^H \right) \mathbf{v}_0, \qquad (15)$$

therefore, optimal spacings with respect to Problem 1 (4) can be written as

$$\hat{\Delta} = \arg\min_{\Delta} \mathcal{P}_1(\Delta, \boldsymbol{\eta}), \qquad (16)$$

where the subscript is a reminder of the rank-1 approximation. The advantage of the rank-1 performance criterion (15) is that it allows to derive an explicit expression as a function of the parameters of interest. In particular, after tedious but straightforward algebra, we get

$$\mathcal{P}_1(\Delta, \boldsymbol{\eta})$$
$$= \sum_{i=1}^{M} \left[\rho_i N + \rho_i \sum_{j=1}^{L} 4S(l_j, \theta_0, \theta_i) + \rho_i \sum_{k=1}^{C} 2S(c_k, \theta_0, \theta_i) \right], \qquad (17)$$

---

[2] Rank-1 approximation for the out-of-cell interferers is quite accurate when considering large reuse factors as the angular spread experienced by the array is reduced by the increased distance of the out-of-cell interferers.

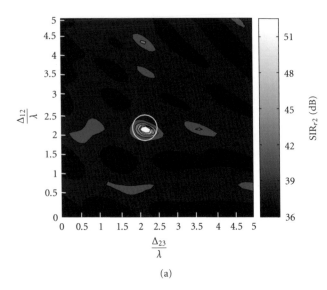 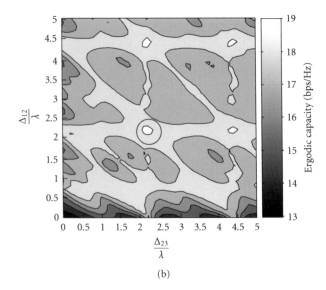

(a)   (b)

FIGURE 6: Setting B: rank-2 approximation of the signal-to-interference ratio, $\text{SIR}_{r2}(\Delta, \overline{\eta})$, (23) versus Δ_{12}/λ and Δ_{23}/λ (a) compared with ergodic capacity $\mathcal{C}(\Delta, \overline{\eta})$ (b) ($r = 50$ m). Circles denote optimal solutions.

where $S(x, \theta_n, \theta_m) = \cos[2\pi x (\sin(\theta_n) - \sin(\theta_m))]$, $L = \left(\binom{N}{2} - N/2 \right)/2$ is the number of "lateral spacings" $l_i = \Delta_{i,j}$ for $i = 1, \ldots, N/2 - 1$, $j = i+1, \ldots, N-i$, $C = N/2$ is the number of "central spacings" $c_i = \Delta_{i,N-i}$ for $i = 1, \ldots, N/2$.

As a remark, notice that if there exists a set of antenna spacing $\widetilde{\Delta}$ such that the user vector $\mathbf{v}_0$ is orthogonal to the M interference vectors $\{\mathbf{v}_i\}_{i=1}^{M}$, then this nulls the interference power, $\mathcal{P}_1(\Delta, \eta) = 0$, and thus implies that $\widetilde{\Delta}$ is a solution to (16) (and therefore to (4)).

4.2. Rank-a ($a > 1$) approximation

In a richer scattering environment, the conditions on the angular spread $\phi_i \ll 1$ that justify the use of rank-1 approximation can not be considered to hold. Therefore, a rank-a approximation with $a > 1$ should be employed (in general) for the spatial correlation matrix of both user and interferers:

$$\mathbf{R}_i \simeq \sum_{k=1}^{a} \rho_i^{(k)} \cdot \mathbf{v}_i^{(k)} \mathbf{v}_i^{(k)H} \qquad (18)$$

for $i = 0, 1, \ldots, M$. The set of vectors $\{\mathbf{v}_i^{(k)}\}_{k=1}^{a}$ in (18) is required to be linearly independent. In this paper, we limit the analysis to the case $a = 2$, which will be shown in Section 5 to account for a wide range of practical environments. The expression of vectors $\mathbf{v}_i^{(k)}$ from (11) with respect to the antenna spacings is not trivial as for the rank-1 case. However, in analogy with (14), we could set

$$\mathbf{v}_i^{(k)} = \left[1 \ \exp\left[-j\omega_i^{(k)} \ \Delta_{12} \right] \ \cdots \ \exp\left[-j\omega_i^{(k)} \ \Delta_{1N} \right] \right]^{T}, \qquad (19)$$

where the wavenumbers $\omega_i = [\omega_i^{(1)}, \omega_i^{(2)}]$ for user and interferers have to be determined according to different criteria. In order to be consistent with the rank-1 case considered in

the previous section, here we minimize the Frobenius norm of approximation error matrix $\| \mathbf{R}_i - \sum_{k=1}^{a} \rho_i^{(k)} \cdot \mathbf{v}_i^{(k)} \mathbf{v}_i^{(k)H} \|^2$ with respect to $\boldsymbol{\omega} = [\omega_i^{(1)}, \omega_i^{(2)}]$ vector and $\boldsymbol{\rho} = [\rho_i^{(1)}, \rho_i^{(2)}]$ vectors. For instance, for a uniform distribution of the scatterers along the ring (i.e., $\kappa = 0$), it can be easily proved that the optimal rank-2 approximation (for $i = 0, \ldots, M$) results in

$$\omega_i^{(1)} = \omega_i(\theta_i) + \varphi_i, \qquad \omega_i^{(2)} = \omega_i(\theta_i) - \varphi_i, \qquad (20)$$

where $\varphi_i = 2\pi/\lambda \cdot \phi_i \cos(\theta_i)$ and $\rho_i^{(1)} = \rho_i^{(2)} = \rho_i/2$.

As for the rank-1 case in (17), after some algebraic manipulations, the performance criterion $\mathcal{P}_2(\Delta, \eta) = E_{\mathbf{h}_0}[\mathbf{h}_0^{H} \mathbf{Q} \mathbf{h}_0]$ admits an explicit expression in terms of the parameters of interest:

$$\mathcal{P}_2(\Delta, \eta) = \sum_{i=1}^{M} \left[\rho_i N + \rho_i \sum_{j=1}^{L} 4S(l_j, \theta_0, \theta_i) \cdot T_i(l_j) \right. \\ \left. + \rho_i \sum_{k=1}^{C} 2S(c_k, \theta_0, \theta_i) T_i(c_k) \right], \qquad (21)$$

where $T_i(x) = \cos(\varphi_0 x) \cos(\varphi_i x)$; notice that in practical environments, the angular spread for the in-cell user, φ_0, is larger than the out-of-cell interferers angular spreads, $\varphi_1, \ldots, \varphi_M$ (see Section 5). Therefore, the optimization problem (4) can be stated as

$$\hat{\Delta} = \arg \min_{\Delta} \mathcal{P}_2(\Delta, \eta). \qquad (22)$$

5. NUMERICAL RESULTS

In this section, numerical results related to the layouts in Figure 1 ($N = 4$, $M = 3$, $F = 3$ for setting A and $F = 7$ for setting B with a cell diameter $\ell = 2$ km) are presented. Both the interference power minimization problems (4), (5)

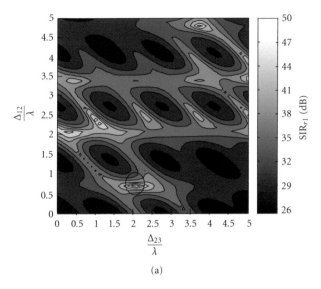
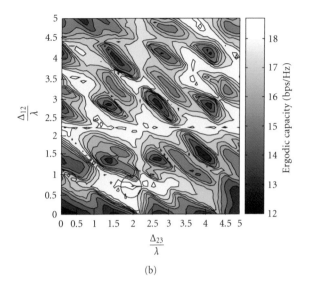

| (a) | (b) |

FIGURE 7: Setting-B: rank-1 approximation (a) of the signal-to-interference ratio $\mathrm{SIR}_{r1}(\Delta, \overline{\eta})$ versus Δ_{12}/λ and Δ_{23}/λ. Circular marker denotes the optimal solution (24) obtained by the rank-1 approximation. As a reference, ergodic capacity is shown (b), for an angular spread approaching zero.

and the ergodic capacity optimization problems (8), (9) for Problems 1 and 2, respectively, are considered and compared for various propagation environments. For Problem 1, user and interferers are located at the center of their respective allowed sectors ($\overline{\eta}$, as in Figure 1), instead, for Problem 2 average system performances are computed over the allowed positions (herein uniformly distributed) of users and interferers.

Exploiting the rank-a-based approximation (rank-1 and rank-2 approximations in (17) and (21), resp.), the interference power (for fixed user and interferers position $\overline{\eta}$ as for Problem 1, or averaged over terminal positions as for Problem 2) is minimized with respect to the array spacings and the resulting optimal solutions are compared to those obtained through maximization of ergodic capacity. Herein, we show that the proposed approach based on interference power minimization is reliable in evaluating the optimal spacings that also maximize the ergodic capacity of the system. Since the number of antenna array is limited to $N = 4$, ergodic capacity optimization can be carried out through an extensive search over the optimization domain.

The channels of user and interferers are assumed to be characterized by the same scatterer radius $r_i = r$ (for the rank-2 case) and $r \to 0$ (for the rank-1 case) with $\kappa = 0$. Furthermore, the path loss exponent is $\alpha = 3.5$. The signal-to-background noise ratio (for the ergodic capacity simulations) is set to $N\rho_0/\sigma^2 = 20$ dB. For the sake of visualization, the rank-a approximation of the interference power is visualized (in dB scale) as the signal-to-interference ratio (SIR):

$$\mathrm{SIR}_{ra}(\Delta, \eta) = \left[\frac{\rho_0}{\mathcal{P}_a(\Delta, \eta)} \right]_{\mathrm{dB}}. \tag{23}$$

5.1. Setting A ($F = 3$)

Assuming at first fixed position $\overline{\eta}$ for user and interferers (Problem 1), Figure 4(b) shows the exact ergodic capacity $\mathcal{C}(\Delta, \overline{\eta})$ for $r = 50$ m (and thus the angular spread is $\phi_0 = 5.75$ deg, $\phi_1 = \phi_3 = 0.87$ deg, $\phi_2 = 0.82$ deg) and Figure 4(a) shows the rank-2 SIR approximation $\mathrm{SIR}_{r2}(\Delta, \overline{\eta})$ (23) versus Δ_{12} and Δ_{23} for setting A. According to both optimization criteria, the optimal array has external spacing $\widehat{\Delta}_{12} \simeq 1.26 \lambda$ and internal spacing $\widehat{\Delta}_{23} \simeq 3.6 \lambda$. It is interesting to compare this result with the case of a line-of-sight channel that is shown in Figure 5. In this latter scenario, the optimal spacings are easily found by solving the rank-1 approximate problem (16) as ($k = 0, 1, \dots$)

$$\widehat{\Delta}_{12} = (2k + 1)\Psi(\theta_1) \quad \text{with any } \widehat{\Delta}_{23} \geq 0 \tag{24a}$$

$$\text{or } \widehat{\Delta}_{12} + \widehat{\Delta}_{23} = (2k + 1)\Psi(\theta_1), \tag{24b}$$

where $\Psi(\theta_1) = \lambda/(2\sin(\theta_1)) \simeq 0.6\lambda$ as $\theta_1 = \theta_2 = 52$ deg. Conditions (24) guarantee that the channel vector of the user is orthogonal to the channel vectors of the first and third interferers (the second is aligned so that mitigation of its interference is not feasible). Moreover, the optimal spacings for the line-of-sight scenario (24) form a grid (see Figure 5(a)) that contains the optimal spacings for the previous case in Figure 4 with larger angular spread. Notice that, for every practical purpose, the solutions to the ergodic capacity maximization (Figure 5(b)) are well approximated by $\mathrm{SIR}_{r1}(\Delta, \overline{\eta})$ maximization in (23). As a remark, we might observe that with line-of-sight channels, there is no advantage of deploying more than two antennas ($\widehat{\Delta}_{12} = 0$ or $\widehat{\Delta}_{23} = 0$ satisfy the optimality conditions (24)) to exploit the interference reduction capability of the array. Instead, for larger angular spread than the line-of-sight case, we can conclude that

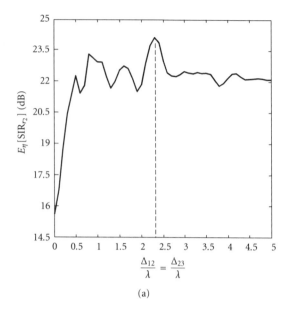

(a)

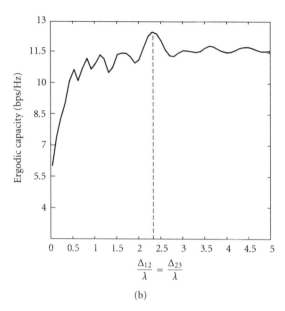

(b)

FIGURE 8: Setting B: rank 2 approximation of the signal-to-interference ratio $E_\eta[\mathrm{SIR}_{r2}(\Delta, \eta)]$ (a) and ergodic capacity $E_\eta[\mathcal{C}(\Delta, \eta)]$ (b) averaged with respect to the position of user and interferers within the corresponding sectors for $\Delta_{12} = \Delta_{23}$.

(i) large enough spacings have to be preferred to accommodate diversity; (ii) contrary to the line-of-sight case, there is great advantage of deploying more than two antennas (approximately 5-6 bit/s/Hz) whereas the benefits of deploying more than three antennas are not as relevant (0.6 bit/s/Hz for an optimally designed three-element array with uniform spacing $3.6\,\lambda$); (iii) compared to the $\lambda/2$-uniformly spaced array, optimizing the antenna spacings leads to a performance gain of approximately 2.5 bit/s/Hz.

Let us now turn to the solution of Problem 2 (9). In this case, the optimal set of spacings $\hat{\Delta}$ should guarantee the best performance on average with respect to the positions of user and interferers within the corresponding sectors. It turns out that the optimal spacings are $\hat{\Delta}_{12} = \hat{\Delta}_{23} \simeq 1.9\,\lambda$ for both optimization criteria (not shown here), and the (average) performance gain with respect to the conventional adaptive arrays with $\hat{\Delta}_{12} = \hat{\Delta}_{23} = \lambda/2$ has decreased to approximately 0.5 bit/s/Hz. This conclusion is substantially different for scenario B as discussed below.

5.2. Setting B (F = 7)

For Problem 1, the exact ergodic capacity $\mathcal{C}(\Delta, \overline{\eta})$ for $r = 50\,\mathrm{m}$ (and angular spread $\phi_0 = 5.75\,\mathrm{deg}$, $\phi_1 = 0.34\,\mathrm{deg}$, $\phi_2 = 0.56\,\mathrm{deg}$, and $\phi_3 = 0.58\,\mathrm{deg}$) and the rank-2 approximation $\mathrm{SIR}_{r2}(\Delta, \overline{\eta})$ (23) are shown versus Δ_{12} and Δ_{23}, in Figure 6, for setting B. In this case, the optimal linear minimum length array consists, as obtained by both optimization criteria, by uniform $2.2\,\lambda$ spaced antennas. Optimal design of linear minimal length array leads to a 2.5 bit/s/Hz capacity gain with respect to the capacity achieved through an array provided with four uniformly $\lambda/2$ spaced antennas. Similarly as before, we compare this result with the case of a line-of-sight channel (Figure 7(a)), where the optimal spacings, solution to the rank-1 approximate problem (16), are

$\hat{\Delta}_{12} = \Psi(\theta_3) \simeq 0.7\,\lambda$ (external spacing) and $\hat{\Delta}_{23} = 3\Psi(\theta_3) \simeq 2.2\,\lambda$ (internal spacing) $\theta_3 = 43.5\,\mathrm{deg}$. In this case, the solutions (confirmed by the ergodic capacity maximization, see Figure 7(b)) guarantee that the channel vector of the user is orthogonal to the channel vector of the third (predominant) interferer (the second is almost aligned so that mitigation of its interference is not feasible, the first one has a minor impact on the overall performances). As pointed out before, a larger angular spread than the line-of-sight case require larger spacings to exploit diversity.

As for Problem 2 (9), in Figure 8, we compare the analytical rank-2 approximation $E_\eta[\mathrm{SIR}_{r2}(\Delta, \eta)]$ averaged over the position of users and interferers with the exact averaged ergodic capacity for a uniform-spaced antenna array. The minimal length optimal solutions turn out again to be $\hat{\Delta}_{12} = \hat{\Delta}_{23} \simeq 2.2\,\lambda$. Moreover, we can conclude that in this interference layout the capacity gain with respect to the capacity achieved through an array provided with four uniformly $\lambda/2$ spaced antennas is 2.5 bit/s/Hz.

5.3. Impact of nonequal power interfering due to shadowing effects

In this section, we investigate the impact of nonequal interfering powers caused by shadowing on the optimal antenna spacings. This amounts to include in the spatial correlation model (12) a log-normal variable for both user and interferers as $\rho_i = (K/d_i^\alpha) \cdot 10^{G_i/10}$ and $G_i \sim \mathcal{N}(0, \sigma_{G_i}^2)$ for $i = 0, 1, \ldots, M$. All shadowing variables $\{G_i\}_{i=0}^M$ affect receiving power levels and are assumed to be independent. Figure 9 shows the ergodic capacity averaged over the shadowing processes for setting B and $r = 50\,\mathrm{m}$ (as in Figure 6), when the standard deviation of the fading processes are $\sigma_{G_0} = 3\,\mathrm{dB}$ for the user (e.g., as for imperfect power control) and $\sigma_{G_i} = 8\,\mathrm{dB}$ $(i = 1, \ldots, M)$ for the interferers. By comparing Figure 9 with

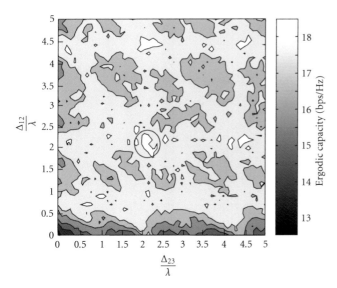

FIGURE 9: Setting B: ergodic capacity $\mathcal{C}(\Delta, \overline{\eta})$ averaged with respect to the distribution of shadowing ($r = 50$ m). Circle denotes the optimal solution.

Figure 6, we see that the overall effect of shadowing is that of reducing the ergodic capacity but not to modify the optimal antenna spacings; similar results can be attained by analyzing the interference power (not shown here).

6. CONCLUSION

In this paper, we tackled the problem of optimal design of linear arrays in a cellular systems under the assumption of Gaussian interference. Two design problems are considered: maximization of the ergodic capacity (through numerical simulations) and minimization of the interference power at the output of the matched filter (by developing a closed form approximation of the performance criterion), for fixed and variable positions of user and interferers. The optimal array deployments obtained according to the two criteria are shown via numerical optimization to coincide for the considered scenarios. The analysis has been validated by studying two scenarios modelling cellular systems with different reuse factors. It is concluded that the advantages of an optimized antenna array as compared to a standard design depend on both the interference layout (i.e., reuse factor) and the propagation environment. For instance, for an hexagonal cellular system with reuse factor 7, the gain can be on average as high as 2.5 bit/s/Hz. As a final remark, it should be highlighted that optimizing the antenna array spacings in such a way to improve the quality of communication (by minimizing the interference power) may render the antenna array unsuitable for other applications where some features of the propagation are of interest, such as localization of transmitters based on the estimation of direction of arrivals.

REFERENCES

[1] J. H. Winters, J. Salz, and R. D. Gitlin, "The impact of antenna diversity on the capacity of wireless communication systems," *IEEE Transactions on Communications*, vol. 42, no. 234, pp. 1740–1751, 1994.

[2] G. J. Foschini and M. J. Gans, "On limits of wireless communications in a fading environment when using multiple antennas," *Wireless Personal Communications*, vol. 6, no. 3, pp. 311–335, 1998.

[3] F. Rashid-Farrokhi, K. J. R. Liu, and L. Tassiulas, "Transmit beamforming and power control for cellular wireless systems," *IEEE Journal on Selected Areas in Communications*, vol. 16, no. 8, pp. 1437–1449, 1998.

[4] J. Saholos, "A solution of the general nonuniformly spaced antenna array," *Proceedings of the IEEE*, vol. 62, no. 9, pp. 1292–1294, 1974.

[5] J.-W. Liang and A. J. Paulraj, "On optimizing base station antenna array topology for coverage extension in cellular radio networks," in *Proceedings of the IEEE 45th Vehicular Technology Conference (VTC '95)*, vol. 2, pp. 866–870, Chicago, Ill, USA, July 1995.

[6] R. Jana and S. Dey, "3G wireless capacity optimization for widely spaced antenna arrays," *IEEE Personal Communications*, vol. 7, no. 6, pp. 32–35, 2000.

[7] A. Abdi and M. Kaveh, "A space-time correlation model for multielement antenna systems in mobile fading channels," *IEEE Journal on Selected Areas in Communications*, vol. 20, no. 3, pp. 550–560, 2002.

[8] P. Zetterberg, "On Base Station antenna array structures for downlink capacity enhancement in cellular mobile radio," Tech. Rep. IR-S3-SB-9622, Department of Signals, Sensors & Systems Signal Processing, Royal Institute of Technology, Stockholm, Sweden, August 1996.

[9] H. L. Van Trees, *Optimum Array Processing*, Wiley-Interscience, New York, NY, USA, 2002.

[10] R. B. Ertel, P. Cardieri, K. W. Sowerby, T. S. Rappaport, and J. H. Reed, "Overview of spatial channel models for antenna array communication systems," *IEEE Personal Communications*, vol. 5, no. 1, pp. 10–22, 1998.

[11] T. Fulghum and K. Molnar, "The Jakes fading model incorporating angular spread for a disk of scatterers," in *Proceedings of the 48th IEEE Vehicular Technology Conference (VTC '98)*, vol. 1, pp. 489–493, Ottawa, Ont, Canada, May 1998.

[12] I. Gupta and A. Ksienski, "Effect of mutual coupling on the performance of adaptive arrays," *IEEE Transactions on Antennas and Propagation*, vol. 31, no. 5, pp. 785–791, 1983.

[13] N. Maleki, E. Karami, and M. Shiva, "Optimization of antenna array structures in mobile handsets," *IEEE Transactions on Vehicular Technology*, vol. 54, no. 4, pp. 1346–1351, 2005.

Hindawi Publishing Corporation
EURASIP Journal on Wireless Communications and Networking
Volume 2007, Article ID 56471, 12 pages
doi:10.1155/2007/56471

Research Article

Capacity Performance of Adaptive Receive Antenna Subarray Formation for MIMO Systems

Panagiotis Theofilakos and Athanasios G. Kanatas

Wireless Communications Laboratory, Department of Technology Education and Digital Systems, University of Piraeus, 80 Karaoli & Dimitriou Street, 18534 Piraeus, Greece

Received 15 November 2006; Accepted 1 August 2007

Recommended by R. W. Heath Jr.

Antenna subarray formation is a novel RF preprocessing technique that reduces the hardware complexity of MIMO systems while alleviating the performance degradations of conventional antenna selection schemes. With this method, each RF chain is not allocated to a single antenna element, but instead to the complex-weighted and combined response of a subarray of elements. In this paper, we derive tight upper bounds on the ergodic capacity of the proposed technique for Rayleigh i.i.d. channels. Furthermore, we study the capacity performance of an analytical algorithm based on a Frobenius norm criterion when applied to both Rayleigh i.i.d. and measured MIMO channels.

1. INTRODUCTION

The interest in multiple-input multiple-output (MIMO) antenna systems has exploded over the last years because of their potential of achieving remarkably high spectral efficiency. However, their practical application has been limited by the increased manufacture cost and energy consumption of the RF chains (performing the frequency transition between microwave and baseband) and analog-to-digital converters, the number of which is proportional to the number of antenna elements.

This high degree of hardware complexity has motivated the introduction of antenna selection schemes, which judiciously choose a subset from all the available antenna elements for processing and thus decrease the number of necessary RF chains. Both analytical [1–11] and stochastic [12] algorithms for antenna selection have been proposed. However, when a limited number of frequency converters are available, antenna selection schemes suffer from severe performance degradations in most fading channels.

In order to alleviate the performance degradations of conventional antenna selection, antenna subarray formation (ASF) has been recently introduced [13]. With this method, each RF chain is not allocated to a single antenna element, but instead to a combined and complex-weighted response of a subarray of antenna elements. Even though additional RF switches (for selecting the antenna elements that participate in each subarray), variable RF phase shifters, or/and variable gain-linear amplifiers (performing the complex-weighting) are required with respect to antenna selection schemes, the proposed method achieves decreased receiver hardware complexity, since less frequency converters and analog-to-digital converters are required with respect to the full system.

Antenna subarray formation actually performs a linear transformation in the RF domain in order to reduce the number of necessary RF chains while taking advantage of the responses of all antenna elements. Since it is a linear preprocessing technique that can be generally applied jointly to both receiver and transmitter, antenna subarray formation can be viewed as a special case of linear precoder-decoder joint designs [14–19]. Indeed, the fundamental mathematical models for both techniques are exactly the same; however, in conventional linear precoding-decoding schemes, preprocessing is performed in the baseband by digital signal processors that are not subject to the practical constraints and hardware nonidealities imposed by the RF components (namely the number of available RF chains, variable phase shifters, or/and variable gain-linear amplifiers) and thus no restrictions on the structure of the preprocessing matrices are required. Instead of decoupling the MIMO channel into independent subchannels (eigenmodes), ASF aims at constructing subchannels (namely, subarrays) that are as mutually independent as possible and deliver the largest receive power gain, under the aforementioned constraints. Note that an RF preprocessing technique for reducing hardware costs has also been introduced in [20], but without grouping antenna elements into subarrays.

Initially, antenna subarray formation was introduced with the restriction that each antenna element participates in one subarray only. For this special case of ASF, the problem of selecting the elements and the weights for the subarray formation has been addressed in [13], where an evolutionary optimization technique is used. In [21], we have introduced an analytical algorithm based on a Frobenius norm criterion. Recognizing that cost-effective analog amplifiers in RF with satisfactory noise figure are practically unavailable, we have also suggested a phase-shift-only design of the technique [22]. Taking into consideration that the performance of ASF may be adversely affected by hardware nonidealities, such as insertion loss, calibration, and phase-shifting errors (which are not an issue in conventional precoder-decoder schemes), we have presented simulation results in [23] that indicate the robustness of ASF to such nonidealities.

In this paper, we elaborate on the capacity performance of ASF and the Frobenius-norm-based algorithm. In particular, we derive a theoretical upper bound on the ergodic capacity of the technique for Rayleigh i.i.d. channels. Moreover, we demonstrate the performance of the technique and the algorithm through extensive computer simulations and application to measured channels.

The rest of the paper is organized as follows: Section 2 explains the proposed technique and its mathematical formulation in more detail, provides capacity calculations for the resulted system and introduces some special ASF schemes. In Section 3, tight theoretical upper bounds on the ergodic capacity of the technique are derived. Section 4 presents an analytical algorithm for ASF and its extensions for several ASF schemes. The capacity performance of the technique and the proposed algorithm is demonstrated in Section 5 through extensive computer simulations. Finally, the paper is concluded with a summary of results.

2. THE ANTENNA SUBARRAY FORMATION TECHNIQUE

In this section, we first present the antenna subarray formation technique and its mathematical formulation. Afterwards, we provide capacity calculations for the resulted system. Finally, some special schemes of ASF are introduced, which are dependent on the number of phase shifters or/and variable gain-linear amplifiers available at the receiver.

2.1. MIMO system model

Consider a flat fading, spatial multiplexing MIMO system with M_T elements at the transmitter and $M_R > M_T$ elements at the receiver. Unless otherwise stated, the $M_R \times M_T$ channel transfer matrix $\mathbf{H}$ is assumed to be perfectly known to the receiver, but unknown to the transmitter.

In spatial multiplexing systems, independent data streams are transmitted simultaneously by each antenna. The received vector for M_R receive elements is given by

$$\mathbf{y} = \mathbf{H}\mathbf{s} + \mathbf{n}, \tag{1}$$

where $\mathbf{n}$ is the zero-mean circularly symmetric complex Gaussian noise vector with covariance matrix $\mathbf{R_n} = N_0 \mathbf{I}_{M_R}$ and $\mathbf{s}$ is the transmitted vector. Assuming that the total transmitter power is P, the covariance matrix for the transmitted vector is constrained as

$$\mathrm{tr}\{E[\mathbf{s}\mathbf{s}^H]\} = P, \tag{2}$$

and the intended average signal-to-noise ratio per antenna at the receiver is

$$\rho = \frac{P}{N_0}. \tag{3}$$

2.2. General mathematical formulation of antenna subarray formation

Antenna Subarray Formation can be applied with any number of RF chains available at the receiver. However, without loss of generality, we assume that the receiver is equipped with exactly M_T RF chains. This assumption is frequently made in antenna selection literature and is justified by the well-known fact that, when the number of receiving RF chains becomes larger than the number of transmit antennas, the number of parallel spatial data pipes that can be opened is constrained by the number of transmit antennas. Thus, the receiver RF chains in excess cannot be exploited to increase the throughput, but can only offer increased diversity order [24]. This assumption is meaningful when the full system channel matrix is of full column rank.

The process of subarray formation, complex weighting and combining at the receiver is linear and thus can be adequately described by the transformation matrix $\mathbf{A}$. In particular, the received vector after antenna subarray formation $\tilde{\mathbf{y}}$ is found by left multiplying the received vector for M_R antenna elements with $\mathbf{A}^H$, that is,

$$\tilde{\mathbf{y}} = \mathbf{A}^H \mathbf{y}. \tag{4}$$

Thus, the response of the jth subarray $\tilde{y}_j$ (i.e., the jth entry of $\tilde{\mathbf{y}}$) is

$$\tilde{y}_j = \boldsymbol{\alpha}_j^H \mathbf{y} = \sum_{i=1}^{M_R} a_{ij}^* y_i, \tag{5}$$

where $\boldsymbol{\alpha}_j$ denotes the jth column of $\mathbf{A}$. Clearly, the response of the jth subarray $\tilde{y}_j$ is a linear combination of the responses of the M_R receiving antenna elements and the conjugated entries of $\boldsymbol{\alpha}_j$ are the corresponding complex weights. Thus, (4) is an adequate mathematical formulation of the subarray formation process, provided that we furthermore enforce the following restriction on the entries of $\mathbf{A}$:

$$a_{ij} = 0, \quad \text{if } i \notin \mathcal{S}_j, \tag{6}$$

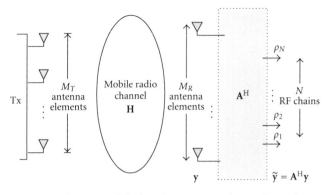

FIGURE 1: System model of receive antenna subarray formation.

with $\mathcal{S}_j$ denoting the set of receive antenna element indices that participate in the jth subarray.

Throughout this paper we assume that the transformation matrix $\mathbf{A}$ is adapted to the instantaneous channel state. Thus, we should have written $\mathbf{A}(\mathbf{H})$, denoting the dependence on the full system channel matrix $\mathbf{H}$. However, to facilitate notation, we just write $\mathbf{A}$ which henceforth implies $\mathbf{A}(\mathbf{H})$.

By substituting (1) into (4), the received vector after subarray formation becomes

$$\widetilde{\mathbf{y}} = \mathbf{A}^{\mathrm{H}}\mathbf{H}\mathbf{s} + \mathbf{A}^{\mathrm{H}}\mathbf{n}. \tag{7}$$

Apparently, the combined effect of the propagation channel and the receive antenna subarrays on the transmitted signal is described by the effective channel matrix

$$\widetilde{\mathbf{H}} = \mathbf{A}^{\mathrm{H}}\mathbf{H}. \tag{8}$$

The effective noise component in (7) is

$$\widetilde{\mathbf{n}} = \mathbf{A}^{\mathrm{H}}\mathbf{n}, \tag{9}$$

which is zero-mean circularly symmetric complex Gaussian vector (ZMCSCGV) [25] with covariance matrix:

$$R_{\widetilde{\mathbf{n}}\widetilde{\mathbf{n}}} = \mathrm{E}[\widetilde{\mathbf{n}}\widetilde{\mathbf{n}}^{\mathrm{H}}] = N_0\mathbf{A}^{\mathrm{H}}\mathbf{A}. \tag{10}$$

The block model of the resulted system is displayed in Figure 1.

2.3. Capacity of receive antenna subarray formation

Depending on the time-variation of the channel, there are different quantities that characterize the capacity of the resulted system. In this paragraph we apply well-known information-theoretic results for MIMO systems to RASF systems and elaborate the capacity of the proposed technique when different assumptions for channel-time variation are made.

2.3.1. Deterministic capacity

Deterministic capacity is a meaningful quantity when the static channel model is adopted, which implies that the channel matrix, despite being random, once chosen it is held fixed

for the whole transmission. In this case, the Shannon capacity of RASF is given in terms of mutual information between the transmitter vector $\mathbf{s}$ and the received vector after subarray formation $\widetilde{\mathbf{y}}$ as

$$C_{\mathrm{RASF}} = \max_{\substack{p(\mathbf{s}) \\ \mathrm{tr}(\mathbf{R_s})=P}} I(\mathbf{s};\widetilde{\mathbf{y}}) = \max_{p(\mathbf{s})}[H(\widetilde{\mathbf{y}}\,|\,\mathbf{H}) - H(\widetilde{\mathbf{y}}\,|\,\mathbf{s},\mathbf{H})], \tag{11}$$

where $H(\mathbf{x})$ is the entropy of $\mathbf{x}$, $p(\mathbf{s})$ denotes the distribution of $\mathbf{s}$ and $\mathrm{tr}(\mathbf{R_s}) = P$ is the power constraint on the transmitter. Recognizing that the transmitted symbols are independent from noise, assuming that $\mathbf{s}$ is ZMCSCGV [25, 26] and taking into account that $\widetilde{\mathbf{n}} \sim \mathcal{N}_C(\mathbf{0}, N_0\mathbf{A}^{\mathrm{H}}\mathbf{A})$, we find that

$$\begin{aligned} C_{\mathrm{RASF}} &= \max_{\substack{p(\mathbf{s}) \\ \mathrm{tr}(\mathbf{R_s})=P}} I(\mathbf{s};\widetilde{\mathbf{y}}) \\ &= \log_2 \det(\pi e \mathbf{R}_{\widetilde{\mathbf{y}}}) - \log_2 \det(\pi e N_0\mathbf{A}^{\mathrm{H}}\mathbf{A}), \end{aligned} \tag{12}$$

where $\mathbf{R}_{\widetilde{\mathbf{y}}} = \mathrm{E}[\widetilde{\mathbf{y}}\widetilde{\mathbf{y}}^{\mathrm{H}}] = \mathbf{A}^{\mathrm{H}}\mathbf{H}\mathbf{R_s}\mathbf{H}^{\mathrm{H}}\mathbf{A} + N_0\mathbf{A}^{\mathrm{H}}\mathbf{A}$ is the covariance matrix of $\widetilde{\mathbf{y}}$. After some mathematical manipulations, (12) becomes

$$C_{\mathrm{RASF}} = \max_{\substack{\mathbf{R_s} \\ \mathrm{tr}(\mathbf{R_s})=P}} \log_2 \det\left[\mathbf{I}_{M_T} + \frac{1}{N_0}\mathbf{R_s}\mathbf{H}^{\mathrm{H}}\mathbf{A}(\mathbf{A}^{\mathrm{H}}\mathbf{A})^{-1}\mathbf{A}^{\mathrm{H}}\mathbf{H}\right]. \tag{13}$$

Since the transmitter does not know the channel and taking into account the power constraint, it is reasonable to assume that

$$\mathbf{R_s} = \frac{P}{M_T}\mathbf{I}_{M_T}. \tag{14}$$

Thus, the Shannon capacity of receive antenna subarray formation with equal power allocation at the transmitter is

$$C_{\mathrm{RASF}} = \log_2 \det\left[\mathbf{I}_{M_T} + \frac{\rho}{M_T}\mathbf{H}^{\mathrm{H}}\mathbf{A}(\mathbf{A}^{\mathrm{H}}\mathbf{A})^{-1}\mathbf{A}^{\mathrm{H}}\mathbf{H}\right]. \tag{15}$$

The capacity of the resulted system is upper bounded by the capacity of the full system, that is

$$C_{\mathrm{RASF}} \le C_{\mathrm{FS}} = \log_2 \det\left(\mathbf{I}_{M_R} + \frac{\rho}{M_T}\mathbf{H}\mathbf{H}^{\mathrm{H}}\right). \tag{16}$$

Proof of this result is given in Appendix A.

2.3.2. Ergodic capacity

In time-varying channels with no delay constraints, ergodic capacity is a meaningful quantity, defined as the probabilistic average of the static channel capacity over the distribution of the channel matrix $\mathbf{H}$. The ergodic capacity for RASF is given by

$$\overline{C}_{\mathrm{RASF}} = \mathrm{E}_{\mathbf{H}}\left[\log_2 \det\left(\mathbf{I}_{M_T} + \frac{\rho}{M_T}\mathbf{H}^{\mathrm{H}}\mathbf{A}(\mathbf{A}^{\mathrm{H}}\mathbf{A})^{-1}\mathbf{A}^{\mathrm{H}}\mathbf{H}\right)\right]. \tag{17}$$

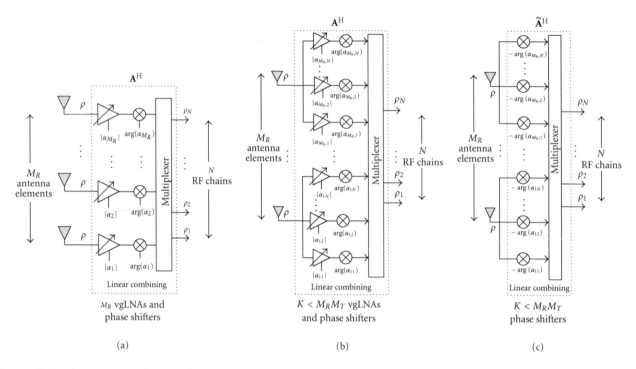

FIGURE 2: Receiver structures for several receive antenna subarray formation (ASF) schemes: (a) strictly-structured ASF (SS-ASF), (b) relaxed-structured ASF (RS-ASF) and (c) reduced hardware complexity ASF (RHC-ASF).

2.3.3. Outage capacity

Outage capacity is a meaningful quantity in slowly varying channels. Assuming a fixed transmission rate R, there is an associated probability P_{out} (bounded away from zero) that the received data will not be received correctly, or equivalently that mutual information will be less than transmission rate R. Outage capacity for RASF is therefore defined as

$$C_{\text{RASF}} = R : \Pr\left\{\log_2 \det\left(\mathbf{I}_{M_T} + \frac{\rho}{M_T}\mathbf{H}^{\text{H}}\mathbf{A}(\mathbf{A}^{\text{H}}\mathbf{A})^{-1}\mathbf{A}^{\text{H}}\mathbf{H}\right) < R\right\}$$

$$= P_{\text{out}}.$$

(18)

2.4. Receive antenna subarray formation schemes

In general, no more constraints on the transformation matrix $\mathbf{A}$ are required. However, depending on the number of available phase shifters or/and variable gain-linear amplifiers (which determine the number of its nonzero entries), further restrictions on matrix $\mathbf{A}$ may be necessary. Motivated by these practical considerations, we have introduced several variations of antenna subarray formation [22], namely, the following.

(1) *Strictly-Structured* ASF (SS-ASF), in which each antenna element is allowed to participate in one subarray only. Thus, each row of the transformation matrix $\mathbf{A}$ may contain only one nonzero element, whereas no restriction is enforced on the columns of $\mathbf{A}$. With this scheme, exactly M_R phase shifters and variable gain-linear amplifiers are required at the receiver.

(2) *Relaxed-Structured* ASF (RS-ASF), where no restrictions on matrix $\mathbf{A}$ are imposed, except for the number of its nonzero entries, which is a fixed system design parameter that determines the number of phase shifters and variable gain-linear amplifiers available to the receiver.

(3) *Reduced Hardware Complexity* ASF (RHC-ASF), which is a phase-shift-only design of the technique. While cost-effective variable gain-linear amplifiers with satisfactory noise figure are not practically available, the economic design and manufacture of variable phase-shifters for the microwave frequency is feasible due to the rapid advances in MMIC technology. Therefore, this scheme reduces even further the hardware complexity of the receiver with negligible capacity loss, as it will be demonstrated in Section 5.

An efficient algorithm for determining the transformation matrix $\mathbf{A}$ for all the aforementioned schemes will be presented in detail in Section 4. Figure 2 presents the receiver architecture for each of the ASF schemes.

3. AN UPPER BOUND ON THE ERGODIC CAPACITY OF ANTENNA SUBARRAY FORMATION FOR I.I.D. RAYLEIGH CHANNELS

In this section, we derive an upper bound on the ergodic capacity of the technique for i.i.d. Rayleigh fading channels, the tightness of which will be verified by extensive computer simulations in Section 5.

A well-known upper bound on the (deterministic) capacity of the full system is given by

$$C_{\text{FS}} \leq \sum_{i=1}^{M_T} \log_2 \left(1 + \frac{\rho}{M_T} \gamma_i \right), \qquad (19)$$

where γ_i are independent chi-squared variates with $2M_R$ degrees of freedom. The equality holds in the "very artificial case" when the transmitted signal vector components "are conveyed over M_T "channels" that are uncoupled and each channel has a separate set of M_R receive antennas" [27]. In other words, when the full MIMO system is consisted of M_T separable and independent parallel SIMO systems, each performing maximum ratio combining (MRC) at the receiver.

In our case, we consider as well that the resulted system is consisted of M_T separable and independent parallel SIMO systems. We suppose that the jth SIMO system is formed by the jth transmit antenna element and the jth receive subarray; thus, for each subarray, only one signal component is received and processed without any interference from the others. Of course, this scheme is practically infeasible; however, it must lead to an upper bound of the resulted system capacity.

A subarray corresponds to an independent SIMO system and is actually formed by choosing a subset of antenna elements, the responses of which are linearly combined and fed to an RF chain. Thus, generalized selection combining (i.e., combining the responses of a subset of antenna elements) is performed in each SIMO system. The maximum SNR (which also achieves maximum capacity) in this case is obtained with the hybrid selection maximum ratio combining scheme (HS/MRC). Furthermore, in this section, we assume that each subarray is formed using a predefined and fixed number of antenna elements (let it be k_j antenna elements for the jth subarray). Therefore, a capacity bound for antenna subarray formation can be obtained by

$$C_{\text{bound}} = \sum_{j=1}^{M_T} \log_2 \left(1 + \xi_j \right). \qquad (20)$$

Assuming that there are no delay constraints, the channel is *ergodic* and therefore it is meaningful to derive an upper bound on *ergodic capacity* as

$$\overline{C}_{\text{bound}} = \sum_{j=1}^{M_T} \text{E} \left[\log_2 \left(1 + \xi_j \right) \right]. \qquad (21)$$

The expectation in (21) can be found [28] by

$$\overline{c}_j \triangleq \text{E} \left[\log_2 \left(1 + \xi_j \right) \right] = \int_0^\infty \log_2 (1 + \xi) \cdot p_{\xi_j}(\xi) d\xi. \qquad (22)$$

Since ξ_j is actually the postprocessing SNR of HS/MRC when k_j out of M_R elements are chosen, its probability density function is [29]

$$p_{\xi_j}(\xi) = \binom{M_R}{k_j} \left[\left(\frac{M_T}{\rho} \right)^{k_j} \frac{\xi^{k_j-1} e^{-(M_T/\rho)\xi}}{(k_j-1)!} \right.$$
$$+ \frac{M_T}{\rho} \sum_{l=1}^{M_R-k_j} (-1)^{k_j+l-1} \binom{M_R-k_j}{l}$$
$$\times \left(\frac{k_j}{l} \right)^{k_j-1} e^{-(M_T/\rho)\xi}$$
$$\times \left(e^{-(M_T l/\rho k_j)\xi} - \sum_{m=0}^{k_j-2} \frac{1}{m!} \left(-\frac{l \cdot M_T}{\rho \cdot k_j} \xi \right)^m \right) \right].$$
$$(23)$$

Substituting (23) into (22) and defining the integral

$$\mathcal{I}_n(x) \triangleq \int_0^\infty t^{n-1} \ln(1+t) e^{-xt} dt \qquad x > 0; \ n = 1, 2, \dots, \qquad (24)$$

we get

$$\overline{c}_j = \frac{1}{\ln 2} \binom{M_R}{k_j} \left[\left(\frac{M_T}{\rho} \right)^{k_j} \frac{\mathcal{I}_{k_j}(M_T/\rho)}{(k_j-1)!} \right.$$
$$+ \frac{M_T}{\rho} \sum_{l=1}^{M_R-k_j} (-1)^{k_j+l-1} \binom{M_R-k_j}{l} \left(\frac{k_j}{l} \right)^{k_j-1}$$
$$\times \left[\mathcal{I}_1 \left(\frac{M_T}{\rho} \left\{ 1 + \frac{l}{k_j} \right\} \right) - \sum_{m=0}^{k_j-2} \frac{1}{m!} \right.$$
$$\times \left(-\frac{l \cdot M_T}{\rho \cdot k_j} \right)^m \mathcal{I}_{m+1}(M_T/\rho) \right] \right],$$
$$(25)$$

which, in fact, is the average channel capacity achieved when employing HS/MRC in a SIMO system with M_R receiving antenna elements and k_j branches.

The integral $\mathcal{I}_n(x)$ can be evaluated by [30]

$$\mathcal{I}_n(x) = (n-1)! \cdot e^x \cdot \sum_{q=1}^n \frac{\Gamma(-n+q, x)}{x^q}, \qquad (26)$$

which for $n = 1$ reduces to

$$\mathcal{I}_1(x) = e^x \frac{E_1(x)}{x}. \qquad (27)$$

Note that $E_1(x)$ is the exponential integral of first-order function defined by

$$E_1(x) = \int_x^\infty \frac{e^{-t}}{t} dt \qquad (28)$$

and $\Gamma(\alpha, x)$ is the complementary incomplete gamma function (or Prym's function) defined as

$$\Gamma(\alpha, x) = \int_x^\infty t^{\alpha-1} e^{-t} dt. \qquad (29)$$

For q positive integer, $\Gamma(-q,x)$ can be calculated by

$$\Gamma(-q,x) = \frac{(-1)^n}{n!}\left[E_1(x) - e^{-x}\sum_{m=0}^{q-1}(-1)^m\frac{m!}{x^{m+1}}\right]. \quad (30)$$

Thus, the ergodic capacity bound for receive antenna subarray formation can be analytically obtained by

$$
\begin{aligned}
C_{\text{bound}} = \frac{1}{\ln 2}\sum_{j=1}^{M_T}\binom{M_R}{k_j}\\
\times\left[\left(\frac{M_T}{\rho}\right)^{k_j}\frac{\mathcal{J}_{k_j}(M_T/\rho)}{(k_j-1)!} + \frac{M_T}{\rho}\sum_{l=1}^{M_R-k_j}(-1)^{k_j+l-1}\right.\\
\times\binom{M_R-k_j}{l}\left(\frac{k_j}{l}\right)^{k_j-1}\\
\times\left[\mathcal{J}_1\left(\frac{M_T}{\rho}\left\{1+\frac{l}{k_j}\right\}\right) - \sum_{m=0}^{k_j-2}\frac{1}{m!}\right.\\
\left.\left.\times\left(-\frac{l\cdot M_T}{\rho\cdot k_j}\right)^m\mathcal{J}_{m+1}(M_T/\rho)\right]\right].
\end{aligned}
$$
$$(31)$$

A simpler expression than (25) can be derived by recognizing that $\log_2(\cdot)$ is a concave function and applying Jensen's inequality to (21),

$$\bar{c}_j = E\left[\log_2\left(1+\xi_j\right)\right] \le \log_2\left(1+E\left[\xi_j\right]\right). \quad (32)$$

It is known for HS/MRC [29] that

$$E[\xi_j] = \frac{\rho}{M_T}k_j\left(1 + \sum_{l=k_j+1}^{M_R}\frac{1}{l}\right). \quad (33)$$

Thus, (21) becomes

$$\overline{C}_{\text{bound}} \le \sum_{j=1}^{M_T}\log_2\left[1 + \frac{\rho}{M_T}k_j\left(1 + \sum_{l=k_j+1}^{M_R}\frac{1}{l}\right)\right], \quad (34)$$

which has a much simpler form than (31) while being almost as tight as computer simulations have demonstrated.

Before concluding this section, we note that analyzing the resulted system into parallel SIMO systems each performing HS/MRC results into capacity bounds of RS-ASF, since HS/MRC requires both phase shifters and variable gain amplifiers. Capacity bounds for RHC-ASF could be derived in a similar manner by considering M_T parallel SIMO systems each performing HS/EGC. Since HS/MRC delivers the best performance amongst all hybrid selection schemes, the upper bound on the ergodic capacity of RS-ASF is also an upper bound on the ergodic capacity of any ASF scheme, including RHC-ASF.

4. ALGORITHM FOR ANTENNA SUBARRAY FORMATION

In this section, we present a novel, analytical algorithm for receive antenna subarray formation, based on a Frobenius norm criterion. We first develop the algorithm for SS-ASF and then provide extensions for RS-ASF and RHC-ASF. The capacity performance of the algorithms will be demonstrated in Section 5.

4.1. Starting point for the algorithm

The starting point for determining the transformation matrix $\mathbf{A}$ will be an optimal solution to the *unconstrained* problem of maximizing the deterministic capacity in (15). As shown in Appendix A, (15) can be maximized when $\mathbf{A}_o = \mathbf{U}$, where the columns of $\mathbf{U}$ are the M_T dominant left singular vectors of the full channel matrix $\mathbf{H}$. Therefore, the entries of the transformation matrix $\mathbf{A}$ will be

$$a_{ij} = \begin{cases} u_{ij} & \text{if } i\in\mathcal{S}_j \\ 0 & \text{otherwise,} \end{cases} \quad (35)$$

with u_{ij} being the (i,j) entry of matrix $\mathbf{U}$. Alternatively,

$$\mathbf{A} = \mathbf{S}\odot\mathbf{U}, \quad (36)$$

where $\odot$ denotes the Hadamard (elementwise) matrix product and the entries of $\mathbf{S}$ are

$$s_{ij} = \begin{cases} 1 & i\in\mathcal{S}_j \\ 0 & \text{otherwise.} \end{cases} \quad (37)$$

4.2. Frobenius norm based algorithm for SS-ASF

We first develop an algorithm for SS-ASF and afterwards extend it for other receive ASF schemes. Due to the additional constraints of SS-ASF, the capacity of the resulted system is given by

$$
\begin{aligned}
C_{\text{RASF}} &= \log_2\det\left(\mathbf{I}_{M_T} + \frac{\rho}{M_T}\mathbf{H}^{\text{H}}\mathbf{A}\mathbf{A}^{\text{H}}\mathbf{H}\right)\\
&= \log_2\det\left(\mathbf{I}_{M_T} + \frac{\rho}{M_T}\tilde{\mathbf{H}}^{\text{H}}\tilde{\mathbf{H}}\right).
\end{aligned}
$$
$$(38)$$

In order to retain the capacity calculations to the intended system SNR measured at the output of every receiver antenna element, $\mathbf{A}$ is now subject to the following normalization:

$$\mathbf{A}^{\text{H}}\mathbf{A} = \mathbf{I}_{M_T}. \quad (39)$$

Intuitively, the desired transformation matrix $\mathbf{A}$ should be such that the distance between the two subspaces defined by $\tilde{\mathbf{H}}_{\text{opt}} = \mathbf{U}^{\text{H}}\mathbf{H}$ (i.e., the effective channel matrix obtained from the optimal solution to the unconstrained problem) and $\tilde{\mathbf{H}} = \mathbf{A}^{\text{H}}\mathbf{H}$ is minimized. As a result, we employ the following minimum distance distortion metric:

$$\varepsilon(\mathbf{A}) = \left\|\tilde{\mathbf{H}}_{\text{opt}} - \tilde{\mathbf{H}}\right\|_{\text{F}}^2 = \left\|(\mathbf{U}-\mathbf{A})^{\text{H}}\mathbf{H}\right\|_{\text{F}}^2. \quad (40)$$

Defining $\mathbf{E} \triangleq \mathbf{U} - \mathbf{A}$ and $\mathbf{F} \triangleq \mathbf{E}^{\text{H}}\mathbf{H}$, (40) can be written as

$$\varepsilon(\mathbf{A}) = \|\mathbf{F}\|_{\text{F}}^2 = \sum_{j=1}^N\left(\sum_{i=1}^{M_T}|f_{ji}|^2\right) = \sum_{j=1}^{M_T}\|\mathbf{f}_j\|^2, \quad (41)$$

TABLE 1: Frobenius-norm-based algorithm for RASF.

Algorithm steps (K, M_R, M_T, and $\mathbf{H}$ are given) (In case of SS-ASF, $K := M_R$)		Complexity
Obtain the SVD of full system channel matrix $\mathbf{H}$.	$\mathbf{H} = \mathbf{U}\boldsymbol{\Sigma}\mathbf{V}^{\mathrm{H}}$	$O(12M_T M_R^2 + 9M_R^3)$

Compute the decision metrics g_{ij} that will determine if the ith antenna element will participate in the jth subarray.	For $i := 1$ to M_R
	    For $j := 1$ to M_T
	        $g_{ij} := U(i,j) \cdot \|\mathbf{H}(i,:)\|^2$
	    end
	end
		$O(M_T^2 M_R)$

Initialize with every $a_{ij} = 0$ and all S_j empty. S_j: set of indices of antenna elements that participate in the jth subarray.	$S_j := \varnothing$ $(\forall j = 1, \ldots, M_T)$ $\mathbf{A} := \mathbf{0}_{M_R \times M_T}$; $n := 0$	

Repeat the following until matrix $\mathbf{A}$ is filled with K nonzero elements:
    (i) let (i_0, j_0) be the indices of the largest g_{ij} element over $1 \leq i \leq M_R$ and $1 \leq j \leq M_T$, provided that $a_{ij} = 0$;
for SS-ASF only, $i \notin \bigcup_j S_j$;
    (ii) set $a_{i_0 j_0} = u_{i_0 j_0}$, that is, the i_0th antenna element participates in the j_0th subarray;
 	
for SS-ASF only, normalize $\mathbf{A}$ so that
 	
$\mathbf{A}^{\mathrm{H}}\mathbf{A} = \mathbf{I}_{M_T}$.
	While $n < K$
	
	    $(i_0, j_0) = \underset{\substack{(i,j) \\ a_{ij}=0}}{\arg\max}\,(g_{ij})$
	    $S_{j_0} := S_{j_0} \cup \{i_0\}$
	    $A(i_0, j_0) := U(i_0, j_0)$
	    $n := n + 1$
	end
	
	For SS-ASF only:
	
	For $j = 1{:}M_T$
	    $\mathbf{A}(:,j) := \mathbf{A}(:,j)/\|\mathbf{A}(:,j)\|$
	end
		$O(KM_R M_T)$

where $\mathbf{f}_j$ denotes the jth row of $\mathbf{F}$, being equal to $\mathbf{f}_j = \mathbf{e}_j^{\mathrm{H}}\mathbf{H}$, and $\mathbf{e}_j$ is the jth column of matrix $\mathbf{E}$.

Recognizing that the ith row of matrix $\mathbf{F}$ can be written as a linear combination of the rows $\mathbf{h}_i$ of the full system channel matrix $\mathbf{H}$ and taking into account that

$$e_{ij} \overset{\wedge}{=} u_{ij} - a_{ij} = \begin{cases} u_{ij} & i \notin \mathcal{S}_j \\ 0 & i \in \mathcal{S}_j, \end{cases} \quad (42)$$

the distortion metric becomes

$$\varepsilon(\mathbf{A}) = \sum_{j=1}^{M_T} \left\| \sum_{i \in \mathcal{S}_j} e_{ij}^* \mathbf{h}_i \right\|^2 = \sum_{j=1}^{M_T} \left\| \sum_{i \notin \mathcal{S}_j} u_{ij}^* \mathbf{h}_i \right\|^2 \leq \sum_{j=1}^{M_T} \sum_{i \notin \mathcal{S}_j} |u_{ij}|^2 \|\mathbf{h}_i\|^2, \quad (43)$$

where the upper bound on the right-hand side follows from the triangular inequality. As a result, the objective is to minimize the upper bound on the distortion metric in (43).

Since the selection of the elements of the transformation matrix $\mathbf{A}$ is based on matrix $\mathbf{U}$, it is trivial to conclude that minimizing the upper bound in (43) is equivalent to maximizing

$$\widetilde{p} = \sum_{j=1}^{M_T} \sum_{i \in \mathcal{S}_j} |u_{ij}|^2 \|\mathbf{h}_i\|^2, \quad (44)$$

which upper-bounds the power of the effective channel matrix $\|\widetilde{\mathbf{H}}\|_{\mathrm{F}}^2$. Indeed, after mathematical manipulations similar to those in (41)–(43), it follows that

$$\|\widetilde{\mathbf{H}}\|_F^2 = \sum_{j=1}^{M_T} \left\| \sum_{i \in \mathcal{S}_j} u_{ij}^* \mathbf{h}_i \right\|^2 \leq \sum_{j=1}^{M_T} \sum_{i \in \mathcal{S}_j} |u_{ij}|^2 \|\mathbf{h}_i\|^2 = \widetilde{p}, \quad (45)$$

where $\widetilde{\mathbf{h}}_j$ denotes the jth row of $\widetilde{\mathbf{H}}$ and $\boldsymbol{\alpha}_j$ is the jth column of matrix $\mathbf{A}$. Consequently, minimizing an upper bound on the minimum distance distortion metric is equivalent to maximizing an upper bound on the power of the effective channel matrix. The latter may not be the optimal way to maximize capacity in spatial multiplexing systems, but it should result into an increased capacity performance, since it is known that [24]

$$C_{\text{SS-ASF}} \geq \log_2 \det\left(1 + \frac{\rho}{M_T} \|\widetilde{\mathbf{H}}\|_{\mathrm{F}}^2\right). \quad (46)$$

The proposed algorithm appoints the receiver antenna elements to the appropriate subarray, so that the metric (44) is maximized. Finally, $\mathbf{A}$ is normalized as in (39). Table 1 presents the algorithm steps in more detail.

4.3. Extension of the algorithm for RS-ASF

The capacity of RS-ASF given by (15) is lower bounded by the capacity formula (38) for SS-ASF, that is,

$$C_{\text{RS-ASF}} \geq \log_2 \det\left(\mathbf{I}_{M_T} + \frac{\rho}{M_T}\mathbf{H}^{\text{H}}\mathbf{A}\mathbf{A}^{\text{H}}\mathbf{H}\right). \quad (47)$$

Proof of this result and indications for the tightness of the bound are provided in Appendix B.

Thus, in the case of RS-ASF we also use the Frobenius norm based algorithm initially developed for SS-ASF. The algorithm terminates when the transformation matrix $\mathbf{A}$ contains exactly K nonzero elements, where $K < M_R M_T$ is a system design parameter that determines the number of variable gain-linear amplifiers and phase shifters available to the receiver.

The computational complexity of the proposed algorithm (see Table 1) is dominated by the initial cost of the singular value decomposition, that is, $O(M_R^3)$ when $M_R \gg M_T$, whereas the complexity of Gorokhov et al. algorithm [4] and of the alternative implementation proposed in [5] for antenna selection is $O(M_T^2 M_R^2)$ and $O(M_T^2 M_R)$, respectively.

4.4. Extention of the algorithm for RHC-ASF

The transformation matrix $\breve{\mathbf{A}}$ for RHC-ASF (a phase-shift-only design of antenna subarray formation) can be obtained from the transformation matrix $\mathbf{A}$ for RS-ASF by applying the following formula to its entries:

$$\breve{a}_{ij} = \begin{cases} \exp(-j \mid \underline{a_{ij}}) & \text{if } i \in \mathcal{S}_j \\ 0 & \text{otherwise.} \end{cases} \quad (48)$$

Intuitively, RHC-ASF follows the notion of equal gain combining. A similar procedure for obtaining a phase-shift-only RF preprocessing technique has been followed in [20].

5. SIMULATION RESULTS

In this section, we present extensive computer simulation results that demonstrate the capacity performance of receive ASF technique, the tightness of the ergodic capacity bounds derived in Section 3, and the performance of the proposed algorithm.

5.1. Upper bound on ergodic capacity for ASF

We first deal with the ergodic capacity bounds of ASF for Rayleigh i.i.d. channels derived in Section 3, namely, (31) and (34). Henceforth, we refer to (34) as "simpler theoretical capacity bound," in order to distinguish it from (31). We consider a flat-fading Rayleigh i.i.d. MIMO channel with $M_R = 8$ receiving and $M_T = 2$ transmitting antenna elements and assume that the receiver is equipped with $N = M_T = 2$ RF chains.

Figure 3 presents the ergodic capacity bounds of RS-ASF over a wide range of SNRs when $K = 8$ variable gain-linear amplifiers and phase shifters are available at the receiver and

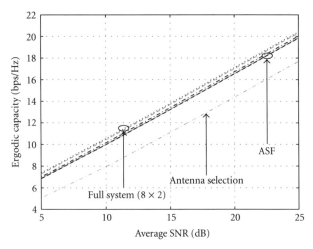

— Exhaustive search ASF
--- Full system (exact capacity)
--- Antenna selection (exact capacity)
--- Theoretical capacity bound of ASF (34)
······ Theoretical capacity bound of full system (34)
-·-· Simpler theoretical capacity bound for ASF (37)

FIGURE 3: Ergodic capacity bounds for ASF and capacity of exhaustive search ASF when $M_R = 8$, $M_T = 2$, and $K = 8$ variable gain-linear amplifiers and phase shifters are available at the receiver (4 antenna elements in each subarray). Results are compared to an ergodic capacity bound and exact ergodic capacity of the full system.

exactly $k \triangleq K/N = 4$ receiving antenna elements participate in each subarray. For purposes of reference, the ergodic capacity of the exhaustive search solution of RS-ASF is also shown. The exhaustive search solution is obtained by considering all the $\binom{M_R}{k}^N$ possible combinations of subarray formation, that is, all possible combinations for the structure of matrix $\mathbf{S}$ as defined in (37), assuming that $\mathbf{A}$ is obtained as in (36). Apparently, both capacity bounds are very tight to the exhaustive search solution.

When each subarray contains M_R antenna elements, the capacity bound of the MIMO system is found by analyzing it into M_T parallel SIMO systems. Each of these parallel systems reduces to a MRC diversity system and therefore the ergodic capacity bound of the full system will be obtained by (31). This observation is verified in Figure 3.

5.2. Frobenius-norm-based algorithm

In this paragraph we demonstrate the capacity performance of the Frobenius-norm-based algorithm for various schemes of receive ASF in terms of outage capacity (when the slowly-varying block fading channel model is adopted) and ergodic capacity (when the channel is assumed ergodic). The proposed algorithm is applied to both Rayleigh i.i.d. and measured MIMO channels.

5.2.1. Rayleigh i.i.d. channels

We consider Rayleigh i.i.d. MIMO channels with $M_T = 2$ elements at the transmitter and assume that the receiver is

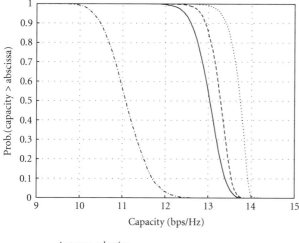

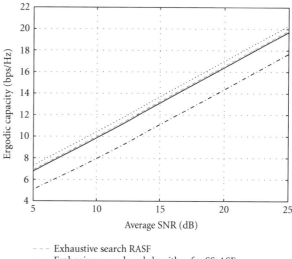

‑·‑ Antenna selection
——— Frobenius norm based algorithm for RASF ($K = 8$)
‑‑‑ Exhaustive search RASF ($K = 8$)
······ Full system (8×2)

FIGURE 4: Empirical complementary cdf of the capacity of the resulted system when the Frobenius-norm-based algorithm for strictly structured receive antenna subarray formation (SS-ASF) is applied to a 8×2 Rayleigh i.i.d. channel with SNR = 15 dB. The performance of the algorithm is compared with the exhaustive search solution for SS-ASF, the full system (8×2), and Gorokhov et al. decremental algorithm for antenna selection.

‑‑‑ Exhaustive search RASF
——— Frobenius norm based algorithm for SS-ASF
······ Full system (8×2)
‑·‑ Antenna selection

FIGURE 5: Performance evaluation of strictly structured ASF (SS-ASF) applied to an 8×2 MIMO Rayleigh i.i.d. channel, in terms of ergodic capacity. The performance of the algorithm is compared to the exhaustive search solution for receive ASF, the full system (8×2), and Gorokhov et al. decremental algorithm for antenna selection.

equipped with $M_T = 8$ elements, $N = M_T = 2$ RF chains, and $K = 8$ phase shifters or/and variable gain-linear amplifiers.

Figure 4 presents the complementary cdf of the capacity of the resulted system for SS-ASF when the SNR is at 15 dB. Clearly, SS-ASF outperforms Gorokhov et al. algorithm for antenna selection [4], which is quasi optimal in terms of capacity performance. Moreover, the performance of the proposed algorithm is very close to the exhaustive search solution. Thus, the SS-ASF technique delivers a significant capacity increase with respect to conventional antenna selection schemes. The same results are verified in Figure 5, where the ergodic capacity of the resulted system over a wide range of SNRs is plotted.

5.2.2. Measured channel

In order to examine the performance in realistic conditions, we have applied the proposed algorithm to measured MIMO channel transfer matrices. Measurements were conducted using a vector channel sounder operating at the center frequency of 5.2 GHz with 120 MHz measurement bandwidth in short-range outdoor environments with LOS propagation conditions. A more detailed description of the measurement setup can be found in [31]. The transmitter has $M_T = 4$ equally spaced antenna elements and the receiver is equipped with $M_R = 16$ receiving elements and $N = M_T = 4$ RF chains. The interelement distance for both the transmitting and receiving antenna arrays is $d = 0, 4\lambda$.

Figure 6 displays the complementary cdf of the capacity of the resulted system when the Frobenius-norm-based algorithm is applied to several schemes of receive ASF and for various values of K (i.e., the number of phase shifters or/and variable gain-linear amplifiers). Clearly, all ASF schemes outperform conventional antenna selection.

Solid black lines correspond to RS-ASF (or SS-ASF for $K = M_R = 16$) and dashed black lines to RHC-ASF. Comparing the solid with the dashed lines for the same value of K, it is evident that RHC-ASF delivers capacity performance very close to RS-ASF. Therefore, the expensive variable gain-linear amplifiers can be abolished from the design of ASF with negligible capacity loss.

For $K = 48$, the capacity performance of RS-ASF and RHC-ASF is very close to the full system, despite the fact that in ASF the receiver is equipped with only $N = M_T = 4$ RF chains (whereas the full system has $M_R = 16$ RF chains). Even when $K = 32$, the capacity loss with respect to the full system is still quite low (10% outage capacity loss of RHC-ASF is less than 1.5 bps/Hz at 15 dB). Similar results are observed for a wide range of signal-to-noise ratios (Figure 7). Consequently, the proposed algorithm can deliver near-optimal capacity performance with respect to the full system while reducing drastically the number of necessary RF chains.

6. CONCLUSIONS

In this paper, we have developed a tight theoretical upper bound on the ergodic capacity of antenna subarray formation and have presented an analytical algorithm for

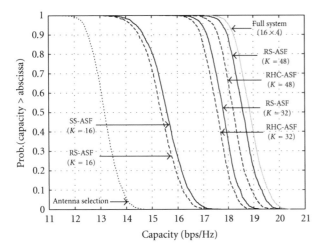

FIGURE 6: Empirical complementary cdf of the capacity of the resulted system when the Frobenius-norm-based algorithm for several schemes of receive antenna subarray formation (ASF) is applied to a 16×4 measured channel with SNR = 15 dB. In particular, the RASF schemes studied are strictly structured ASF (SS-ASF), relaxed-structured ASF (RS-ASF), and reduced hardware complexity ASF (RHC-ASF). K denotes the number of phase shifters or/and variable gain-linear amplifiers available to the receiver. The performance of the algorithm is compared to the full system (16×4) and Gorokhov et al. decremental algorithm for antenna selection.

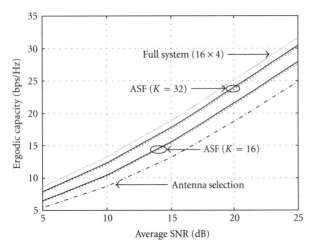

FIGURE 7: Performance evaluation of Frobenius-norm-based algorithm for several schemes of receive antenna subarray formation (RASF) applied to a 16×4 MIMO measured channel, in terms of ergodic capacity. In particular, the RASF schemes studied are strictly structured ASF (SS-ASF), relaxed-structured ASF (RS-ASF) (solid lines), and reduced hardware complexity ASF (RHC-ASF) (dotted lines). K denotes the number of phase shifters or/and variable gain-linear amplifiers available to the receiver. The performance of the algorithm is compared to the full system (16×4) and Gorokhov et al. decremental algorithm for antenna selection.

adaptively grouping receive array elements to subarrays. Application in Rayleigh i.i.d. and measured channels demonstrates significant capacity performance, which can become near optimal with respect to the full system, depending on

the number of available phase shifters or/and variable gain-linear amplifiers. Furthermore, it has been shown that a phase-shift-only design of the technique is feasible with negligible performance penalty. Thus, it has been established that antenna subarray formation is a promising RF preprocessing technique that reduces hardware costs while achieving incredible performance enhancement with respect to conventional antenna selection schemes.

APPENDICES

A.

Let $\mathbf{A} = \mathbf{U}_A \boldsymbol{\Sigma}_A \mathbf{V}_A^H$ be a singular value decomposition [32] of matrix $\mathbf{A}$. We get

$$
\begin{aligned}
\mathbf{A}(\mathbf{A}^H \mathbf{A})^{-1} \mathbf{A}^H &= \mathbf{U}_A \boldsymbol{\Sigma}_A \mathbf{V}_A^H (\mathbf{V}_A \boldsymbol{\Sigma}_A^2 \mathbf{V}_A^H)^{-1} \mathbf{V}_A \boldsymbol{\Sigma}_A \mathbf{U}_A^H \\
&= \mathbf{U}_A \boldsymbol{\Sigma}_A \mathbf{V}_A^H \mathbf{V}_A \boldsymbol{\Sigma}_A^{-2} \mathbf{V}_A^H \mathbf{V}_A \boldsymbol{\Sigma}_A \mathbf{U}_A^H \qquad \text{(A.1)} \\
&= \mathbf{U}_A \mathbf{U}_A^H.
\end{aligned}
$$

Thus, the capacity formula in (15) becomes

$$
C_{\text{RASF}} = \log_2 \det\left(\mathbf{I}_{M_T} + \frac{\rho}{M_T} \mathbf{H}^H \mathbf{U}_A \mathbf{U}_A^H \mathbf{H} \right). \qquad \text{(A.2)}
$$

Applying the known formula for determinants [32]

$$
\det(\mathbf{I} + \mathbf{A}\mathbf{B}) = \det(\mathbf{I} + \mathbf{B}\mathbf{A}) \qquad \text{(A.3)}
$$

to (A.2), we get

$$
C_{\text{RASF}} = \log_2 \det\left(\mathbf{I}_{M_T} + \frac{\rho}{M_T} \mathbf{U}_A^H \mathbf{H}\mathbf{H}^H \mathbf{U}_A \right) \qquad \text{(A.4)}
$$

which can be written as

$$
C_{\text{RASF}} = \sum_{m=1}^{M_T} \log_2\left(1 + \frac{\rho}{M_T} \lambda_m\left(\mathbf{U}_A^H \mathbf{H}\mathbf{H}^H \mathbf{U}_A \right) \right), \qquad \text{(A.5)}
$$

where $\lambda_m(\mathbf{X})$ denotes the mth eigenvalue of square matrix $\mathbf{X}$ in descending order. Poincare separation theorem [32] states that

$$
\lambda_m\left(\mathbf{U}_A^H \mathbf{H}\mathbf{H}^H \mathbf{U}_A \right) \le \lambda_m\left(\mathbf{H}\mathbf{H}^H \right) \qquad \text{(A.6)}
$$

with equality occurring when the columns of $\mathbf{U}_A$ are the M_T dominant left singular vectors of $\mathbf{H}$. Thus,

$$
\begin{aligned}
C_{\text{RASF}} &\le \sum_{k=1}^{M_T} \log_2\left(1 + \frac{\rho}{M_T} \lambda_k\left(\mathbf{H}\mathbf{H}^H \right) \right) \\
&= \log_2 \det\left(\mathbf{I}_{M_R} + \frac{\rho}{M_T} \mathbf{H}\mathbf{H}^H \right) = C_{\text{FS}},
\end{aligned} \qquad \text{(A.7)}
$$

where equality occurs when

$$\mathbf{U}_A = \begin{bmatrix} \mathbf{u}_1 & \mathbf{u}_2 & \cdots & \mathbf{u}_{M_T} \end{bmatrix} \tag{A.8}$$

and $\mathbf{u}_k$ is the kth dominant singular vector of $\mathbf{H}$. Therefore, an *optimal* solution to the *unconstrained* (i.e., without the subarray formation constraints in (6) capacity maximization problem is

$$\mathbf{A}_o = \begin{bmatrix} \mathbf{u}_1 & \mathbf{u}_2 & \cdots & \mathbf{u}_{M_T} \end{bmatrix} \mathbf{Q}, \tag{A.9}$$

where $\mathbf{Q} = \Sigma_A \mathbf{V}_A^H$ is a matrix with orthogonal rows and columns.

B.

Let $\mathbf{A} = \mathbf{U}_A \Sigma_A \mathbf{V}_A^H$ be a singular value decomposition of the transformation matrix $\mathbf{A}$. Exploiting Hadarmard's inequality for determinants [32] and after some trivial mathematical manipulations, it follows that

$$\det\left(\Sigma_A^2\right) = \det\left(\mathbf{V}_A \Sigma_A^2 \mathbf{V}_A^H\right) = \det\left(\mathbf{A}^H \mathbf{A}\right) \le \prod_{k=1}^{M_T} \left[\mathbf{A}^H \mathbf{A}\right]_{kk}$$

$$= \prod_{k=1}^{M_T} \mathbf{a}_k^H \mathbf{a}_k = \prod_{k=1}^{M_T} \|\mathbf{a}_k\|^2 \le 1,$$

$$\tag{B.1}$$

where $\mathbf{a}_k$ denotes the kth column of the transformation matrix $\mathbf{A}$. The last inequality in (B.1) follows from $\|\mathbf{a}_k\| \le \|\mathbf{u}_k\| = 1$, with $\mathbf{u}_k$ being the kth left singular vector of the full system channel matrix, and it is justified by the fact that the entries of matrix $\mathbf{A}$ are obtained as in (35).

In the high SNR regime, after substituting for $\mathbf{A} = \mathbf{U}_A \Sigma_A \mathbf{V}_A^H$ and taking into account (B.1), it is valid to write

$$\det\left(\mathbf{I}_{M_T} + \frac{\rho}{M_T}\mathbf{H}^H \mathbf{A}\mathbf{A}^H \mathbf{H}\right) \approx \det\left(\frac{\rho}{M_T}\mathbf{H}^H \mathbf{U}_A \Sigma_A^2 \mathbf{U}_A^H \mathbf{H}\right)$$

$$= \det\left(\Sigma_A^2\right)\det\left(\frac{\rho}{M_T}\mathbf{H}^H \mathbf{U}_A \mathbf{U}_A^H \mathbf{H}\right)$$

$$\le \det\left(\frac{\rho}{M_T}\mathbf{H}^H \mathbf{U}_A \mathbf{U}_A^H \mathbf{H}\right). \tag{B.2}$$

Recognizing that the right-hand side of (B.2) is an approximation of (A.2), that is, the capacity of the RASF system, in the high SNR regime, the validity of the bound in (47) is proven.

Note that the same approximation for the capacity of MIMO systems at high SNR has been widely used (see, e.g., [24]). Simulation results in Figure 8 demonstrate that the bound is quite tight.

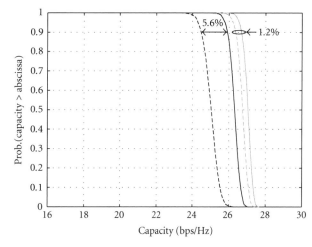

FIGURE 8: Comparison between capacity bound (47) for relaxed structured ASF and true capacity (15) of the resulted system in terms of empirical complementary cdf, when applied to a 16×4 MIMO Rayleigh i.i.d. channel with SNR = 15 dB. Proof of this bound can be found in Appendix B.

ACKNOWLEDGMENT

This work has been partially funded by Antenna Centre of Excellence (ACE2) research programme, under the EU 6th Framework Programme.

REFERENCES

[1] D. A. Gore, R. U. Nabar, and A. J. Paulraj, "Selecting an optimal set of transmit antennas for a low rank matrix channel," in *Proceedings of IEEE Interntional Conference on Acoustics, Speech, and Signal Processing (ICASSP '00)*, vol. 5, pp. 2785–2788, Istanbul, Turkey, June 2000.

[2] R. S. Blum and J. H. Winters, "On optimum MIMO with antenna selection," *IEEE Communications Letters*, vol. 6, no. 8, pp. 322–324, 2002.

[3] A. F. Molisch, M. Z. Win, Y.-S. Choi, and J. H. Winters, "Capacity of MIMO systems with antenna selection," *IEEE Transactions on Wireless Communications*, vol. 4, no. 4, pp. 1759–1772, 2005.

[4] A. Gorokhov, D. A. Gore, and A. J. Paulraj, "Receive antenna selection for MIMO spatial multiplexing: theory and algorithms," *IEEE Transactions on Signal Processing*, vol. 51, no. 11, pp. 2796–2807, 2003.

[5] M. Gharavi-Alkhansari and A. B. Gershman, "Fast antenna subset selection in MIMO systems," *IEEE Transactions on Signal Processing*, vol. 52, no. 2, pp. 339–347, 2004.

[6] D. A. Gore and A. J. Paulraj, "MIMO antenna subset selection with space-time coding," *IEEE Transactions on Signal Processing*, vol. 50, no. 10, pp. 2580–2588, 2002.

[7] R. W. Heath Jr., S. Sandhu, and A. J. Paulraj, "Antenna selection for spatial multiplexing systems with linear receivers," *IEEE Communications Letters*, vol. 5, no. 4, pp. 142–144, 2001.

[8] D. A. Gore, R. W. Heath Jr., and A. J. Paulraj, "Transmit se-
     lection in spatial multiplexing systems," *IEEE Communications
     Letters*, vol. 6, no. 11, pp. 491–493, 2002.

[9] M. A. Jensen and M. L. Morris, "Efficient capacity-based an-
     tenna selection for MIMO Systems," *IEEE Transactions on Ve-
     hicular Technology*, vol. 54, no. 1, pp. 110–116, 2005.

[10] A. F. Molisch, M. Z. Win, and J. H. Winter, "Reduced-
     complexity transmit/receive-diversity systems," *IEEE Transac-
     tions on Signal Processing*, vol. 51, no. 11, pp. 2729–2738, 2003.

[11] L. Dai, S. Sfar, and K. B. Letaief, "Receive antenna selection for
     MIMO systems in correlated channels," in *Proceedings of the
     IEEE International Conference on Communications (ICC '04)*,
     vol. 5, pp. 2944–2948, Paris, France, June 2004.

[12] P. D. Karamalis, N. D. Skentos, and A. G. Kanatas, "Selecting
     array configurations for MIMO systems: an evolutionary com-
     putation approach," *IEEE Transactions on Wireless Communi-
     cations*, vol. 3, no. 6, pp. 1994–1998, 2004.

[13] P. D. Karamalis, N. D. Skentos, and A. G. Kanatas, "Adaptive
     antenna subarray formation for MIMO systems," *IEEE Trans-
     actions on Wireless Communications*, vol. 5, no. 11, pp. 2977–
     2982, 2006.

[14] G. G. Raleigh and J. M. Cioffi, "Spatio-temporal coding for
     wireless communication," *IEEE Transactions on Communica-
     tions*, vol. 46, no. 3, pp. 357–366, 1998.

[15] A. Scaglione, G. B. Giannakis, and S. Barbarossa, "Redundant
     filterbank precoders and equalizers—I: unification and opti-
     mal designs," *IEEE Transactions on Signal Processing*, vol. 47,
     no. 7, pp. 1988–2006, 1999.

[16] H. Sampath, P. Stoica, and A. J. Paulraj, "Generalized linear
     precoder and decoder design for MIMO channels using the
     weighted MMSE criterion," *IEEE Transactions on Communi-
     cations*, vol. 49, no. 12, pp. 2198–2206, 2001.

[17] A. Scaglione, P. Stoica, S. Barbarossa, G. B. Giannakis, and
     H. Sampath, "Optimal designs for space-time linear precoders
     and decoders," *IEEE Transactions on Signal Processing*, vol. 50,
     no. 5, pp. 1051–1064, 2002.

[18] D. P. Palomar, J. M. Cioffi, and M. A. Lagunas, "Joint Tx-Rx
     beamforming design for multicarrier MIMO channels: a uni-
     fied framework for convex optimization," *IEEE Transactions on
     Signal Processing*, vol. 51, no. 9, pp. 2381–2401, 2003.

[19] C. Mun, J.-K. Han, and D.-H. Kim, "Quantized principal com-
     ponent selection precoding for limited feedback spatial multi-
     plexing," in *Proceedings of the IEEE International Conference on
     Communications (ICC '06)*, pp. 4149–4154, Istanbul, Turkey,
     June 2006.

[20] X. Zhang, A. F. Molisch, and S.-Y. Kung, "Variable-phase-shift-
     based RF-baseband codesign for MIMO antenna selection,"
     IEEE Transactions on Signal Processing, vol. 53, no. 11, pp.
     4091–4103, 2005.

[21] P. Theofilakos and A. G. Kanatas, "Frobenius norm based re-
     ceive antenna subarray formation for MIMO systems," in *Pro-
     ceedings of the 1st European Conference on Antennas and Propa-
     gation (EuCAP '06)*, vol. 626, Nice, France, November 2006.

[22] P. Theofilakos and A. G. Kanatas, "Reduced hardware com-
     plexity receive antenna subarray formation for MIMO systems
     based on frobenius norm criterion," in *Proceedings of the 3rd
     International Symposium on Wireless Communication Systems
     (ISWCS '06)*, Valencia, Spain, September 2006.

[23] P. Theofilakos and A. G. Kanatas, "Robustness of receive
     antenna subarray formation to hardware and signal non-
     idealities," in *Proceedings of the 65th IEEE Vehicular Technol-
     ogy Conference (VTC '07)*, pp. 324–328, Dublin, Ireland, April
     2007.

[24] O. Oyman, R. U. Nabar, H. Bölcskei, and A. J. Paulraj, "Char-
     acterizing the statistical properties of mutual information
     in MIMO channels," *IEEE Transactions on Signal Processing*,
     vol. 51, no. 11, pp. 2784–2795, 2003.

[25] F. D. Neeser and J. L. Massey, "Proper complex random pro-
     cesses with applications to information theory," *IEEE Trans-
     actions on Information Theory*, vol. 39, no. 4, pp. 1293–1302,
     1993.

[26] T. M. Cover and J. A. Thomas, *Elements of Information Theory*,
     John Wiley & Sons, New York, NY, USA, 1991.

[27] G. J. Foschini and M. J. Gans, "On limits of wireless commu-
     nications in a fading environment when using multiple an-
     tennas," *Wireless Personal Communications*, vol. 6, no. 3, pp.
     311–335, 1998.

[28] A. Papoulis and S. U. Pillai, *Probability, Random Variables
     and Stochastic Processes*, McGraw-Hill, New York, NY, USA,
     4th edition, 2002.

[29] M. K. Simon and M.-S. Alouini, *Digital Communication over
     Fading Channels*, John Wiley & Sons, New York, NY, USA,
     1st edition, 2000.

[30] M.-S. Alouini and A. J. Goldsmith, "Capacity of Rayleigh
     fading channels under different adaptive transmission and
     diversity-combining techniques," *IEEE Transactions on Vehic-
     ular Technology*, vol. 48, no. 4, pp. 1165–1181, 1999.

[31] N. D. Skentos, A. G. Kanatas, P. I. Dallas, and P. Constantinou,
     "MIMO channel characterization for short range fixed wire-
     less propagation environments," *Wireless Personal Communi-
     cations*, vol. 36, no. 4, pp. 339–361, 2006.

[32] R. A. Horn and C. R. Johnson, *Matrix Analysis*, Cambridge
     University Press, Cambridge, UK, 1985.

Hindawi Publishing Corporation
EURASIP Journal on Wireless Communications and Networking
Volume 2007, Article ID 32460, 12 pages
doi:10.1155/2007/32460

Research Article
Capacity of MIMO-OFDM with Pilot-Aided Channel Estimation

Ivan Cosovic and Gunther Auer

DoCoMo Euro-Labs, Landsberger Straße 312, 80687 München, Germany

Received 31 October 2006; Revised 9 July 2007; Accepted 4 October 2007

Recommended by A. Alexiou

An analytical framework is established to dimension the pilot grid for MIMO-OFDM operating in time-variant frequency selective channels. The optimum placement of pilot symbols in terms of overhead and power allocation is identified that maximizes the training-based capacity for MIMO-OFDM schemes without channel knowledge at the transmitter. For pilot-aided channel estimation (PACE) with perfect interpolation, we show that the maximum capacity is achieved by placing pilots with maximum equidistant spacing given by the sampling theorem, if pilots are appropriately boosted. Allowing for realizable and possibly suboptimum estimators where interpolation is not perfect, we present a semianalytical method which finds the best pilot allocation strategy for the particular estimator.

1. INTRODUCTION

Systems employing multiple transmit and receive antennas, known as multiple-input multiple-output (MIMO) systems, promise significant gains in channel capacity [1–3]. Together with orthogonal frequency division multiplexing (OFDM), MIMO-OFDM is selected for the wireless local area network (WLAN) standard IEEE 802.11n [4], and for beyond 3rd generation (B3G) mobile communication systems [5].

As multiple signals are transmitted from different transmit antennas simultaneously, coherent detection requires accurate channel estimates of all transmit antennas' signals at the receiver. The most common technique to obtain channel state information is via pilot-aided channel estimation (PACE) where known training symbols termed pilots are multiplexed with data. For PACE, channel estimates are exclusively generated by means of pilot symbols, and these estimates are then processed for the detection of data symbols as if they were the true channel response. A sophisticated pilot design should strike a balance between the attainable accuracy of the channel estimate and the resources consumed by pilot symbols. An appropriate means to optimize this trade-off is to maximize the channel capacity of pilot-aided schemes. As the training overhead grows proportionally to the number of transmitted spatial streams [6], the attainable capacity gains for MIMO-OFDM are traded with the bandwidth and energy consumed by a growing number of pilot symbols to estimate the MIMO channels.

A lower bound for the attainable capacity for multiple antenna systems with pilot-aided channel estimation for the block fading channel was derived in [7]. The capacity lower bound was then used to optimize the energy allocation and the fraction of resources consumed by pilots. The capacity lower bound of [7] was extended to single carrier systems operating in frequency-selective channels [8–10], and to spatially correlated MIMO channels [11]. Furthermore, in [8, 12] the capacity achieving pilot design for MIMO-OFDM over frequency-selective channels was studied. For OFDM, pilot symbols inserted in the frequency domain (before OFDM modulation) sample the channel, allowing to recover the channel response for data bearing subcarriers by means of interpolation. This implies that, besides pilot overhead and power allocation, also the placement of pilots is to be optimized. It was found that equidistant placement of pilot symbols not only minimizes the mean squared error (MSE) of the channel estimates [13], but also maximizes the capacity [8].

All previous work deriving the capacity for training-based schemes for single-carrier [7–11] as well as for multi-carrier systems [8, 12] considered time-invariant channels, where the channel was assumed static for the block of transmitted symbols. Furthermore, work on OFDM was limited to perfect interpolation [8, 12]. That is, additive white Gaussian noise (AWGN) is the only source of channel estimation errors. This implies that, in the absence of noise channel estimates perfectly match the true channel response. However,

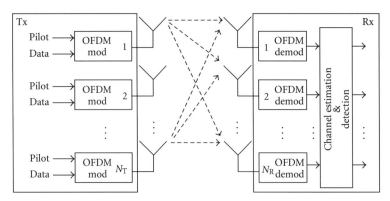

FIGURE 1: MIMO-OFDM system employing N_T transmit and N_R receive antennas.

from the sampling theorem it is well known that an infinite number of pilots is necessary for perfect interpolation [14]. In the context of OFDM, perfect interpolation is observed only if the channel model is sample spaced [15], that is, all channel taps are integer multiples of the sampling duration. In practice, however, pilot sequences are finite while channel taps are nonsample spaced, ultimately leading to reduced capacity bounds. Whereas in related work for OFDM only time-invariant channels and perfect interpolation are considered, in this work real world problems are taken into account, and a more realistic capacity bound is achieved.

Furthermore, previous work on maximizing the training based capacity was exclusively dedicated to minimum mean squared error (MMSE) channel estimation [7–12]. Applied to PACE the MMSE criterion finds the best tradeoff between the attainable interpolation accuracy and the mitigation of noise [16]. While MMSE estimates are favorable in terms of performance, knowledge of the 2nd order statistics is required for implementation, together with a computationally expensive matrix inversion [16]. Unlike previous work, our objective is *not* to find the optimum estimator that achieves capacity, rather we aim to identify the optimum pilot design that maximizes capacity for a given and possibly suboptimum estimator.

The present paper addresses the above mentioned limitations of previous work; its main contributions are summarized as follows.

(i) The work from [7], conducted for block fading channels, is extended to time-variant frequency-selective channels. Assuming perfect interpolation and placing pilot symbols with maximum possible distance in time and frequency still satisfying the sampling theorem, the capacity is shown to approach that for the block fading channel, provided that the size of the block is chosen according to the maximum pilot spacing imposed by the sampling theorem.

(ii) Previous work in [7, 8, 12] is extended to arbitrary linear estimators. Capacity-achieving pilot design for realizable, and possibly suboptimum, channel estimation schemes is therefore possible, as for example, interpolation by finite impulse response (FIR) filtering [16–18], discrete Fourier transform (DFT) based

interpolation [19, 20], or linear interpolation [21]. We derive a closed form expression of the training based-capacity for perfect interpolation, and propose a semianalytical procedure for practical estimation techniques.

(iii) For a particular class of estimators, namely FIR interpolation filters, we demonstrate that the pilot grid that maximizes capacity is mostly independent of the chosen channel model, as long as the maximum channel delay and the maximum Doppler frequency are within a certain range. This is an appealing property, as a sophisticated pilot design should be valid for an as wide as possible range of channel conditions.

The remainder of this paper is structured as follows. In Section 2, the system model is introduced. In Section 3, the estimation error model is established and analyzed, whereas bounds on the achievable capacity of the optimized pilot design are derived in Section 4. Numerical examples in Section 5 verify the developed framework in terms of pilot boost and overhead, as well as number of transmit antennas.

2. SYSTEM MODEL

Consider a MIMO-OFDM system with N_T transmit and N_R receive antennas as illustrated in Figure 1. We assume that N_T spatial streams are transmitted and that channel knowledge is not available at the transmitter. Denote with N_c the number of used subcarriers, and with L the number of OFDM symbols per frame. OFDM modulation is performed by N_DFT-point ($N_\mathrm{DFT} \geq N_\mathrm{c}$) inverse DFT (IDFT), followed by insertion of a cyclic prefix (CP) of N_CP samples. Assuming perfect orthogonality in time and frequency, the received signal of subcarrier n of the ℓth OFDM symbol block and νth receive antenna is given by

$$Y_{n,\ell}^{(\nu)} = \sum_{\mu=1}^{N_\mathrm{T}} \sqrt{\frac{E_\mathrm{d}}{N_\mathrm{T}}} X_{n,\ell}^{(\mu)} H_{n,\ell}^{(\mu,\nu)} + Z_{n,\ell}^{(\nu)}, \tag{1}$$
$$0 \leq n < N_\mathrm{c}, \quad 0 \leq \ell < L, \quad 0 \leq \nu < N_\mathrm{R}.$$

In (1), $X_{n,\ell}^{(\mu)}$, $H_{n,\ell}^{(\mu,\nu)}$, and $Z_{n,\ell}^{(\nu)}$, denote the normalized transmitted symbol over transmit antenna μ with $E\{|X_{n,\ell}^{(\mu)}|^2\} = 1$, the channel transfer function (CTF) between transmit antenna μ and receive antenna ν, and AWGN at the νth receive antenna with zero mean and variance N_0, respectively. An energy per transmitted data symbol of E_d/N_T and a normalized average channel gain, $E\{|H_{n,\ell}^{(\mu,\nu)}|^2\} = \sigma_H^2 = 1$ is assumed.

The discrete CTF $H_{n,\ell}^{(\mu,\nu)}$, is obtained by sampling $H^{(\mu,\nu)}(f, t)$ at frequency $f = n/T$ and time $t = \ell T_{sym}$, where $T_{sym} = (N_c + N_{CP})T_{spl}$ and $T = N_c T_{spl}$ represent the OFDM symbol duration with and without the cyclic prefix, and T_{spl} is the sample duration. Considering a frequency selective time-variant channel, modeled by a tapped delay line with Q_0 nonzero taps with channel impulse response (CIR), $h^{(\mu,\nu)}(\tau, t) = \sum_{q=1}^{Q_0} h_q^{(\mu,\nu)}(t)\delta(t - \tau_q^{(\mu,\nu)})$, the CTF is described by

$$H_{n,\ell}^{(\mu,\nu)} = H^{(\mu,\nu)}\left(\frac{n}{T}, \ell T_{sym}\right) = \sum_{q=1}^{Q_0} h_{q,\ell}^{(\mu,\nu)} \exp\left(-j2\pi\tau_q^{(\mu,\nu)}\frac{n}{T}\right),$$

(2)

where $h_{q,\ell}^{(\mu,\nu)} = h_q^{(\mu,\nu)}(\ell T_{sym})$ denotes the complex valued channel tap q, assumed to be constant over one OFDM symbol block, with associated tap delay $\tau_q^{(\mu,\nu)}$. Intersymbol interference is avoided by ensuring that $N_{CP}T_{spl} \geq \tau_{max}$, where τ_{max} denotes the maximum delay of the CIR. Then an arbitrary CIR is supported for which all channel taps $h_{q,\ell}^{(\mu,\nu)}$ are contained within the range $0 \leq \tau_q^{(\mu,\nu)} \leq \tau_{max}$, and the received signal is given by (1).

The 2nd-order statistics are determined by the two-dimensional (2D) correlation function $R^{(\mu,\nu)}[\Delta_n, \Delta_\ell] = E\{H_{n,\ell}^{(\mu,\nu)}(H_{n+\Delta_n,\ell+\Delta_\ell}^{(\mu,\nu)})^*\}$, composed of two independent correlation functions in frequency and time, $R^{(\mu,\nu)}[\Delta_n, \Delta_\ell] = R_f^{(\mu,\nu)}[\Delta_n]R_t^{(\mu,\nu)}[\Delta_\ell]$. Both $R_f^{(\mu,\nu)}[\Delta_n] = E\{H_{n,\ell}^{(\mu,\nu)}(H_{n+\Delta_n,\ell}^{(\mu,\nu)})^*\}$ and $R_t^{(\mu,\nu)}[\Delta_\ell] = E\{H_{n,\ell}^{(\mu,\nu)}(H_{n,\ell+\Delta_\ell}^{(\mu,\nu)})^*\}$ are strictly band-limited [18]. That is, the inverse Fourier transform of $R_f^{(\mu,\nu)}[\Delta_n]$ described by the power delay profile is essentially nonzero in the range $[0, \tau_{max}]$, where τ_{max} is the maximum channel delay. Likewise, the Fourier transform of $R_t^{(\mu,\nu)}[\Delta_\ell]$ describing time variations due to mobile velocities is given by the Doppler power spectrum, nonzero within $[-f_{D,max}, f_{D,max}]$, where $f_{D,max}$ is the maximum Doppler frequency. No further assumptions regarding the distribution of $h^{(\mu,\nu)}(\tau, t)$ are imposed. To this end, the CIR may possibly be nonsample spaced, that is, tap delays $\tau_q^{(\mu,\nu)}$ in (2) may *not* be placed at integer multiples of the sampling duration.

In order to recover the transmitted information, pilot symbols are commonly used for channel estimation. Channel estimation schemes for MIMO-OFDM based on the least squares (LS) and MMSE criterion are studied in [22, 23] and [6, 24, 25], respectively. We assume that channel state information about all $N_T \times N_R$ channels is required at the receiver. To enable this, pilots belonging to different transmit antennas are orthogonally separated in time and/or frequency. Thus, the problem of MIMO-OFDM channel estimation breaks down to estimating the channel of a single antenna OFDM system. Note, there are other possibilities to orthogonally separate the pilots, but they lead to higher

complexity and/or at least the same pilot overhead [6]. Furthermore, pilots belonging to the same transmit antenna are equidistantly spaced in time and frequency within the OFDM frame [17]. This is motivated by the findings in [8], where it is shown that equidistant placement of pilots minimizes the harmonic mean of the MSE of channel estimates over all subcarriers and thus maximizes the capacity. Figure 2 illustrates the resulting placement of pilots for four spatial streams, arranged in a rectangular shaped pattern. Extension to other regular pilot patters, such as a diamond shaped grid [26], is straightforward.

Resources are constraint to bandwidth and energy. Unlike the orthogonally separated pilots, data symbols are spatially multiplexed. One frame is assigned $N_p^{(\mu)}$ pilot and N_d data symbols per spatial dimension which amounts to (cf., Figure 2)

$$N_c L = N_d + \sum_{\mu=1}^{N_T} N_p^{(\mu)}.$$

(3)

The resulting pilot overhead of the μth antenna is defined by

$$\Omega_p^{(\mu)} = \frac{N_p^{(\mu)}}{N_c L}.$$

(4)

With an energy per transmitted data symbol of E_d/N_T, the total transmit energy over all N_T antennas equals

$$E_{tot} = E_d N_d + \sum_{\mu=1}^{N_T} E_p^{(\mu)} N_p^{(\mu)},$$

(5)

where $E_p^{(\mu)}$ is the energy per pilot symbol of the μth transmit antenna.

The accuracy of the channel estimates may be improved by a pilot boost $S_p^{(\mu)}$. With an energy per transmitted pilot set to $E_p^{(\mu)} = S_p^{(\mu)} E_d$, the signal-to-noise ratio (SNR) at the input of the channel estimation unit is improved by a factor of $S_p^{(\mu)}$. On the other hand, the useful transmit energy of the payload information is reduced, if the overall transmit energy in (5) is kept constant. The energy dedicated to pilot symbols is determined by the pilot overhead per antenna $\Omega_p^{(\mu)}$, and the pilot boost per antenna $S_p^{(\mu)}$. Including the pilot overhead, the ratio of the energy per symbol E_d of a system with pilots, to the energy per symbol E_0 of an equivalent system with the same frame size $N_c \times L$, same transmit energy E_{tot}, but without pilot symbols is

$$\frac{E_d}{E_0} = \frac{1}{1 + \sum_{\mu=1}^{N_T} \Omega_p^{(\mu)}\left(S_p^{(\mu)} - 1\right)}.$$

(6)

The ratio E_d/E_0 is a measure for the pilot insertion loss relative to a reference system assuming no overhead due to pilots. Note, (6) is obtained exploiting (4), (5), and the constraint $E_{tot} = N_c L E_0$.

In the following, we assume that for each transmit antenna the same number of pilot symbols and the same boosting level are used, that is, $N_p = N_p^{(\mu)}$, $\Omega_p = \Omega_p^{(\mu)}$, and $S_p = S_p^{(\mu)}$,

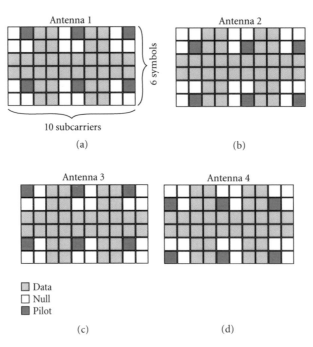

FIGURE 2: Example for placing orthogonal pilots over antenna-specific OFDM subframes, $N_T = 4$, $N_p^{(1)} = N_p^{(2)} = N_p^{(3)} = N_p^{(4)} = 6$, and $N_d = 36$.

$\mu = \{1, \ldots, N_T\}$. Now, the total pilot overhead in (4) amounts to $N_T \Omega_p$ and (6) simplifies to

$$\frac{E_d}{E_0} = \frac{1}{1 + N_T \Omega_p (S_p - 1)}. \tag{7}$$

3. ESTIMATION ERROR MODELING

The channel estimation unit outputs an estimate of the CTF, $H_{n,\ell}^{(\mu,\nu)}$, denoted by $\hat{H}_{n,\ell}^{(\mu,\nu)} = \mathbf{w}^H \widetilde{\mathbf{y}}^{(\mu,\nu)}$. Let M_f and M_t denote the number of pilot symbols in frequency and time used to generate $\hat{H}_{n,\ell}^{(\mu,\nu)}$. The $M_t M_f \times 1$ column vector $\widetilde{\mathbf{y}}^{(\mu,\nu)}$ contains the received pilots from transmit antenna μ to receive antenna ν. The $M_t M_f \times 1$ column vector $\mathbf{w}$ represents an arbitrary linear estimator.

3.1. Parametrization of the MSE

Channel estimation impairments are quantified by the MSE of the estimation error $\varepsilon_{n,\ell}^{(\mu,\nu)} = H_{n,\ell}^{(\mu,\nu)} - \hat{H}_{n,\ell}^{(\mu,\nu)}$. With the assumption that all transmit and receive antennas are mutually uncorrelated, the MSE is independent of μ and ν, denoted by $\sigma_\varepsilon^2[n,\ell] = E[|\varepsilon_{n,\ell}^{(\mu,\nu)}|^2]$. The MSE of an arbitrary 2D pilot-aided scheme is given by

$$\sigma_\varepsilon^2[n,\ell] = E\Big[|\varepsilon_{n,\ell}^{(\mu,\nu)}|^2\Big] = E\Big[|H_{n,\ell}^{(\mu,\nu)} - \hat{H}_{n,\ell}^{(\mu,\nu)}|^2\Big]$$
$$= E\Big[|H_{n,\ell}^{(\mu,\nu)}|^2\Big] - 2\Re\Big\{\mathbf{w}^H \mathbf{r}_{\widetilde{y}H}^{(\mu,\nu)}[n,\ell]\Big\} + \mathbf{w}^H \mathbf{R}_{\widetilde{y}\widetilde{y}}^{(\mu,\nu)} \mathbf{w}. \tag{8}$$

The 2D correlation functions $\mathbf{r}_{\widetilde{y}H}^{(\mu,\nu)}[n,\ell] = E\{\widetilde{\mathbf{y}}^{(\mu,\nu)}(H_{n,\ell}^{(\mu,\nu)})^*\}$ and $\mathbf{R}_{\widetilde{y}\widetilde{y}}^{(\mu,\nu)} = E\{\widetilde{\mathbf{y}}^{(\mu,\nu)}(\widetilde{\mathbf{y}}^{(\mu,\nu)})^H\}$ represent the cross-correlation between $\widetilde{\mathbf{y}}^{(\mu,\nu)}$ and the desired response $H_{n,\ell}^{(\mu,\nu)}$, and the autocorrelation matrix of the received pilots, $\widetilde{\mathbf{y}}^{(\mu,\nu)}$, respectively, [16]. The autocorrelation matrix is composed of $\mathbf{R}_{\widetilde{y}\widetilde{y}}^{(\mu,\nu)} = \mathbf{R}_{\widetilde{h}\widetilde{h}}^{(\mu,\nu)} + \mathbf{I}/\gamma_p$, where $\mathbf{R}_{\widetilde{h}\widetilde{h}}^{(\mu,\nu)} = E\{\widetilde{\mathbf{h}}^{(\mu,\nu)}(\widetilde{\mathbf{h}}^{(\mu,\nu)})^H\}$ is the autocorrelation matrix of the CTF at pilot positions excluding the AWGN term, and $\mathbf{I}$ denotes the identity matrix, all of dimension $M_f M_t \times M_f M_t$. With the pilot insertion loss of (7), the SNR at pilot positions amounts to $\gamma_p = S_p E_d / N_0 = \gamma_0 S_p / (1 + N_T \Omega_p (S_p - 1))$, where $\gamma_0 = E_0 / N_0$ denotes the SNR of a reference system assuming perfect channel knowledge and no overhead due to pilots.

The MSE in (8) is dependent on n and ℓ. In order to allow for a tractable model, we choose to average the MSE over the entire sequence, so $\sigma_\varepsilon^2[n,\ell] \to \sigma_\varepsilon^2$.

The channel estimates generated by a linear estimator $\mathbf{w}$ can be decomposed into a signal and noise part, denoted by $\hat{H}_{n,\ell}^{(\mu,\nu)} = \mathbf{w}^H \widetilde{\mathbf{h}}^{(\mu,\nu)} + \mathbf{w}^H \widetilde{\mathbf{z}}^{(\nu)}$, where $\widetilde{\mathbf{h}}^{(\mu,\nu)}$ and $\widetilde{\mathbf{z}}^{(\nu)}$ account for CTF and AWGN vectors at pilot positions. Likewise, the estimation error $\varepsilon_{n,\ell}^{(\mu,\nu)}$ can be separated into an interpolation error $H_{n,\ell}^{(\mu,\nu)} - \mathbf{w}^H \widetilde{\mathbf{h}}^{(\mu,\nu)}$ and a noise error $\mathbf{w}^H \widetilde{\mathbf{z}}^{(\nu)}$. Assuming that CTF $H_{n,\ell}^{(\mu,\nu)}$ and AWGN $Z_{n,\ell}^{(\nu)}$ are uncorrelated, the MSE also separates into a noise and interpolation error:

$$\sigma_\varepsilon^2 = E\Big[|H_{n,\ell}^{(\mu,\nu)} - \mathbf{w}^H \widetilde{\mathbf{h}}^{(\mu,\nu)} - \mathbf{w}^H \widetilde{\mathbf{z}}^{(\nu)}|^2\Big]$$
$$= E\Big[|H_{n,\ell}^{(\mu,\nu)} - \mathbf{w}^H \widetilde{\mathbf{h}}^{(\mu,\nu)}|^2\Big] + E\Big[|\mathbf{w}^H \widetilde{\mathbf{z}}^{(\nu)}|^2\Big]. \tag{9}$$

We note that this separation of the MSE is possible for any linear estimator. The noise part $\sigma_n^2 = E[|\mathbf{w}^H \widetilde{\mathbf{z}}^{(\nu)}|^2]$ is inversely proportional to the SNR and is given by

$$\sigma_n^2 = \frac{\mathbf{w}^H \mathbf{w}}{\gamma_p} = \frac{1}{G_n \gamma_0} \cdot \frac{1 + N_T \Omega_p (S_p - 1)}{S_p}, \quad (10)$$

where $G_n = 1/(\mathbf{w}^H \mathbf{w})$ defines the estimator gain. According to (8) the variance of the interpolation error is determined by

$$\sigma_i^2 = E\left[\,|H_{n,\ell}^{(\mu,\nu)} - \mathbf{w}^H \widetilde{\boldsymbol{h}}^{(\mu,\nu)}|^2\,\right]$$
$$= E\left[\,|H_{n,\ell}^{(\mu,\nu)}|^2\,\right] - 2\Re\left\{\mathbf{w}^H \mathbf{r}_{\widetilde{h}H}^{(\mu,\nu)}\right\} + \mathbf{w}^H \mathbf{R}_{\widetilde{h}\widetilde{h}}^{(\mu,\nu)} \mathbf{w}. \quad (11)$$

3.2. Equivalent system model and effective SNR

In order to derive a model taking into account channel estimation errors, we assume a receiver that processes the channel estimates $\hat{H}_{n,\ell}^{(\mu,\nu)}$ as if these were the true CTF. The effect of channel estimation errors on the received signal in (1) is described by the equivalent system model:

$$Y_{n,\ell}^{(\mu,\nu)} = \sum_{\mu=1}^{N_T} \sqrt{\frac{E_d}{N_T}} X_{n,\ell}^{(\mu)} \breve{H}_{n,\ell}^{(\mu,\nu)} + \underbrace{\sum_{\mu=1}^{N_T} \sqrt{\frac{E_d}{N_T}} X_{n,\ell}^{(\mu)} \varepsilon_{n,\ell}^{(\mu,\nu)} + Z_{n,\ell}^{(\nu)}}_{\eta_{n,\ell}^{(\mu,\nu)}}, \quad (12)$$

where $\eta_{n,\ell}^{(\mu,\nu)}$ denotes the effective noise term with zero mean and variance $\sigma_\eta^2 = N_0 + E_d \sigma_\varepsilon^2$. Apart from the increased noise term $\eta_{n,\ell}^{(\mu,\nu)}$, channel estimation impairments affect the equivalent system model (12) by distortions in the signal part of $\breve{H}_{n,\ell}^{(\mu,\nu)} = \mathbf{w}^H \breve{\boldsymbol{h}}^{(\mu,\nu)}$. The useful signal energy observed at the receiver is given by

$$\sigma_{\breve{H}}^2 = E\left[\,|\mathbf{w}^H \breve{\boldsymbol{h}}^{(\mu,\nu)}|^2\,\right] = \mathbf{w}^H \mathbf{R}_{\breve{h}\breve{h}} \mathbf{w}. \quad (13)$$

Now the effective SNR including channel estimation of the equivalent system model (12) yields

$$\gamma = E_d \frac{\sigma_{\breve{H}}^2}{\sigma_\eta^2} = \frac{E_d \sigma_{\breve{H}}^2}{N_0 + E_d \sigma_\varepsilon^2}. \quad (14)$$

An important difference to the equivalent system model devised by [7] is the definition of $\sigma_{\breve{H}}^2$ in (13). In [7] the estimate $\hat{H}_{n,\ell}^{(\mu,\nu)} = \mathbf{w}^H \widetilde{\boldsymbol{y}}^{(\mu,\nu)}$ replaces $\breve{H}_{n,\ell}^{(\mu,\nu)} = \mathbf{w}^H \breve{\boldsymbol{h}}^{(\mu,\nu)}$ in (12), thus containing contributions from the CTF and noise, so that $\sigma_{\hat{H}}^2 = \sigma_{\breve{H}}^2 + N_0/E_d$. Instead, our model with $\sigma_{\breve{H}}^2$ in (13) exclusively captures the signal part of the channel estimate. As the model of [7] was tailored for an MMSE estimator with $\sigma_{\hat{H}}^2 = 1 - \sigma_\varepsilon^2$, meaningful results are produced. However, the model of [7] implies that the noise term contained in $\sigma_{\hat{H}}^2$ contributes to the useful signal energy of the effective SNR γ in (14). This becomes problematic at low SNR, when the equiv-

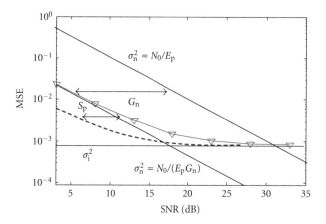

FIGURE 3: Parametrization of an MSE estimation curve.

alent system model is to be applied to other than MMSE estimators. For instance, consider an unbiased estimator with $\mathbf{w}^H \mathbf{R}_{\widetilde{h}\widetilde{h}} \mathbf{w} = 1$ and unitary estimator gain $\mathbf{w}^H \mathbf{w} = 1$. Then with the model of [7]: $\sigma_{\hat{H}}^2 = 1 + N_0/E_d$, so for low SNR, $N_0 \to \infty$, the effective SNR in (14) approaches $\gamma = 1/2$, which is clearly a contradiction. On the other hand, by using (13) we get $\sigma_{\breve{H}}^2 = 1$ and $\gamma \to 0$ as $N_0 \to \infty$, so our model in (12) produces meaningful results for low SNR. In any case, in the high SNR regime with $N_0 \ll E_d$ both models converge and we get $\sigma_{\breve{H}}^2 \approx \sigma_{\hat{H}}^2$.

Inserting (7) into the effective SNR γ in (14) and computing the ratio γ_0/γ quantifies the SNR degradation due to channel estimation errors, given by

$$\Delta\gamma = \frac{\gamma_0}{\gamma} = \frac{1}{\sigma_{\breve{H}}^2} \left(1 + N_T \Omega_p (S_p - 1) + \sigma_\varepsilon^2 \gamma_0\right). \quad (15)$$

Substituting the MSE $\sigma_\varepsilon^2 = \sigma_n^2 + \sigma_i^2$ into (15), with σ_n^2 being expressed in the parametrized form of (10), the loss in SNR due to channel estimation can be transformed to

$$\Delta\gamma = \frac{1}{\sigma_{\breve{H}}^2} \left(1 + N_T \Omega_p (S_p - 1)\right) \cdot \left(1 + \frac{1}{G_n S_p}\right) + \frac{\sigma_i^2}{\sigma_{\breve{H}}^2} \gamma_0. \quad (16)$$

In the following, a fixed estimator $\mathbf{w}$ is considered where the estimator coefficients are computed once and are not adapted for changing channel conditions. Then, according to (16) the performance penalty due to channel estimation is fully determined by two SNR independent parameters, the estimator gain G_n and the interpolation error σ_i^2. On the other hand, allowing for an SNR dependent estimator, $\mathbf{w} = \mathbf{w}(\gamma)$, the parameters G_n and σ_i^2 would be strictly speaking only valid for one particular SNR value γ. A prominent example for an SNR dependent estimator is the MMSE estimator, known as Wiener filter [27]. In this case, the SNR for which G_n and σ_i^2 lead to a maximum $\Delta\gamma$ in (16) should be used, so to maintain a certain performance under worst case conditions.

The MSE of a fixed estimator $\mathbf{w}$ is plotted in Figure 3. At low SNR, the MSE is dominated by the noise error σ_n^2.

Hence, the MSE linearly decreases with the SNR. At high SNR the MSE experiences an error floor caused by the SNR independent interpolation error. A pilot boost is only effective to reduce the noise part of the MSE in (10), while the interpolation error, σ_i^2 in (11), remains unaffected. This is shown in Figure 3, where a pilot boost shifts the MSE S_p dB to the left.

4. CAPACITY ANALYSIS

The ergodic channel capacity that includes channel estimation and pilot insertion losses when the channel is not known at transmitter can be lower bounded by [7]

$$C \geq (1 - N_T \Omega_p) E \left[\log_2 \det \left(\mathbf{I}_{N_R} + \frac{\mathbf{H}_{n,\ell} \mathbf{H}_{n,\ell}^H}{N_T} \frac{\gamma_0}{\Delta \gamma} \right) \right], \quad (17)$$

where $\mathbf{I}_{N_R}$ is the $N_R \times N_R$ identity matrix and the CTF is defined by

$$\mathbf{H}_{n,\ell} = \begin{pmatrix} H_{n,\ell}^{(1,1)} & \cdots & H_{n,\ell}^{(N_T,1)} \\ \vdots & \ddots & \vdots \\ H_{n,\ell}^{(1,N_R)} & \cdots & H_{n,\ell}^{(N_T,N_R)} \end{pmatrix}. \quad (18)$$

In (17), the expectation is taken over the frequency and time dimension of $\mathbf{H}_{n,\ell}$, that is, over indices n and ℓ. The capacity penalty due to the pilot-aided channel estimation is characterized by two factors: the SNR loss due to estimation errors, $\Delta \gamma$ from (15) or (16), and the loss in spectral efficiency due to resources consumed by pilot symbols, $N_T \Omega_p$. Inserting (16) into (17) we obtain

$$C \geq (1 - N_T \Omega_p) E \left[\log_2 \det \left(\mathbf{I}_{N_R} + \frac{\mathbf{H}_{n,\ell} \mathbf{H}_{n,\ell}^H}{N_T} \right. \right.$$
$$\left. \left. \cdot \frac{\gamma_0 \sigma_{\tilde{H}}^2}{(1 + N_T \Omega_p (S_p - 1)) \cdot (1 + 1/(G_n S_p)) + \sigma_i^2 \gamma_0} \right) \right]. \quad (19)$$

An important requirement for the capacity lower bound to become tight is that the signal and noise terms in the equivalent system model of (12) are uncorrelated [7]. Unfortunately, for arbitrary linear estimators $\mathbf{w}$ this may not be the case, as the interpolation error $H_{n,\ell}^{(\mu,\nu)} - \breve{H}_{n,\ell}^{(\mu,\nu)}$, with $\breve{H}_{n,\ell}^{(\mu,\nu)} = \mathbf{w}^H \tilde{\mathbf{h}}^{(\mu,\nu)}$, introduces correlations between the effective noise term $\eta_{n,\ell}^{(\mu,\nu)}$ and the CTF $H_{n,\ell}^{(\mu,\nu)}$. On the other hand, the noise part of the estimation error $\mathbf{w}^H \tilde{\mathbf{z}}^{(\nu)}$ is statistically independent of both $H_{n,\ell}^{(\mu,\nu)}$ and $\breve{H}_{n,\ell}^{(\mu,\nu)}$. Therefore, for perfect interpolation ($\breve{H}_{n,\ell}^{(\mu,\nu)} = H_{n,\ell}^{(\mu,\nu)}$ and $\sigma_i^2 = 0$), the signal part $\breve{H}_{n,\ell}^{(\mu,\nu)}$ and the effective noise term $\eta_{n,\ell}^{(\mu,\nu)}$ in (12) become statistically independent. To this end, one condition for a tight capacity bound is $\sigma_n^2 \gg \sigma_i^2$, so to ensure that $\breve{H}_{n,\ell}^{(\mu,\nu)}$ and $\eta_{n,\ell}^{(\mu,\nu)}$ are sufficiently decorrelated.

In the following, we focus on the problem of capacity maximization. By doing so, we consider

(i) pilot boost S_p,

(ii) pilot overhead Ω_p,

(iii) number of transmit antennas N_T

as optimization parameters such that the capacity is maximized.

4.1. Optimum pilot boost

The effect of a pilot boost is twofold: first, the estimation error decreases; second, the energy dedicated to pilot symbols increases. So, there clearly exists an optimum pilot boost S_p which minimizes the loss in SNR due to channel estimation, $\Delta \gamma$ in (16), and thus maximizes the system capacity in (19).

The optimum pilot boost for the parametrized estimation error model is obtained by differentiating $\Delta \gamma$ from (16) or C from (19) with respect to S_p and setting the result to zero. This results in

$$S_{p,opt} = \sqrt{\frac{1 - N_T \Omega_p}{N_T \Omega_p G_n}}. \quad (20)$$

The optimum pilot boost $S_{p,opt}$ is seen to increase if less pilots are used and/or less transmit antennas are used, but decreases with growing estimator gain, G_n. In any case, a pilot boost is only effective to reduce the noise part of the MSE σ_n^2 in (10), while the interpolation error σ_i^2 in (11) remains unaffected. Hence, the attainable gains of a pilot boost diminish with growing σ_i^2 and SNR γ_0, as deduced from $\Delta \gamma$ in (16) or C in (19), although the optimum pilot boost $S_{p,opt}$ in (20) is independent of both SNR and σ_i^2.

The loss in SNR for the optimally chosen pilot boost (20) yields

$$\Delta \gamma \Big|_{S_p = S_{p,opt}} = \frac{1}{\sigma_{\tilde{H}}^2} \left(\sqrt{1 - N_T \Omega_p} + \sqrt{\frac{N_T \Omega_p}{G_n}} \right)^2 + \frac{\sigma_i^2}{\sigma_{\tilde{H}}^2} \gamma_0, \quad (21)$$

whereas the capacity becomes

$$C \Big|_{S_p = S_{p,opt}} = (1 - N_T \Omega_p) E \left[\log_2 \det \left(\mathbf{I}_{N_R} + \frac{\mathbf{H}_{n,\ell} \mathbf{H}_{n,\ell}^H}{N_T} \right. \right.$$
$$\left. \left. \cdot \frac{\gamma_0 \sigma_{\tilde{H}}^2}{\left(\sqrt{1 - N_T \Omega_p} + \sqrt{N_T \Omega_p / G_n} \right)^2 + \sigma_i^2 \gamma_0} \right) \right]. \quad (22)$$

As both G_n and σ_i^2 depend on Ω_p, an analytical solution for Ω_p that maximizes (22) is a task of formidable complexity which is not pursued here. Instead, for the special case of perfect interpolation ($\sigma_i^2 = 0$), we derive the optimum pilot overhead Ω_p in closed form, whereas we propose a semianalytical procedure for the general case.

4.2. Ideal lowpass interpolation filter (LPIF)

Motivated by previous work where it was shown that equidistant placement of pilot symbols minimizes the MSE [13],

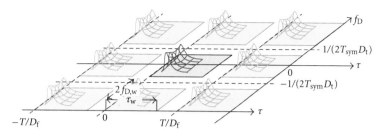

FIGURE 4: Filter transfer function of an ideal 2D low-pass interpolation filter.

as well as maximizes the capacity [8], we focus on channel estimation by interpolation with equispaced pilots in time and frequency. By using a scattered pilot grid the received OFDM frame is sampled in two dimensions, with rate D_f/T and $D_t T_{sym}$ in frequency and time, respectively.

An ideal lowpass interpolation filter (LPIF) is characterized by the 2D rectangular shaped filter transfer function

$$W(\tau, f_D) = \sum_{n=-\infty}^{\infty} \sum_{\ell=-\infty}^{\infty} w_{n,\ell} \, e^{-j2\pi(n\tau/T + \ell f_D T_{sym})}$$

$$= \begin{cases} 1, & 0 \leq \tau \leq \tau_w, \; |f_D| \leq f_{D,w}, \\ 0, & \tau_w < \tau \leq \dfrac{T}{D_f}, f_{D,w} < |f_D| \leq \dfrac{1}{2T_{sym}D_t}, \end{cases}$$

$$(23)$$

where $w_{n,\ell}$ denotes the filter coefficient of pilot subcarrier n and OFDM symbol ℓ. The filter parameters τ_w and $f_{D,w}$ specify the cut-off region of the filter. The transformation to the time (delay) and Doppler domains is described by a discrete time Fourier transform (DTFT) [14], between the variable pairs $n \to \tau$ and $\ell \to f_D$. Due to sampling, $W(\tau, f_D)$ is periodically repeated at intervals $[0, T/D_f]$ and $[-1/(2T_{sym}D_t), 1/(2T_{sym}D_t)]$. This is illustrated in Figure 4 where the filter transfer function of an ideal LPIF is drawn in the 2D plane.

Applied to PACE the LPIF is to be designed so that the spectral components of the CTF $H_{n,\ell}^{(\mu,\nu)}$ which are nonzero within the range $[0, \tau_{max}]$ and $[-f_{D,max}, f_{D,max}]$ pass the filter undistorted, while spectral components outside this range are blocked. Furthermore, the pilot spacings D_f and D_t must be sufficiently small to prevent spectral overlap between the filter passband and its aliases. Hence, in order to reconstruct the signal the sampling theorem requires that [18]

$$\tau_{max} \leq \tau_w < \frac{T}{D_f}, \qquad f_{D,max} \leq f_{D,w} < \frac{1}{2D_t T_{sym}}. \quad (24)$$

The filter parameters τ_w and $f_{D,w}$ represent the maximum assumed delay of the channel and the maximum assumed Doppler frequency, according to worst case channel conditions, as indicated in Figure 4.

Applied to the MSE analysis in Section 3.1, an ideal LPIF has some appealing properties as follows.

(i) Provided that (24) is satisfied, the interpolation error diminishes, $\sigma_i^2 = 0$; that is, the LPIF resembles perfect interpolation. Consequently, the MSE is equivalent to

$\sigma_\varepsilon^2 = \sigma_n^2 = \mathbf{w}^H\mathbf{w}/\gamma_p$ in (10). Furthermore, the MSE becomes independent of the subcarrier and OFDM symbol indices n and ℓ. Hence, no deviation over n and ℓ is observed, $\sigma_\varepsilon^2[n, \ell] = \sigma_\varepsilon^2$. As opposed to the general case of linear estimators in Section 3.1, averaging over n and ℓ is not required.

(ii) Due to perfect interpolation, the useful signal power in (13) becomes $\sigma_{\hat{H}}^2 = \sigma_H^2 = 1$, that is, the LPIF produces unbiased estimates.

(iii) As the number of filter coefficients in frequency and time approach infinity, $\{M_f, M_t\} \to \infty$, the pilot overhead becomes $\Omega_p = 1/(D_f D_t)$.

(iv) In the high SNR regime, the performance of an ideal LPIF asymptotically approaches the MMSE [28]. In general, however, the MSE of the ideal LPIF is strictly larger than the MMSE.

By invoking Parseval's theorem [14], the MSE can be transformed to

$$\sigma_\varepsilon^2 = \frac{\mathbf{w}^H\mathbf{w}}{\gamma_p}$$
$$= \frac{D_f D_t T_{sym}}{\gamma_p T} \int_0^{T/D_f} \int_{-1/(2D_t T_{sym})}^{1/(2D_t T_{sym})} |W(\tau, f_D)|^2 \, d\tau df_D. \quad (25)$$

The MSE, σ_ε^2, is determined by the fraction of the AWGN suppressed by the filter. Inserting (23) and solving (25) yields

$$\sigma_\varepsilon^2 = \frac{1}{\gamma_p G_n} = \frac{1}{\gamma_p \beta_f \beta_t} = \frac{c_w^2}{\gamma_p \Omega_p} \quad (26)$$

with $c_w = \sqrt{2\tau_w f_{D,w} T_{sym}/T}$. The factors $\beta_f = T/(D_f \tau_w)$ and $\beta_t = 1/(2D_t f_{D,w} T_{sym})$ are a measure for the amount of oversampling in frequency and time, with respect to minimum sampling rates T/D_f and $1/(2D_t T_{sym})$, required by the sampling theorem in (24). The MSE is inversely proportional to the oversampling factors β_f and β_t, as well as the pilot overhead $\Omega_p = 1/(D_f D_t)$. Hence, increasing the pilot overhead directly improves the MSE.

4.2.1. Capacity of PACE with perfect 2D interpolation

The expression for the MSE in (26) establishes a relation between estimator gain and pilot overhead, $G_n = \Omega_p/c_w^2$, that allows to maximize the channel capacity in closed form. Moreover, the effective signal and noise terms in the equivalent system model (12) become statistically independent, ensuring a tight capacity bound in (22).

For an ideal LPIF, the optimum pilot boost (20) can be conveniently expressed as

$$S_{\mathrm{p,opt}} = c_{\mathrm{w}} \frac{\sqrt{1 - N_{\mathrm{T}} \Omega_{\mathrm{p}}}}{\sqrt{N_{\mathrm{T}} \Omega_{\mathrm{p}}}}. \tag{27}$$

Inserting $G_{\mathrm{n}} = \Omega_{\mathrm{p}}/c_{\mathrm{w}}^2$ into (21) and after some algebraic transformations the SNR loss for $S_{\mathrm{p}} = S_{\mathrm{p,opt}}$ becomes

$$\Delta\gamma|_{S_{\mathrm{p}}=S_{\mathrm{p,opt}}} = \left(\sqrt{1 - N_{\mathrm{T}} \Omega_{\mathrm{p}}} + c_{\mathrm{w}} \sqrt{N_{\mathrm{T}}} \right)^2. \tag{28}$$

This means that $\Delta\gamma|_{S_{\mathrm{p}}=S_{\mathrm{p,opt}}}$ is minimized by the maximum pilot overhead Ω_{p}, that is, all transmitted symbols are dedicated to pilots $\Omega_{\mathrm{p}} = 1/N_{\mathrm{T}}$. However, in this case the capacity becomes zero. In fact, the capacity is maximized by selecting the smallest pilot overhead $\Omega_{\mathrm{p,min}}$ which still satisfies the sampling theorem

$$C_{\max} = (1 - N_{\mathrm{T}} \Omega_{\mathrm{p,min}}) E \left[\log_2 \det \left(\mathbf{I}_{N_{\mathrm{R}}} + \frac{\mathbf{H}_{n,\ell} \mathbf{H}_{n,\ell}^H}{N_{\mathrm{T}}} \right. \right.$$
$$\left. \left. \cdot \frac{\gamma_0}{\left(\sqrt{1 - N_{\mathrm{T}} \Omega_{\mathrm{p,min}}} + c_{\mathrm{w}} \sqrt{N_{\mathrm{T}}} \right)^2} \right) \right], \tag{29}$$

where $\Omega_{\mathrm{p,min}} = 1/(D_{\mathrm{f,max}} D_{\mathrm{t,max}})$ is attained by the maximum pilot spacings which satisfy (24), $D_{\mathrm{f,max}} = \lfloor T/\tau_{\mathrm{w}} \rfloor$ and $D_{\mathrm{t,max}} = \lfloor 1/(2 f_{\mathrm{D,w}} T_{\mathrm{sym}}) \rfloor$, where $\lfloor x \rfloor$ is the largest integer equal or smaller than x. To prove (29) it can be easily checked that $C_{\max}$ is a monotonically decreasing function with respect to Ω_{p}, with the global maximum at $\Omega_{\mathrm{p}} = 0$. Hence, (29) is maximized by $\Omega_{\mathrm{p,min}}$, since $\Delta\gamma|_{S_{\mathrm{p}}=S_{\mathrm{p,opt}}}$ is only valid for pilot grids which satisfy the sampling theorem in (24).

By ignoring the rounding effects and thus approximating $D_{\mathrm{f,max}} D_{\mathrm{t,max}} \approx c_{\mathrm{w}}^2$, we obtain $\Omega_{\mathrm{p,min}} \approx c_{\mathrm{w}}^2$, that is, the effects of channel estimation errors for PACE are completely described by $\Omega_{\mathrm{p,min}}$. Interestingly, the estimator gain now approaches unity, $G_{\mathrm{n}} = 1$ and the SNR loss becomes

$$\Delta\gamma_{\min} = \left(\sqrt{1 - N_{\mathrm{T}} \Omega_{\mathrm{p,min}}} + \sqrt{N_{\mathrm{T}} \Omega_{\mathrm{p,min}}} \right)^2. \tag{30}$$

Furthermore, (27) and (29) can be approximated by

$$S_{\mathrm{p,opt}} = \frac{\sqrt{1 - N_{\mathrm{T}} \Omega_{\mathrm{p}}}}{\sqrt{N_{\mathrm{T}} \Omega_{\mathrm{p}}}}, \tag{31}$$

$$C_{\max} \approx (1 - N_{\mathrm{T}} \Omega_{\mathrm{p,min}}) E \left[\log_2 \det \left(\mathbf{I}_{N_{\mathrm{R}}} + \frac{\mathbf{H}_{n,\ell} \mathbf{H}_{n,\ell}^H}{N_{\mathrm{T}}} \right. \right.$$
$$\left. \left. \cdot \frac{\gamma_0}{\left(\sqrt{1 - N_{\mathrm{T}} \Omega_{\mathrm{p,min}}} + \sqrt{N_{\mathrm{T}} \Omega_{\mathrm{p,min}}} \right)^2} \right) \right]. \tag{32}$$

Finally, it turns out that the fraction of energy dedicated to data relative to the overall transmit energy becomes

$$\frac{E_{\mathrm{d}} N_{\mathrm{d}}}{E_0 N_{\mathrm{c}} L} = \frac{E_{\mathrm{d}}}{E_0} (1 - N_{\mathrm{T}} \Omega_{\mathrm{p,min}})$$
$$\approx \frac{\sqrt{1 - N_{\mathrm{T}} \Omega_{\mathrm{p,min}}}}{\sqrt{1 - N_{\mathrm{T}} \Omega_{\mathrm{p,min}}} + \sqrt{N_{\mathrm{T}} \Omega_{\mathrm{p,min}}}}, \tag{33}$$

where the ratio E_{d}/E_0 is defined in (6). Interestingly, (30) and (33) are equivalent to the results obtained for the block fading channel (see [7, equation (34)]). Applied to MIMO-OFDM, the block fading assumption translates to a time/frequency area the channel is assumed constant. Then, the same results apply given that the *interval, the channel is constant* for the block fading assumption in [7], is replaced by the *maximum pilot spacing that satisfies the sampling theorem*, $D_{\mathrm{f,max}} D_{\mathrm{t,max}} = 1/\Omega_{\mathrm{p,min}}$. This means that the capacity lower bound of [7] is extended to the more general case of time-variant frequency selective channels.

An interesting observation can be made by setting

$$N_{\mathrm{T}} \Omega_{\mathrm{p,min}} = \frac{1}{2}. \tag{34}$$

By devoting half of the resources to pilot symbols we obtain from (31) and (33), respectively,

$$S_{\mathrm{p,opt}} = 1, \qquad \frac{E_{\mathrm{d}} N_{\mathrm{d}}}{E_0 N_{\mathrm{c}} L} = \frac{1}{2}. \tag{35}$$

In case half of the frame is devoted to training purposes, in order to maximize capacity, pilots should not be boosted, and consequently half of the transmit energy is invested on pilots. A similar conclusion is also provided in [7] assuming a block fading channel.

4.2.2. Number of transmit antennas

Suppose that N_{T}' out of the N_{T} transmit antennas are used for communication. Inserting N_{T}' for N_{T} in the capacity expression for MIMO-OFDM with optimum pilot grid, $C_{\max}$ in (32), the number of transmit antennas that maximizes channel capacity $C_{\max}$ is given by [7, 29]

$$N_{\mathrm{T}}' = \min \left\{ N_{\mathrm{T}}, N_{\mathrm{R}}, \frac{1}{2 \Omega_{\mathrm{p,min}}} \right\}. \tag{36}$$

Several important conclusions with respect to the capacity maximization in MIMO-OFDM can be drawn from (36).

(i) If $N_{\mathrm{T}} = N_{\mathrm{R}} = 1/(2\Omega_{\mathrm{p,min}})$, from (34) and (35) it follows that pilots should not be boosted, that is, they should be of equal energy as the data symbols..

(ii) The amount of training should not exceed half of the OFDM frame.

(iii) The number of transmit and receive antennas should be equal.

4.3. Semianalytical approach for pilot grid design

The optimal pilot grid that maximizes the channel capacity is derived for an ideal LPIF in Section 4.2. For realizable estimators, we propose the following procedure to obtain the optimum pilot grid.

(i) Specify the filter parameters τ_{w} and $f_{\mathrm{D,w}}$ so that the relation in (24) is satisfied.

(ii) Choose maximum possible pilot spacings and estimator dimensions M_{f} and M_{t}, that maintain a certain interpolation error σ_{i}^2. This determines the minimum pilot overhead $\Omega_{\mathrm{p,min}}$, and the estimator gain $G_{\mathrm{n}} = 1/(\mathbf{w}^H\mathbf{w})$.

(iii) Determine the optimum pilot boost $S_{\mathrm{p,opt}}$ using (20).

(iv) Calculate the optimum number of transmit antennas using (36).

Considering (i), in a well-designed OFDM system the maximum channel delay $\tau_{\max}$ should not exceed the cyclic prefix. Therefore, it is reasonable to assume $\tau_{\mathrm{w}} = T_{\mathrm{CP}}$. In addition, $f_{\mathrm{D,w}}$ is set according to the maximum Doppler frequency expected in a certain propagation scenario.

Considering (ii), this condition is imposed to keep σ_{i}^2 sufficiently low. The impact of σ_{i}^2 on the SNR penalty in (21) becomes negligible if $\sigma_{\mathrm{i}}^2 < \epsilon_{\mathrm{th}}/\gamma_{\mathrm{w}}$, where ϵ_{th} is a small positive constant and γ_{w} denotes the largest expected SNR. This condition effectively enforces a sufficient degree of oversampling. That is, $\Omega_{\mathrm{p,min}}$ is required to be larger than the theoretical minimum.

The condition on the interpolation error σ_{i}^2 in step (ii) serves another important requirement. The fact that σ_{i}^2 is negligible in the SNR region of interest ensures that the useful signal part and the estimation error in the equivalent signal model in (12) become uncorrelated, so that the capacity bound in (22) becomes tight.

Considering (iii), this step ensures that the capacity in (22) is maximized, given that in step (ii) $\Omega_{\mathrm{p,min}}$ is appropriately chosen, so that the interpolation error σ_{i}^2 is sufficiently small. As a formal proof appears difficult, we verify through a numerical example in Section 5 that the proposed semianalytical procedure indeed maximizes capacity.

5. NUMERICAL RESULTS

An OFDM system with $N_{\mathrm{c}} = 512$ subcarriers, and a cyclic prefix of duration $T_{\mathrm{CP}} = 64 \cdot T_{\mathrm{spl}}$, is employed. One frame consists of $L = 65$ OFDM symbols, and it is assumed that channel estimation is carried out, after all pilots of one frame have been received. The signal bandwidth is 20 MHz, which translates to a sampling duration of $T_{\mathrm{spl}} = 50$ ns. This results in an OFDM symbol duration of $T_{\mathrm{sym}} = 35.97\,\mu s$ of which the cyclic prefix is $T_{\mathrm{CP}} = 3.2\,\mu s$. A high mobility scenario is considered with velocities up to 300 km/h. At 5 GHz carrier frequency this translates to a normalized maximum Doppler frequency of $f_{\mathrm{D,max}}T_{\mathrm{sym}} \le 0.04$.

Channel estimation unit

Since pilots belonging to different transmit antennas are orthogonally separated in time and/or frequency, the problem of MIMO-OFDM channel estimation breaks down to estimating the channel of a single antenna OFDM system. A cas-

TABLE 1: Power delay profile of WINNER channel model C2 [31].

Delay (ns)	0	5	135	160	215	260	385
Power (dB)	−0.5	0	−3.4	−2.8	−4.6	−0.9	−6.7
Delay (ns)	400	530	540	650	670	720	750
Power (dB)	−4.5	−9.0	−7.8	−7.4	−8.4	−11	−9.0
Delay (ns)	800	945	1035	1185	1390	1470	
Power (dB)	−5.1	−6.7	−12.1	−13.2	−13.7	−19.8	

caded channel estimator consisting of two one-dimensional (1D) estimators termed 2×1D PACE is implemented. 2×1D PACE performs only slightly worse than optimal 2D PACE, while being significantly less complex [16].

The estimator was implemented by a Wiener interpolation filter (WIF) with model mismatch [16]. The filter coefficients in frequency and time are generated assuming a uniformly distributed power delay profile and Doppler power spectrum, nonzero within the range $[0, \tau_{\mathrm{w}}]$ and $[-f_{\mathrm{D,w}}, f_{\mathrm{D,w}}]$. Furthermore, the average SNR at the filter input, γ_{w}, is required, which should be equal or larger than actual average SNR, so $\gamma_{\mathrm{w}} \ge \gamma_0$. To generate the filter coefficients, we set $\tau_{\mathrm{w}} = T_{\mathrm{CP}}$, $f_{\mathrm{D,w}} = 0.04T_{\mathrm{spl}}$ and $\gamma_{\mathrm{w}} = 30$ dB. With these parameters, the sampling theorem in (24) requires for the pilot spacings in frequency and time $D_{\mathrm{f}} < 8$ and $D_{\mathrm{t}} \le 12$.

The WIF with model mismatch is closely related to an LPIF, and therefore inherits many of its properties—signals with spectral components within the range $[0, \tau_w]$ and $[0, f_{\mathrm{D,max}}]$ pass the filter undistorted, while spectral components outside this range are blocked. In fact, it was shown in [30] that for an infinite number of coefficients, $\{M_{\mathrm{f}}, M_{\mathrm{t}}\} \to \infty$, the mismatched WIF approaches an ideal LPIF.

Results

The performance of a channel estimation unit generally depends on the chosen channel model. On the other hand, the optimum pilot grid and the associated channel estimation unit is expected to operated in a wide variety of channel conditions. Hence, it is important to test the performance of the considered estimators for various channel models. In Figure 5 the channel estimation MSE determined by (8) is plotted for the pilot grid $D_{\mathrm{f}} = 6$, $D_{\mathrm{t}} = 8$, and filter orders $M_{\mathrm{f}} = 16$, $M_{\mathrm{t}} = 9$. The following channel models are considered:

Chn A: IST-WINNER channel model C2 for typical urban propagation environments [31]; the power delay profile (PDP) is shown in Table 1;

Chn B: flat fading channel with PDP $\rho(\tau) = \delta(\tau - T_{\mathrm{CP}}/2)$;

Chn C: 2-tap channel with PDP $\rho(\tau) = (\delta(\tau)+\delta(\tau - T_{\mathrm{CP}}))/2$;

Chn D: uniformly distributed PDP nonzero within the range $[0, T_{\mathrm{CP}}]$. This is the channel used to generate the WIF coefficients, that is, the WIF is matched to Chn D.

For all models, the independent fading taps are generated using Jakes' model [32] with $f_{\mathrm{D,max}}T_{\mathrm{sym}} = 0.033$, corresponding to a velocity of 250 km/h at 5 GHz carrier frequency. It is seen in Figure 5 that the MSE is virtually independent

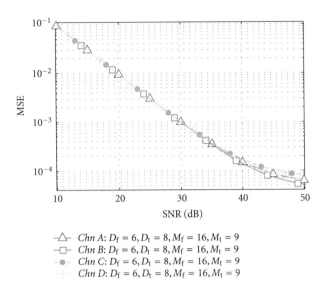

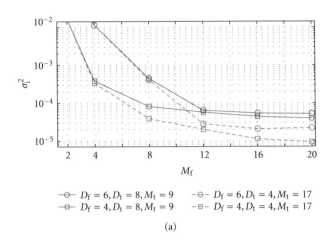

(a)

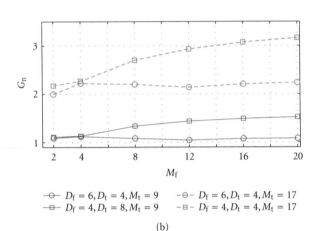

(b)

FIGURE 5: MSE versus SNR of $2 \times$ 1D-PACE for various channel models, $D_f = 6$, $D_t = 8$, $M_f = 16$, $M_t = 9$.

FIGURE 6: Interpolation error σ_i^2 and estimator gain G_n at an SNR of $\gamma_0 = 30$ dB against the filter order in frequency M_f.

of the particular channel model, although the PDPs of the considered channels cover an extensive range of possible propagation scenarios. Note that *Chn C* is the worst-case channel, since its two taps are placed at the closest position with respect to the cutoff regions of the WIF (compare with Figure 4). Likewise, *Chn B* is the best-case channel, as its single tap is located right in the center of the filter passband.

The optimum pilot grid for the considered MIMO-OFDM system is assembled in the following as discussed in Section 4.3. All results are plotted for *Chn A* with normalized maximum Doppler $f_{D,max} T_{sym} = 0.033$. We note that results in Figure 5 suggest that the identified optimum pilot grid is valid for any channel model with $\tau_{max} \leq \tau_w$, $f_{D,max} \leq f_{D,w}$ and $\gamma_0 \leq \gamma_w$.

From the set of allowable D_f and D_t, the following candidate grids are selected: $D_f = \{4, 6\}$ and $D_t = \{4, 8\}$, which translates to the oversampling factors $\beta_f = \{2, 1.25\}$ and $\beta_t = \{3, 1.5\}$. The filter order in time direction was set equal to the number of pilots per frame, so $M_t = \{17, 9\}$. In frequency direction, on the other hand, the number of pilots is $N_c/D_f = 85$ and 128, respectively, allowing for much higher filter orders M_f.

Figure 6 shows the interpolation error σ_i^2 (in Figure 6(a)) by computing (11) and the estimator gain $G_n = 1/(\mathbf{w}^H \mathbf{w})$ (in Figure 6(b)) against the filter order in frequency M_f for $M_t = \{9, 17\}$ and various pilot grids (parameters D_f and D_t). Provided that $M_f \geq 12$ we observe that for all pilot grids, $\gamma_w \sigma_i^2 < 0.1$ with $\gamma_0 \leq \gamma_w = 30$ dB. Therefore, the impact of σ_i^2 on $\Delta\gamma$ is negligible (less than 0.5 dB). By setting $M_f = 16$ in the following, none of the considered grids can be ruled out at this point.

In Figure 7, the channel capacity versus pilot boost S_p of an 8×8 MIMO-OFDM system with different pilot grids is depicted at SNR $\gamma_0 = 10$ dB. The plots are obtained by inserting σ_i^2 and $G_n = 1/(\mathbf{w}^H \mathbf{w})$ obtained in Figure 6 into the capacity expression (19) assuming different pilot grids. It is

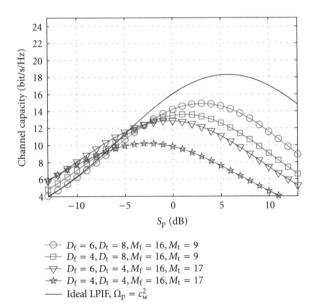

FIGURE 7: Capacity versus pilot boost for 8×8 MIMO-OFDM system with different pilot grids at an SNR of $\gamma_0 = 10$ dB.

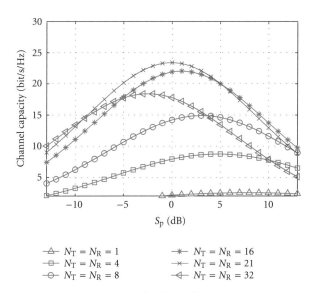

FIGURE 8: Capacity versus pilot boost for $N \times N$ MIMO-OFDM system for different number of antennas N, SNR $\gamma_0 = 10\,\text{dB}$, $D_f = 6$, $D_t = 8$, $M_f = 16$, $M_t = 9$.

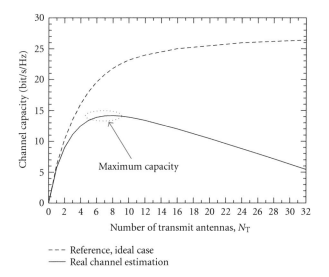

FIGURE 9: Capacity versus number of transmit antennas N_T for MIMO-OFDM system with $N_R = 8$ receive antennas, SNR $\gamma_0 = 10\,\text{dB}$, $D_f = 6$, $D_t = 8$, $M_f = 16$, $M_t = 9$.

seen that the most bandwidth efficient grid ($D_f = 6, D_t = 8$) maximizes capacity. Furthermore, maximum capacity C_{max} for all grids is achieved for those S_p that satisfy (20). This plot confirms the proposed semianalytical procedure described in Section 4.3. As reference, the capacity assuming an ideal LPIF is also plotted in Figure 7. A significant gap in capacity between the ideal LPIF relative to the realizable estimators is visible. This is mainly due to the fact that a realizable filter does not exhibit a rectangular filter transfer function. This inevitably requires a higher pilot overhead and also reduces the attainable estimator gain.

The channel capacity versus pilot boost, S_p, of an $N \times N$ MIMO-OFDM system with pilot grid $D_f = 6$, $D_t = 8$, filter orders $M_f = 16$, $M_t = 9$ and for different number of trans-

mit/receive antennas, $N = N_T = N_R$, is depicted in Figure 8. Again, the SNR is set to $\gamma_0 = 10\,\text{dB}$. Maximum capacity is achieved for $N_T = 21$ yielding $N_T \Omega_p \approx 1/2$ and $S_p = 0\,\text{dB}$. This confirms the analytical results from Section 4.2.2 as well as the proposed semianalytical procedure from Section 4.3. The significance of these results lie in the fact that results of [7] are generalized to MIMO-OFDM operating in dynamic channels and to arbitrary linear estimators.

The channel capacity versus the number of transmit antennas N_T of an $N_T \times 8$ MIMO-OFDM system for grid $D_f = 6$, $D_t = 8$, filter orders $M_f = 16$, $M_t = 9$, at SNR $\gamma_0 = 10\,\text{dB}$ is shown in Figure 9. The plots were generated using the capacity expression in (22) including the optimum pilot boost $S_{p,opt}$ according to (20). As a reference, capacity of the corresponding system assuming perfect channel estimation and no loss due to pilots is shown. It can be observed that for $N_T \approx 8$ maximum capacity is achieved, corresponding to the conclusion from Section 4.2.2. For higher values the reduction in available bandwidth due to the pilot insertion dominates, lowering the achievable capacity.

6. CONCLUSIONS

In this paper, a framework for pilot grid design in MIMO-OFDM was developed and used to determine the pilot spacing and boost, so to maximize the capacity of the target MIMO-OFDM system, including channel estimation errors and pilot overhead. The analysis show that the previously derived capacity lower bound for a block fading channel is also valid for MIMO-OFDM over time-varying frequency-selective channels. The derived bound applies to perfect interpolation, which essentially requires infinitely long pilot sequences and filter coefficients. Furthermore, a semianalytical procedure was proposed to maximize the capacity for realizable and possibly suboptimum channel estimation schemes.

ACKNOWLEDGMENTS

This work has been performed in the framework of the IST project IST-4-027756 WINNER (World Wireless Initiative New Radio), which is partly funded by the European Union. This paper was presented in part at the IEEE International Conference on Communications (ICC-07), Glasgow, UK, June 2007.

REFERENCES

[1] E. Telatar, "Capacity of multi-antenna Gaussian channels," *European Transactions on Telecommunications*, vol. 10, no. 6, pp. 585–595, 1999.

[2] G. J. Foschini and M. J. Gans, "On limits of wireless communications in a fading environment when using multiple antennas," *Wireless Personal Communications*, vol. 6, no. 3, pp. 311–335, 1998.

[3] H. Bolcskei, D. Gesbert, and A. J. Paulraj, "On the capacity of OFDM-based spatial multiplexing systems," *IEEE Transactions on Communications*, vol. 50, no. 2, pp. 225–234, 2002.

[4] R. Van Nee, V. K. Jones, G. Awater, A. Van Zelst, J. Gardner, and G. Steele, "The 802.11n MIMO-OFDM standard for wireless LAN and beyond," *Wireless Personal Communications*, vol. 37, no. 3-4, pp. 445–453, 2006.

[5] 3rd Generation Partnership Project; Technical Specification Group Radio Access Network, "Physical layer aspects for evolved Universal Terrestrial Radio Access (UTRA)," June 2006.

[6] G. Auer, "Analysis of pilot-symbol aided channel estimation for OFDM systems with multiple transmit antennas," in *Proceedings of IEEE International Conference on Communications (ICC '04)*, vol. 6, pp. 3221–3225, Paris, France, June 2004.

[7] B. Hassibi and B. M. Hochwald, "How much training is needed in multiple-antenna wireless links?" *IEEE Transactions on Information Theory*, vol. 49, no. 4, pp. 951–963, 2003.

[8] S. Adireddy, L. Tong, and H. Viswanathan, "Optimal placement of training for frequency-selective block-fading channels," *IEEE Transactions on Information Theory*, vol. 48, no. 8, pp. 2338–2353, 2002.

[9] S. Adireddy and L. Tong, "Optimal placement of known symbols for slowly varying frequency-selective channels," *IEEE Transactions on Wireless Communications*, vol. 4, no. 4, pp. 1292–1296, 2005.

[10] H. Vikalo, B. Hassibi, B. Hochwald, and T. Kailath, "On the capacity of frequency-selective channels in training-based transmission schemes," *IEEE Transactions on Signal Processing*, vol. 52, no. 9, pp. 2572–2583, 2004.

[11] O. Simeone and U. Spagnolini, "Lower bound on training-based channel estimation error for frequency-selective block-fading Rayleigh MIMO channels," *IEEE Transactions on Signal Processing*, vol. 52, no. 11, pp. 3265–3277, 2004.

[12] S. Ohno and G. B. Giannakis, "Capacity maximizing MMSE-optimal pilots for wireless OFDM over frequency-selective block Rayleigh-fading channels," *IEEE Transactions on Information Theory*, vol. 50, no. 9, pp. 2138–2145, 2004.

[13] R. Negi and J. Cioffi, "Pilot tone selection for channel estimation in a mobile OFDM system," *IEEE Transactions on Consumer Electronics*, vol. 44, no. 3, pp. 1122–1128, 1998.

[14] A. V. Oppenheim and R. W. Schafer, *Discrete-Time Signal Processing*, Prentice Hall, Englewood Cliffs, NJ, USA, 2nd edition, 1999.

[15] J.-J. van de Beek, O. Edfors, M. Sandell, S. Wilson, and P. Börjesson, "On channel estimation in OFDM systems," in *Proceedings of the 45th IEEE Vehicular Technology Conference (VTC '95)*, vol. 2, pp. 815–819, Chicago, Ill, USA, July 1995.

[16] P. Höher, S. Kaiser, and P. Robertson, "Pilot-symbol-aided channel estimation in time and frequency," in *Proceedings of the IEEE Global Telecommunications Conference (Globecom '97)*, vol. 4, pp. 90–96, Phoenix, Ariz, USA, November 1997.

[17] P. Höher, "TCM on frequency selective land-mobile radio channels," in *Proceedings of the 5th Tirrenia International Workshop on Digital Communications*, pp. 317–328, Tirrenia, Italy, September 1991.

[18] F. Sanzi and J. Speidel, "An adaptive two-dimensional channel estimator for wireless OFDM with application to mobile DVB-T," *IEEE Transactions on Broadcasting*, vol. 46, no. 2, pp. 128–133, 2000.

[19] Y. Li, "Pilot-symbol-aided channel estimation for OFDM in wireless systems," *IEEE Transactions on Vehicular Technology*, vol. 49, no. 4, pp. 1207–1215, 2000.

[20] Y.-H. Yeh and S.-G. Chen, "DCT-based channel estimation for OFDM systems," in *Proceedings of IEEE International Conference on Communications (ICC '04)*, vol. 4, pp. 2442–2446, Paris, France, June 2004.

[21] M.-H. Hsieh and C.-H. Wei, "Channel estimation for OFDM systems based on comb-type pilot arrangement in frequency selective fading channels," *IEEE Transactions on Consumer Electronics*, vol. 44, no. 1, pp. 217–225, 1998.

[22] Y. Li, N. Seshadri, and S. Ariyavisitakul, "Channel estimation for OFDM systems with transmitter diversity in mobile wireless channels," *IEEE Journal on Selected Areas in Communications*, vol. 17, no. 3, pp. 461–471, 1999.

[23] Y. Li, "Simplified channel estimation for OFDM systems with multiple transmit antennas," *IEEE Transactions on Wireless Communications*, vol. 1, no. 1, pp. 67–75, 2002.

[24] W. G. Jeon, K. H. Paik, and Y. S. Cho, "Two-dimensional MMSE channel estimation for OFDM systems with transmitter diversity," in *Proceedings of the 54th IEEE Vehicular Technology Conference (VTC '01 fall)*, vol. 3, pp. 1682–1685, Atlantic City, NJ, USA, October 2001.

[25] M. Speth, "LMMSE channel estimation for MIMO OFDM," in *Proceedings of the 8th International OFDM Workshop (InOWo '03)*, Hamburg, Germany, September 2003.

[26] J.-W. Choi and Y.-H. Lee, "Optimum pilot pattern for channel estimation in OFDM systems," *IEEE Transactions on Wireless Communications*, vol. 4, no. 5, pp. 2083–2088, 2005.

[27] S. M. Kay, *Fundamentals of Statistical Signal Processing: Estimation Theory*, Prentice Hall, Englewood Cliffs, NJ, USA, 1993.

[28] H. Meyr, M. Moeneclaey, and S. A. Fechtel, *Digital Communication Receivers*, John Wiley & Sons, New York, NY, USA, 2nd edition, 1998.

[29] L. Zheng and D. N. C. Tse, "Communication on the Grassmann manifold: a geometric approach to the noncoherent multiple-antenna channel," *IEEE Transactions on Information Theory*, vol. 48, no. 2, pp. 359–383, 2002.

[30] G. Auer and E. Karipidis, "Pilot aided channel estimation for OFDM: a separated approach for smoothing and interpolation," in *Proceedings of IEEE International Conference on Communications (ICC '05)*, vol. 4, pp. 2173–2178, Seoul, Korea, May 2005.

[31] IST-4-027756 WINNER II, "D1.1.2 WINNER II channel models," September 2007.

[32] W. C. Jakes, *Microwave Mobile Communications*, John Wiley & Sons, New York, NY, USA, 1974.

Hindawi Publishing Corporation
EURASIP Journal on Wireless Communications and Networking
Volume 2007, Article ID 21093, 12 pages
doi:10.1155/2007/21093

Research Article

Distributed Antenna Channels with Regenerative Relaying: Relay Selection and Asymptotic Capacity

Aitor del Coso and Christian Ibars

Centre Tecnològic de Telecomunicacions de Catalunya (CTTC), Av. Canal Olímpic, Castelldefels, Spain

Received 15 November 2006; Accepted 3 September 2007

Recommended by Monica Navarro

Multiple-input-multiple-output (MIMO) techniques have been widely proposed as a means to improve capacity and reliability of wireless channels, and have become the most promising technology for next generation networks. However, their practical deployment in current wireless devices is severely affected by antenna correlation, which reduces their impact on performance. One approach to solve this limitation is *relaying diversity*. In relay channels, a set of N wireless nodes aids a source-destination communication by relaying the source data, thus creating a distributed antenna array with uncorrelated path gains. In this paper, we study this multiple relay channel (MRC) following a *decode-and-forward* (D&F) strategy (*i.e.*, regenerative forwarding), and derive its achievable rate under AWGN. A half-duplex constraint on relays is assumed, as well as distributed channel knowledge at both transmitter and receiver sides of the communication. For this channel, we obtain the optimum relay selection algorithm and the optimum power allocation within the network so that the transmission rate is maximized. Likewise, we bound the ergodic performance of the achievable rate and derive its asymptotic behavior in the number of relays. Results show that the achievable rate of regenerative MRC grows as the logarithm of the Lambert W function of the total number of relays, that is, $\mathcal{C} = \log_2(W_0(N))$. Therefore, D&F relaying, cannot achieve the capacity of actual MISO channels.

1. INTRODUCTION

Current wireless applications demand an ever-increasing transmission capacity and highly reliable communications. Voice transmission, video broadcasting, and web browsing require wire-like channel conditions that the wireless medium still cannot support. In particular, channel impairments, namely, path loss and multipath fading do not allow wireless channels to reach the necessary rate and robustness expected for next generation systems. Recently, a wide range of multiple antenna techniques have been proposed to overcome these channel limitations [1–4]; however, the deployment of multiple transmit and/or receive antennas on the wireless nodes is not always possible or worthwhile. For these cases, the most suitable technique to take advantage of spatial diversity is *node cooperation* and *relay channels* [5, 6].

Relay channels consist of single source-destination pairs aided in their communications by a set of wireless relay nodes that creates a distributed antenna array (see Figure 1). The relay nodes can be either infrastructure nodes, placed by the service provider in order to enhance coverage and rate [7], or a set of network users that cooperate with the source, while having own data to transmit [8]. Relay-based architectures have been shown to improve capacity, diversity, and delay of wireless channels when properly allocating network resources, and have become a key technique for the evolution of wireless communications [9].

Background

The use of relays to increase the achievable rate of point-to-point transmissions was initially proposed by Cover and El Gamal in [10]. Motivated by this work, many relaying techniques have been recently studied, which can be classified, based on their forwarding strategy and required processing at the relay nodes, as *regenerative relaying* and *nonregenerative relaying* [5, 11]. The former assumes that relay nodes decode the source information, prior to reencoding and sending it to destination [12, 13]. On the other hand, with the latter, relay nodes transform and retransmit their received signals but do not decode them [14–16].

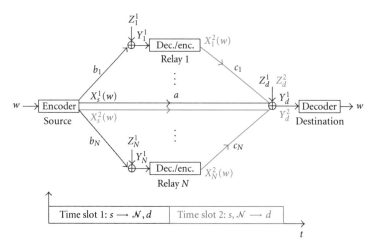

FIGURE 1: Half-duplex regenerative multiple relay channel with N parallel relays.

Regenerative relaying was initially presented in [10, Theorem 1] for a single-relay channel, and consists of relay nodes decoding the source data and transmitting it to destination, ideally without errors. Such signal regeneration allows for cooperative coherent transmissions. Therefore, source and relays can operate as a distributed antenna array and implement multiple-input single-output (MISO) beamforming. We distinguish two techniques: *decode-and-forward* (D&F), presented in [10], and *partial decoding* (PD), analyzed in [17]. D&F requires the relay nodes to fully decode the source message before retransmitting it. Thus, it penalizes the achievable rate when poor source-to-relay channel conditions occur. Nevertheless, for poor source-to-destination channels (e.g., degraded relay channels), it was shown to be the capacity achieving technique [10]. On the other hand, with PD the relay nodes only partially decode the source message. Part of the transmitted message is sent directly to the destination without being relayed [18]. PD is specifically appropriate when the source node can adapt the amount of information transmitted through relays to the network channel conditions; otherwise it does not improve the D&F scheme [19]. The diversity analysis of regenerative multiple relay networks was carried out by Laneman and Wornell in [20], showing that signal regeneration achieves full transmit diversity of the system. However, regenerative relaying has some drawbacks as well: first, decoding errors at the relay nodes generate error propagation; second, synchronization among relays (specifically in the low SNR regime) may complicate its implementation, and finally, the processing capabilities required at the relays increase their cost [5].

The two previously mentioned techniques are well known for the single-relay channel. However, the only significant extensions to the multiple relay setup are found in [6, 21, 22]. In these works, they were applied to physical-layer multihop networks and to the multiple relay channel with orthogonal components, respectively.

Contributions

This paper studies the point-to-point Gaussian channel with N parallel relays that use *decode-and-forward* relaying. On the relays, a half duplex constraint is considered, that is, the relay nodes cannot transmit and receive simultaneously in the same frequency band. The communication is arranged into two consecutive, identical time slots, as shown in Figure 1. The source uses the first time slot to transmit the message to the set of relays and to the destination. Then, during time slot 2, the set of nodes who have successfully decoded the message, and the source, transmit extra parity bits to the destination node, which uses its received signal during the two slots to decode the message. Transmit and receive channel state information (CSI) are available at both transmitter and receiver sides, and channel conditions are assumed not to vary during the two slots of the communication. Additionally, we consider that the source knows all relay-to-destination channels, so that it can implement a relay selection algorithm. Finally, the overall transmitted power during the two time slots is constrained to a constant, and we maximize the achievable rate through power allocation on the two slots of the communication, and on the useful relays.

The contributions of this paper are as follows.

(i) First, the instantaneous achievable rate of the proposed communication is derived in Proposition 1; then the optimum power allocation on the two slots is obtained in Proposition 2. Results show that the achievable rate is maximized through an optimum relay selection algorithm and through power allocation on the two slots, referred to as constrained temporal *waterfilling*.

(ii) Second, we analyze the ergodic performance of the instantaneous achievable rate derived in Proposition 2, assuming independent, identically distributed (i.i.d.) random channel fading and i.i.d. random relay positions. We assume that the source node transmits over several concatenated two-slot transmissions. The channel is invariant during the two slots, and uncorrelated from one two-slot transmission to the next (see Figure 2). Thus, the source transmits with an effective rate equal to the ergodic achievable rate of the link, which is lower- and upper-bounded in this paper.

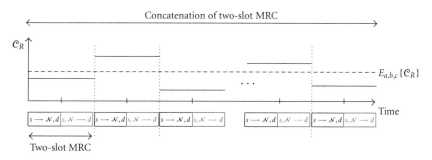

FIGURE 2: Ergodic capacity: concatenation in time of half-duplex multiple relay channels.

(iii) Finally, we study the asymptotic performance (in the number of relays) of the instantaneous achievable rate, and we show that it grows asymptotically with the logarithm of the branch 0 of the Lambert W function[1] of the total number of relays, that is, $C = \log_2(W_0(N))$.

The remainder of the paper is organized as follows: in Section 2, we introduce the channel and signal model; in Section 3, the instantaneous achievable of the D&F MRC is derived and the optimum relay selection and power allocation are obtained. In Section 4, the ergodic achievable rate is upper- and lower-bounded, and Section 5 analyzes the asymptotic achievable rate of the channel. Finally, Section 6 contains simulation results and Section 7 summarizes conclusions.

Notation

We define $\mathbf{X}_{1:n}^{(2)} = [X_1^{(2)}, \ldots, X_n^{(2)}]^T$ with $n \in \{1, \ldots, N\}$. Moreover, in the paper, $I(A; B)$ denotes mutual information between random variables A and B, $C(x) = \log_2(1 + x)$, $\mathbf{b}^{\dagger}$ denotes the conjugate transpose of vector $\mathbf{b}$, and b^* denotes the conjugate of b.

2. CHANNEL MODEL

We consider a wireless multiple-relay channel (MRC) with a source node s, a destination node d, and a set of parallel relays $\mathcal{N} = \{1, \ldots, N\}$ (see Figure 1). Wireless channels among network nodes are *frequency-flat, memoryless,* and modelled with a complex, Gaussian-distributed coefficient; $a \sim \mathcal{CN}(0, 1)$ denotes the unitary power, Rayleigh distributed channel between source and destination, and $c_i \sim \mathcal{CN}(0, 1)$ the complex channel from relay i to destination. In the system, b_i is modelled as a superposition of path loss (with exponent α) and Rayleigh distributed fading, in order to account for the different transmission distances from the source to relays, d_i, $i = 1, \ldots, N$, and from source to destina-

tion d_o (used as reference), that is,

$$b_i \sim \mathcal{CN}\left(0, \left(\frac{d_o}{d_i}\right)^{\alpha}\right). \tag{1}$$

We assume invariant channels during the two-slot communication.

As mentioned, the communication is arranged in two consecutive time slots of equal duration (see Figure 1). During the first slot, a single-input multiple-output (SIMO) transmission from the source node to the set of relays and destination takes place. The second slot is then used by relays and source to retransmit data to destination via a distributed MISO channel. In both slots, the transmitted signals are received under additive white Gaussian noise (AWGN), and destination attemps to decode making use of the signal received during the two phases. The complex signals transmitted by the source during slot $t = \{1, 2\}$, and by relay i during phase 2, are denoted by $X_s^{(t)}$ and $X_i^{(2)}$, respectively. Therefore, considering memoryless channels, the received signal at the relay nodes during time slot 1 is given by

$$Y_i^{(1)} = b_i \cdot X_s^{(1)} + Z_i^{(1)} \quad \text{for } i \in \mathcal{N}, \tag{2}$$

where $Z_i^{(1)} \sim \mathcal{CN}(0, 1)$ is normalized AWGN at relay i. Likewise, considering the channel definition in Figure 1, the received signal at the destination node d during time slots 1 and 2 is written as

$$Y_d^{(1)} = a \cdot X_s^{(1)} + Z_d^{(1)},$$
$$Y_d^{(2)} = a \cdot X_s^{(2)} + \sum_{i=1}^{N} c_i \cdot X_i^{(2)} + Z_d^{(2)}, \tag{3}$$

where, as previously said, $Z_d^{(t)} \sim \mathcal{CN}(0, 1)$ is AWGN. Notice that, due to half-duplex limitations, the relay nodes do not transmit during time slot 1 and do not receive during time slot 2. The overall transmitted power during the two time slots is constrained to $2P$; thus, defining $\gamma_1 = \mathrm{E}\{X_s^{(1)}(X_s^{(1)})^*\}$ and $\gamma_2 = \mathrm{E}\{X_s^{(2)}(X_s^{(2)})^*\} + \sum_{i=1}^{N}\mathrm{E}\{X_i^{(2)}(X_i^{(2)})^*\}$ as the

---

[1] The branch 0 of the Lambert W function, $W_0(N)$, is defined as the function satisfying $W_0(N)e^{W_0(N)} = N$, with $W_0(N) \in R^+$ [23].

transmitted power[2] during slots 1 and 2, respectively, we enforce the following two-slot power constraint:

$$\gamma_1 + \gamma_2 = 2P. \tag{4}$$

3. ACHIEVABLE RATE IN AWGN

In order to determine the achievable rate of the channel, we consider updated transmitter and receiver channel state information (CSI) at all nodes, and assume symbol and phase synchronization among transmitters. The achievable rate with D&F is given in the following proposition.

Proposition 1. *In a half-duplex multiple-relay channel with decode-and-forward relaying and N parallel relays, the rate*

$$\mathcal{C}_{\mathrm{D\&F}} = \max_{1 \le n \le N} \left\{ \max_{p(\mathbf{X}_s, \mathbf{X}_{1:n}^{(2)}):\gamma_1 + \gamma_2 = 2P} \frac{1}{2} \cdot I\left(X_s^{(1)}; Y_d^{(1)}\right) \right.$$
$$\left. + \frac{1}{2} \cdot I\left(X_s^{(2)}, \mathbf{X}_{1:n}^{(2)}; Y_d^{(2)}\right) \right\}$$
$$s.t. \quad I\left(X_s^{(1)}; Y_n^{(1)}\right) \ge I\left(X_s^{(1)}; Y_d^{(1)}\right) + I\left(X_s^{(2)}, \mathbf{X}_{1:n}^{(2)}; Y_d^{(2)}\right) \tag{5}$$

is achievable. Source-relay path gains have been ordered as

$$|b_1| \ge \cdots \ge |b_n| \ge \cdots \ge |b_N|. \tag{6}$$

Remark 1. Factor 1/2 comes from time division signalling. Variable n in the maximization represents the number of active relays; hence, the relay selection is carried out through the maximization in (5), considering (6).

Proof. Let the N relays in Figure 1 be ordered as in (6), and assume that only the subset $\mathcal{R}_n = \{1, \ldots, n\} \subseteq \mathcal{N}$ is active, with $n \le N$. The source node selects message $\omega \in [1, \ldots, 2^{mR}]$ for transmission (with m the total number of transmitted symbols during the two slots, and R the transmission rate) and maps it into two codebooks $\mathcal{X}_1, \mathcal{X}_2 \in \mathcal{C}^{m/2}$, using two independent encoding functions,[3] $x_1 : \{1, \ldots, 2^{mR}\} \to \mathcal{X}_1$ and $x_2 : \{1, \ldots, 2^{mR}\} \to \mathcal{X}_2$. The codeword $x_1(\omega)$ is then transmitted by the source during time slot 1, that is, $X_s^{(i)} = x_1(\omega)$. At the end of this slot, all relay nodes belonging to $\mathcal{R}_n$ are able to decode the transmitted message with arbitrarily small error probability if and only if the transmission rate satisfies [24]:

$$R \le \frac{1}{2} \cdot \min_{i \in \mathcal{R}_n} \left\{ I\left(X_s^{(1)}; Y_i^{(1)}\right) \right\}$$
$$= \frac{1}{2} \cdot I\left(X_s^{(1)}; Y_n^{(1)}\right), \tag{7}$$

where equality follows from (6), taking into account that all noises are i.i.d. Later, once decoded ω and knowing the codebook $\mathcal{X}_2$ and its associated encoding function, nodes in $\mathcal{R}_n$

(and also the source) calculate $x_2(\omega)$ and transmit it during phase 2. Hence, considering *memoryless* time-division channels with uncorrelated signalling between the two phases, the destination is able to decode ω if

$$R \le \frac{1}{2} \cdot I\left(X_s^{(1)}; Y_d^{(1)}\right) + \frac{1}{2} \cdot I\left(X_s^{(2)}, \mathbf{X}_{1:n}^{(2)}; Y_d^{(2)}\right). \tag{8}$$

Therefore, the maximum source-to-destination transmission rate for the MRC is given by (8) with equality, subject to (7) being satisfied. Finally, noting that the set of active relay nodes $\mathcal{R}_n$ can be chosen out of $\{\mathcal{R}_1, \ldots, \mathcal{R}_N\}$ concludes the proof. □

As previously mentioned, we consider all receiver nodes under unitary power AWGN. The evaluation of Proposition 1 for faded Gaussian channels is established in Proposition 2. Previously, from an intuitive view of (5), some conclusions can be inferred: first, we note that the relay nodes which have successfully decoded during phase 1 transmit during phase 2 using a distributed MISO channel to destination. Assuming transmit CSI and phase synchronization among them, the performance of such a distributed MISO is equal to that of the actual MISO channel. Therefore, the optimum power allocation on the relays will also be the optimum beamforming [1]. For the power allocation over the two time slots, we also notice the following tradeoff: the higher the power allocated during time slot 1 is, the more the relays belong to the decoding set, but the less power they have during time slot 2 to transmit. Both considerations are discussed in Proposition 2.

Proposition 2. *In a Gaussian, half-duplex, multiple relay channel with decode-and-forward relaying and N parallel relays, the rate*

$$\mathcal{C}_{\mathrm{D\&F}} = \max_{1 \le n \le N} \frac{1}{2} \cdot \mathcal{C}(\gamma_{1n}\lambda_1) + \frac{1}{2} \cdot \mathcal{C}(\gamma_{2n}\lambda_{2n}) \tag{9}$$

is achievable, where

$$\lambda_1 = |a|^2, \qquad \lambda_{2n} = |a|^2 + \sum_{i=1}^{n} |c_i|^2 \tag{10}$$

are the beamforming gains during time slots 1 and 2, respectively, and the power allocation is computed from

$$\gamma_{1n} = \max\left\{ \left(\frac{1}{\mu_n} - \frac{1}{\lambda_1}\right), \gamma_n^c \right\},$$
$$\gamma_{2n} = \min\left\{ \left(\frac{1}{\mu_n} - \frac{1}{\lambda_{2n}}\right), 2P - \gamma_n^c \right\} \tag{11}$$

subject to $(\mu_n^{-1} - \lambda_1^{-1}) + (\mu_n^{-1} - \lambda_{2n}^{-1}) = 2P$, and

$$\gamma_n^c = \phi_n + \sqrt{\phi_n^2 + \frac{2P}{\lambda_1}},$$
$$\phi_n = \left(\frac{1}{\mu_n} - \frac{1}{\lambda_1}\right) - \frac{|b_n|^2}{2\lambda_1\lambda_{2n}}. \tag{12}$$

Source-relay path gains have been ordered as

$$|b_1| \ge \cdots \ge |b_n| \ge \cdots \ge |b_N|, \tag{13}$$

---

[2] $E\{\cdot\}$ denotes expectation.
[3] Codewords in $\mathcal{X}_1, \mathcal{X}_2$ have length $m/2$ since each one is transmitted in one time slot, respectively.

Remark 2. As previously, maximization over n selects the optimum number of relays. The optimum power allocation γ_{1n}, γ_{2n} results in a constrained temporal *water-filling* over the two slots of the communication. Furthermore, γ_n^c is the minimum power allocation during time slot 1 that satisfies simultaneously, for a given set of active relays $\mathcal{R}_n = \{1, \ldots, n\}$, the power constraint (4) and the constraint in (5).

Proof. To derive expression (9), we independently solve the optimization problems in (5):

$$\max_{p(\mathbf{X}_s, \mathbf{X}_{1:n}^{(2)}):\gamma_1+\gamma_2=2P} \frac{1}{2} \cdot I\left(X_s^{(1)}; Y_d^{(1)}\right) + \frac{1}{2} \cdot I\left(X_s^{(2)}, \mathbf{X}_{1:n}^{(2)}; Y_d^{(2)}\right)$$

$$\text{s.t.} \quad I\left(X_s^{(1)}; Y_n^{(1)}\right) \geq I\left(X_s^{(1)}; Y_d^{(1)}\right) + I\left(X_s^{(2)}, \mathbf{X}_{1:n}^{(2)}; Y_d^{(2)}\right)$$

$$\tag{14}$$

for every $n \in \{1, \ldots, N\}$. First, we notice that for AWGN and *memoryless* channels, the optimum input signal during the two slots is i.i.d. with Gaussian distribution. Hence, the mutual information in (14) are given by

$$I\left(X_s^{(1)}; Y_d^{(1)}\right) = \mathcal{C}(\gamma_1 \lambda_1),$$

$$I\left(X_s^{(2)}, \mathbf{X}_{1:n}^{(2)}; Y_d^{(2)}\right) = \mathcal{C}(\gamma_2 \lambda_{2n}), \tag{15}$$

$$I\left(X_s^{(1)}; Y_n^{(1)}\right) = \mathcal{C}(\gamma_1 |b_n|^2),$$

with λ_1 and λ_{2n} defined in (10), and γ_1 and γ_2 the transmitted powers during time slot 1 and 2, respectively. Then maximization (14) reduces to

$$\max_{\gamma_1, \gamma_2:\gamma_1+\gamma_2=2P} \frac{1}{2} \cdot \mathcal{C}(\gamma_1 \lambda_1) + \frac{1}{2} \cdot \mathcal{C}(\gamma_2 \lambda_{2n})$$

$$\text{s.t.} \quad \mathcal{C}\left(\gamma_1 |b_n|^2\right) \geq \mathcal{C}(\gamma_1 \lambda_1) + \mathcal{C}(\gamma_2 \lambda_{2n}). \tag{16}$$

The optimization above is solved in Appendix A yielding (9), with γ_{1n} and γ_{2n} the optimum power allocation on each slot for a given value n. Maximization over n results in the optimum relay selection. □

4. ERGODIC ACHIEVABLE RATE

In this section, we analyze the ergodic behavior of the instantaneous achievable rate obtained in Proposition 2. We assume that the source transmits over several, concatenated two-slot multiple relay transmissions, with uncorrelated channel conditions (see Figure 2). Thus, it achieves an effective rate equal to the expectation (on the channel distribution) of the achievable rate defined in Proposition 2, that is, it achieves a rate equal to the ergodic achievable rate. Throughout the paper, we assume random channel fading and random i.i.d. relay positions, invariant during the two-phase transmission but independent between transmissions.

Accordingly, considering the result in (9), we define the ergodic achievable rate[4] of the half-duplex MRC as

$$\begin{aligned} \mathcal{C}_{\text{D\&F}}^e &= \mathrm{E}_{\mathbf{a},\mathbf{b},\mathbf{c}}\{\mathcal{C}_{\text{D\&F}}\} \\ &= \mathrm{E}_{\mathbf{a},\mathbf{b},\mathbf{c}}\left\{\max_{1 \leq n \leq N} \mathcal{C}_n\right\}, \end{aligned} \tag{17}$$

where $\mathbf{a} = |a|^2$ is the source-to-destination channel; $\mathbf{c} = [|c_1|^2, \ldots, |c_N|^2]$ the relay-to-destination channels, and $\mathbf{b} = [|b_1|^2, \ldots, |b_N|^2]$ the source-to-relay channels ordered as (6). Notice that all elements in $\mathbf{c}$ are i.i.d. while, due to ordering, elements in $\mathbf{b}$ are mutually dependent. Finally, $\mathcal{C}_n$ in (17) is defined from Proposition 2 as

$$\mathcal{C}_n = \frac{1}{2} \cdot \mathcal{C}(\gamma_{1n} \lambda_1) + \frac{1}{2} \cdot \mathcal{C}(\gamma_{2n} \lambda_{2n}). \tag{18}$$

There is no closed-form expression for the ergodic capacity of the multiple-relay channel in (17); capacities $\mathcal{C}_1, \ldots, \mathcal{C}_N$ are mutually dependent, therefore closed-form expression for the cumulative density function (cdf) of $\max_{1 \leq n \leq N} \mathcal{C}_n$ cannot be obtained. Hence, we turn our attention to obtaining upper and lower bounds.

4.1. Lower bound

A lower bound can be derived using Jensen's inequality, taking into account the convexity of the pointwise maximum function:

$$\begin{aligned} \mathcal{C}_{\text{D\&F}}^e &= \mathrm{E}_{\mathbf{a},\mathbf{b},\mathbf{c}}\left\{\max_{1 \leq n \leq N} \mathcal{C}_n\right\} \\ &\geq \max_{1 \leq n \leq N} \mathrm{E}_{\mathbf{a},\mathbf{b},\mathbf{c}}\{\mathcal{C}_n\}. \end{aligned} \tag{19}$$

The interpretation of such bound is as follows: the inequality shows that the ergodic capacities achieved assuming a fixed number of active relays are, obviously, always lower than the ergodic capacity achieved with instantaneous optimal relay selection. Analyzing (19) carefully, we notice that $\mathcal{C}_n$ does not depend upon entire vector $\mathbf{b}$ but only upon $|b_n|^2$. Furthermore, we have seen that $\mathcal{C}_n$ depends on fading between source and destination, and between relays and destination just in terms of beamforming gains $\lambda_1 = |a|^2$ and $\lambda_{2n} = |a|^2 + \sum_{i=1}^{n} |c_i|^2$; therefore, renaming $\delta = |a|^2$ and $\beta_n = \sum_{i=1}^{n} |c_i|^2$, expression (19) simplifies to

$$\mathcal{C}_{\text{D\&F}}^e \geq \max_{1 \leq n \leq N} \mathrm{E}_{\delta,\beta_n,|b_n|^2}\{\mathcal{C}_n\}, \tag{20}$$

where δ is a unitary-mean, exponential random variable describing the square of the fading coefficient between source and destination. Likewise, β_n describes the relay beamforming gain assuming only the set of relays $\mathcal{R}_n = \{1, \ldots, n\}$ to be active. It is obtained as the sum of n exponentially distributed, unitary mean random variables, and hence it is distributed as a chi-squared random variable with $2n$ degrees

---

[4] Notice that, due to the power constraint (4), the ergodic achievable rate is directly computed as the expectation of the instantaneous achievable rate of the link.

of freedom. Both variables are described by their probability density functions (pdf) as

$$f_\delta(\delta) = e^{-\delta},$$
$$f_{\beta_n}(\beta) = \frac{\beta^{(n-1)}e^{-\beta}}{(n-1)!}. \qquad (21)$$

The study of $|b_n|^2$ is more involved; b_n, as defined previously, is the nth better channel from source to relays, following the ordering in (13). As stated earlier, source-to-relay channels in (1) are i.i.d. with complex Gaussian distribution and power $(d_o/d)^\alpha$; d is the random source-to-relay distance, assumed i.i.d. for all relays and with a generic pdf $f_d(d)$, $d \in [0, d^+]$. Hence, defining $\xi \sim \mathcal{CN}(0, (d_o/d)^\alpha)$, we make use of ordered statistics to obtain the pdf of $|b_n|^2$ as [25]

$$f_{|b_n|^2}(b) = \frac{N!}{(N-n)!1!(n-1)!} f_{|\xi|^2}(b) P[|\xi|^2 \le b]^{N-n}$$
$$\times P[|\xi|^2 \ge b]^{n-1}, \qquad (22)$$

where cumulative density function $P[|\xi|^2 \le b]$ may be derived as

$$P[|\xi|^2 \le b] = 1 - \int_0^{d^+} e^{-b(x/d_o)^\alpha} f_d(x) dx, \qquad (23)$$

and probability density function $f_{|\xi|^2}(b)$ is computed as the first derivative of (23) respect to b:

$$f_{|\xi|^2}(b) = \int_0^{d^+} \left(\frac{x}{d_o}\right)^\alpha e^{-b(x/d_o)^\alpha} f_d(x) dx. \qquad (24)$$

Therefore, proceeding from (20),

$$\mathcal{C}_{D\&F}^e \ge \max_{1 \le n \le N} \iint_0^\infty E_{|b_n|^2}\{\mathcal{C}_n \mid \delta, \beta_n\} f_\delta(\delta) f_{\beta_n}(\beta) db\, d\beta, \qquad (25)$$

where $E_{|b_n|^2}\{\mathcal{C}_n \mid \delta, \beta\}$ is the mean of $\mathcal{C}_n$ over $|b_n|^2$ conditioned on beamforming gains δ and $\beta_n = \beta$. This mean may be readily obtained using the pdf (22) and power allocation defined in (10):

$$E_{|b_n|^2}\{\mathcal{C}_n \mid \delta, \beta\} = \frac{1}{2} \int_0^\infty (\mathcal{C}(\gamma_{1n}\delta) + \mathcal{C}(\gamma_{2n}(\delta+\beta)))$$
$$\times f_{|b_n|^2}(b) db. \qquad (26)$$

Notice that

$$[\gamma_{1n}, \gamma_{2n}] = \begin{cases} \left[\left(\frac{1}{\mu_n} - \frac{1}{\delta}\right), \left(\frac{1}{\mu_n} - \frac{1}{\delta+\beta}\right)\right], & b \ge \psi(\delta, \beta), \\ [\gamma_n^c, 2P - \gamma_n^c], & b < \psi(\delta, \beta), \end{cases} \qquad (27)$$

where

$$\psi(\delta, \beta) = \left[\left(\frac{1}{\mu_n} - \frac{1}{\delta}\right) + \frac{2P}{\delta}\left(\frac{1}{\mu_n} - \frac{1}{\delta}\right)\right]^{-1} \delta(\delta+\beta). \qquad (28)$$

4.2. Upper bound

To upper bound the ergodic achievable rate use, once again, Jensen's inequality. Nevertheless, in this case, we focus on the concavity of functions $\mathcal{C}_n$ in (18). As previously mentioned, the capacity $\mathcal{C}_n$ only depends on 3 variables: the random source-to-user channel $|a|^2$, the relays-to-destination beamforming gain $\sum_{i=1}^n |c_i|^2$, and the random path gain $|b_n|^2$. Obviously, it also depends on the power allocation and the power constraint, but notice that power allocation is a direct function of those three variables and that the power constraint is assumed constant.

The concavity of $\mathcal{C}_n$ over the three random variables is shown in Appendix B, and obtained applying properties of the composition of concave functions [26]. This result allows us to conclude that $\mathcal{C}_{D\&F}$, being defined as the maximum of a set concave functions (9), is also concave over the variables that define $\mathcal{C}_n$. Therefore, the capacity of regenerative MRC is concave over variable $\mathbf{a}$ and vectors $\mathbf{b}$ and $\mathbf{c}$, and thus we may define the following upper bound:

$$\mathcal{C}_{D\&F}^e = E_{a,b,c}\left\{\max_{1 \le n \le N} \mathcal{C}_n\right\}$$
$$\le \max_{1 \le n \le N} \mathcal{C}_n(\bar{\mathbf{a}}, \bar{\mathbf{b}}, \bar{\mathbf{c}}), \qquad (29)$$

where $\bar{\mathbf{a}} = E_a\{a\} = 1$, $\bar{\mathbf{c}} = E_c\{c\} = [1, \ldots, 1]$, and $\bar{\mathbf{b}} = E_b\{\mathbf{b}\} = [|\bar{b}_1|^2, \ldots, |\bar{b}_n|^2, \ldots, |\bar{b}_N|^2]$ are the mean squared source-to-destination, relay-to-destination, and source-to-relay channels, respectively. Notice that $|\bar{b}_n|^2 = \int_0^\infty b f_{|b_n|^2}(b) db$ is computed by using the pdf in (22). Therefore, considering the capacity derivation in Proposition 2, we obtain

$$\mathcal{C}_n(\bar{\mathbf{a}}, \bar{\mathbf{b}}, \bar{\mathbf{c}}) = \frac{1}{2}\log_2(1+\rho_{1n}) + \frac{1}{2}\log_2(1+\rho_{2n}\cdot(n+1)), \qquad (30)$$

where

$$\rho_{1n} = \max\left\{\left(\frac{1}{\mu_n} - 1\right), \gamma_n^c\right\}$$
$$\rho_{2n} = \min\left\{\left(\frac{1}{\mu_n} - \frac{1}{n+1}\right), (2P - \gamma_n^c)\right\},$$
$$\gamma_n^c = \left(\left(\frac{1}{\mu_n} - 1\right) - \frac{|\bar{b}_n|^2}{2(n+1)}\right)$$
$$+ \sqrt{\left(\left(\frac{1}{\mu_n} - 1\right) - \frac{|\bar{b}_1|^2}{2(n+1)}\right)^2 + 2P}. \qquad (31)$$

Hence, the upper bound on the ergodic capacity of MRC is

$$\mathcal{C}_{D\&F}^e \le \max_{1 \le n \le N} \frac{1}{2}\log_2(1+\rho_{1n}) + \frac{1}{2}\log_2(1+\rho_{2n}\cdot(n+1)). \qquad (32)$$

The interpretation of this upper bound leads to the comparison of faded and nonfaded channels: from (29) we conclude that the capacity of the MRC with nonfaded channels is always higher than the ergodic capacity of the MRC with unitary-mean Rayleigh-faded channels.

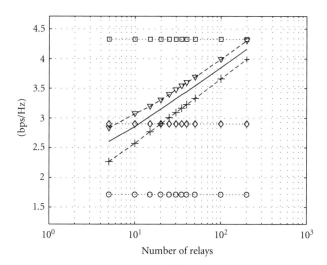

FIGURE 3: Ergodic achievable rate in [bps/Hz] of a Gaussian multiple relay channel with transmit SNR = 5 dB, under Rayleigh fading. The upper and lower bounds proposed in the paper are shown, and the ergodic capacity of a direct link plotted as reference.

5. ASYMPTOTIC ACHIEVABLE RATE

In previous sections, we analyzed the instantaneous and ergodic achievable rate of multiple-relay channels with full CSI, assuming a finite number of potential relays N. Results suggest (as it can be shown in Figure 3) a growth of the spectral efficiency with the total number relays. Nevertheless, neither the result in Proposition 2 nor the bounds (25) and (32) are tractable enough to infer the asymptotic behavior. In this section, we introduce the necessary approximations to simplify the problem and to analyze the asymptotic achievable rate of the MRC. We show that capacity grows with the logarithm of the branch zero of the Lambert W function of the total number of parallel relays.

Prior to the analysis, in the asymptotic domain ($N \to \infty$), we rename variable n in maximization (9) as $n = \kappa \cdot N$ with $\kappa \in [0, 1]$ (see [25, page 71]), and we introduce four key approximations.

(1) For a large number of network nodes, we consider capacities $\mathcal{C}_n$ in (18) defined only by the second slot mutual information,[5] that is,

$$\mathcal{C}_{\kappa \cdot N} = \frac{1}{2}\mathcal{C}(\gamma_{1\kappa \cdot N}\lambda_1) + \frac{1}{2}\mathcal{C}(\gamma_{2\kappa \cdot N}\lambda_{2\kappa \cdot N}) \approx \frac{1}{2}\mathcal{C}(\gamma_{2\kappa \cdot N}\lambda_{2\kappa \cdot N}). \tag{33}$$

---

[5] The proposed approximation is also a lower bound. Thus, the asymptotic performance of the lower bound is valid to lower bound the asymptotic performance of the achievable rate.

The proposed approximation is justified by the large beamforming gain obtained during time slot 2 when the number of relays grows to ∞ (as shown in approximation 2). As a consequence, $\gamma^c_{\kappa \cdot N}$ computed in Appendix A is recalculated as

$$\gamma^c_{\kappa \cdot N} = 2P \frac{\lambda_{2\kappa \cdot N}}{|b_{\kappa \cdot N}|^2 + \lambda_{2\kappa \cdot N}}. \tag{34}$$

To derive (34), we recall that $\gamma^c_{\kappa \cdot N}$ is defined in (A.5) as the power allocation during slot 1 that simultaneously satisfies $\sum_{i=1}^2 \gamma_i = 2P$ and $\mathcal{C}(\gamma_1|b_{\kappa \cdot N}|^2) = \mathcal{C}(\gamma_1\lambda_1) + \mathcal{C}(\gamma_2\lambda_{2\kappa \cdot N})$ (i.e., $\gamma^c_{\kappa \cdot N} = \{\gamma_1 : \mathcal{C}(\gamma_1|b_{\kappa \cdot N}|^2) = \mathcal{C}(\gamma_1\lambda_1) + \mathcal{C}((2P - \gamma_1)\lambda_{2\kappa \cdot N})\}$). Hence, neglecting the factor $\mathcal{C}(\gamma_1\lambda_1)$, then (34) is obtained.

(2) From the Law of Large Numbers, $\lambda_{2\kappa \cdot N}$ in (10) is approximated as $\lambda_{2\kappa \cdot N} \approx \kappa \cdot N$.

(3) From [25, pages 255–258], the pdf of the ordered random variable $|b_{\kappa \cdot N}|^2$ asymptotically satisfies $\text{pdf}_{|b_{\kappa \cdot N}|^2} = \mathcal{N}(Q(1 - \kappa), \varepsilon \cdot N^{-1})$ as $N \to \infty$ (with ε a fixed constant). $Q(\kappa) : [0, 1] \to R^+$ is the inverse function of the cdf of the squared modulus of the nonordered source-to-relay channel defined in (1), that is , $Q(\Pr\{|b|^2 < \tilde{b}\}) = \tilde{b}$ with $b \sim \mathcal{CN}(0, (d_o/d)^\alpha)$ and d the source-to-relay random distance. From the asymptotic pdf, the following convergence in probability holds:

$$|b_{\kappa \cdot N}|^2 \xrightarrow{P} Q(1 - \kappa). \tag{35}$$

(4) We consider high-transmitted power, so that $\mu_{\kappa \cdot N} \approx P^{-1}$ is in the power allocation (11).

Making use of those four approximations, we may apply (9) to define the asymptotic instantaneous capacity as

$$\begin{aligned}
\mathcal{C}^a_{\text{D\&F}} &= \frac{1}{2} \lim_{N \to \infty} \max_{\kappa \in [0,1]} \mathcal{C}_{\kappa \cdot N} \\
&\approx \frac{1}{2} \lim_{N \to \infty} \max_{\kappa \in [0,1]} \mathcal{C}(\gamma_{2\kappa \cdot N}\lambda_{2\kappa \cdot N}) \\
&= \frac{1}{2} \lim_{N \to \infty} \max_{\kappa \in [0,1]} \min\left\{\mathcal{C}\left(\left(\frac{1}{\mu_{\kappa \cdot N}} - \frac{1}{\kappa \cdot N}\right)\kappa \cdot N\right), \right. \\
&\qquad\qquad\qquad\qquad \left. \mathcal{C}\left((2P - \gamma^c_{\kappa \cdot N})\kappa \cdot N\right)\right\} \\
&= \frac{1}{2} \lim_{N \to \infty} \max_{\kappa \in [0,1]} \min\left\{\mathcal{C}(P \cdot \kappa \cdot N - 1), \right. \\
&\qquad\qquad\qquad\qquad \left. \mathcal{C}\left(2P \frac{Q(1 - \kappa)\kappa \cdot N}{Q(1 - \kappa) + \kappa \cdot N}\right)\right\},
\end{aligned} \tag{36}$$

where first equality follows from Proposition 2, and second equality from approximation 1; third equality comes from the power allocation $\gamma_{2\kappa \cdot N}$ in (11) and considering $\lambda_{2\kappa \cdot N} = 2\kappa \cdot N$ as approximation 2. Finally, forth equality is obtained making use of approximation 4, and introducing the asymptotic convergence of $|b_{\kappa \cdot N}|^2$ in (34).

Let us focus now on the last equality in (36). We notice that (i) $\mathcal{C}(P \cdot \kappa \cdot N - 1)$ is an increasing function in $\kappa \in [0, 1]$,

(ii) $Q(1 - \kappa)$ is a decreasing function in the same interval, (iii) therefore, $\mathcal{C}(2P(Q(1 - \kappa)\kappa \cdot N)/(Q(1 - \kappa) + \kappa \cdot N))$ is asymptotically a decreasing function in $\kappa \in [0, 1]$. Hence, in the limit, the maximum in κ of the minimum of an increasing and a decreasing functions would be given at the intersection of the two curves. As derived in Appendix C, the intersection point[6] $\kappa_o(N)$ satisfies

$$\kappa_o(N) \geq \frac{W_0(\rho N)}{\rho N} \qquad (37)$$

with ρ a fixed constant in $(0, 1)$, and with equality whenever the relay positions are not random but deterministic. As mentioned earlier, $W_0(N)$ is the branch zero of the Lambert W function evaluated at N [23].

Finally, applying the forth equality in (36), we derive

$$\begin{aligned}
\mathcal{C}_{\text{D\&F}}^a &= \frac{1}{2} \lim_{N \to \infty} \mathcal{C}\left(P \cdot \kappa_o(N) \cdot N - 1\right) \\
&\geq \frac{1}{2} \lim_{N \to \infty} \log_2\left(P \cdot \frac{W_0(\rho N)}{\rho}\right).
\end{aligned} \qquad (38)$$

This result shows that, for any random distribution of relays, the capacity of MRC with channel knowledge grows asymptotically with the logarithm of the Lambert W function of the total number relays. However, due to approximations 2 and 3, our proof only demonstrates asymptotic performance in probability.

6. NUMERICAL RESULTS

In this section, we evaluate the lower and upper bounds described in (25) and (32), respectively, and compare them with the ergodic achievable rate of the link, obtained through Monte Carlo simulation.

As previously pointed out, we assume i.i.d., unitary mean, Rayleigh-distributed fading from all transmitter nodes to destination, while source-to-relay channels are modelled as a superposition of path loss and unitary mean Rayleigh fading. Likewise, source and destination are fixed nodes, while the position of the N relays is i.i.d. throughout a square, limited at its diagonal by the point-to-point source-to-destination link. As mentioned earlier, the position of relays is invariant during the two-slot communication but variant and uncorrelated from one transmission to the other. To deal with propagation effects, we defined a simplified exponential indoor propagation model with path loss exponent $\alpha = 4$. Finally, we consider normalized distances, defining distance between source and destination equal to 1, and source-relay random distance $d_i \in [0, 1]$.

Taking into account the considerations above, we focus the analysis on the number of relay nodes and the transmitted SNR, that is, P/σ_0^2. Figure 3 depicts the ergodic bounds computed for transmit SNR equal to 5 dB for an MRC with the number of relay nodes ranging from 5 to 200. Likewise,

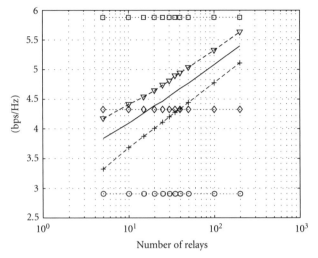

-▽- Ergodic upper bound, SNR = 10 dB
— Ergodic achievable rate
-+- Ergodic lower bound, SNR = 10 dB
··○·· Direct link ergodic capacity, SNR = 10 dB
··◇·· Direct link ergodic capacity, SNR = 15 dB
··□·· Direct link ergodic capacity, SNR = 20 dB

FIGURE 4: Ergodic achievable rate in [bps/Hz] of a Gaussian multiple relay channel with transmit SNR = 10 dB, under Rayleigh fading. The upper and lower bounds proposed in the paper are shown, and the ergodic capacity of a direct link plotted as reference.

Figures 4 and 5 plot results for transmit SNR equal to 10 dB and 20 dB, respectively. Firstly, we clearly note that, for all plots, ergodic bounds and simulated result increase with the number of users, as we have previously demonstrated in the asymptotic capacity section.

Moreover, the comparison of the three plots shows that the advantage of relaying diminishes as the transmitted power increases. In such a way, it can be seen that for transmit SNR = 5 dB only $N = 20$ parallel relay nodes are needed to double the noncooperative capacity, while for SNR = 10 dB more than $N = 200$ nodes would be necessary to obtain twice the spectral efficiency. Furthermore, we may see that for SNR = 5 dB with only 10 relays, it is possible to obtain the same ergodic capacity as a Rayleigh-faded direct link with SNR = 10 dB, while to obtain the same power saving for MRC with SNR = 20 dB, 50 nodes are needed. Finally, plots show that the accuracy of the presented bounds grows as the transmit SNR diminishes, which may be interpreted in terms of the meaning of such bounds: for decreasing transmitted power, the effect of instantaneous relay selection and the effect of Rayleigh fading over the cooperative links lose significance.

Figures 6–8 show results on the mean number of active relays versus the total number of relay nodes. Recall that the optimum number of relay nodes is calculated from maximization over n in Proposition 2. Specifically, Figure 6 depicts results for SNR = 5 dB while Figures 7 and 8 show cooperating nodes for SNR = 10 dB and SNR = 20 dB. In all three, the number of active nodes n that maximizes the lower and upper bounds, (25) and (32), respectively, is

---

[6] For a fixed number of relays N, a fixed intersection point κ_o is derived. Thus, $\kappa_o = \kappa_o(N)$.

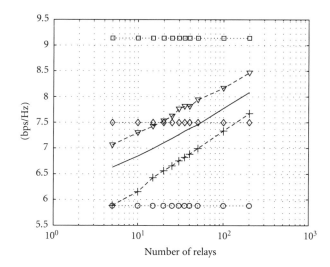

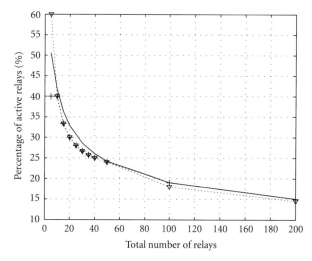

-▽- Ergodic upper bound, SNR = 15 dB
—— Ergodic achievable rate
-+- Ergodic lower bound, SNR = 15 dB
·○· Direct link ergodic capacity, SNR = 15 dB
·◇· Direct link ergodic capacity, SNR = 20 dB
·□· Direct link ergodic capacity, SNR = 25 dB

FIGURE 5: Ergodic achievable rate in [bps/Hz] of a Gaussian multiple relay channel with transmit SNR = 15 dB, under Rayleigh fading. The upper and lower bounds proposed in the paper are shown, and the ergodic capacity of a direct link plotted as reference.

·▽· Active relays with the upper bound, SNR = 5 dB
—— Active relays, SNR = 5 dB
·+· Active relays with the lower bound, SNR = 5 dB

FIGURE 6: Expected number of active relays (in %) of a multiple relay channel with transmit SNR = 5 dB, under Rayleigh fading. The number of relays that optimizes the upper and lower bounds are shown for comparison.

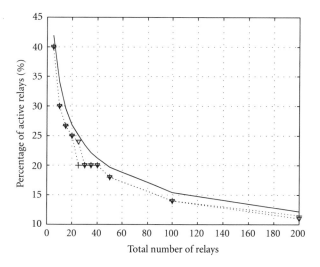

·▽·· Active relays with the upper bound
—— Active relays
·+·· Active relays with the lower bound

FIGURE 7: Expected number of active relays (in %) of a multiple relay channel with transmit SNR = 10 dB, under Rayleigh fading. The number of relays that optimizes the upper and lower bounds are shown for comparison.

also plotted; hence, it allows for comparison between the mean number of relays with capacity achieving relaying and the optimum number of relays with no instantaneous relay selection (25) and with no fading channels (32), respectively. Firstly, results show that the simulated mean number of relays is close to the number of relays maximizing the upper and lower bounds, being closer for the low SNR regime. Finally, we notice that, as the transmit SNR increases, the percentage of relays cooperating with the source decreases. Therefore, we conclude that regenerative relaying is, as previously mentioned, more powerful in the low SNR regime.

7. CONCLUSIONS

In this paper, we examined the achievable rate of a decode-and-forward (D&F) multiple-relay channel with half-duplex constraint and transmitter and receiver channel state information. The transmission was arranged in two phases: during the first phase, the source transmits its message to relays and destination. During the second phase, the relays and the source are configured as a distributed antenna array to transmit extra parity bits. The instantaneous achievable rate for the optimum relay selection and power allocation was obtained. Furthermore, we studied and bounded the ergodic performance of the achievable rate for Rayleigh-faded channels. We also found the asymptotic performance of the achievable rate in number of relays. Results show that

(i) $\mathcal{C}_{D\&F} \propto \log(W_0(N))$ as $N \to \infty$; (ii) with regenerative relaying, higher capacity is obtained for low signal-to-noise ratio, (iii) the percentage of active relays (i.e., the number of nodes who can decode the source message) decreases for increasing N, and (iv) this percentage is low, even at low SNR, due to the regenerative constraint.

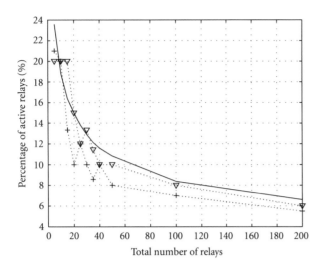

··▽·· Active relays with the upper bound, SNR = 15 dB
—— Active relays, SNR = 15 dB
··+·· Active relays with the lower bound, SNR = 15 dB

FIGURE 8: Expected number of active relays (in %) of a multiple relay channel with transmit SNR = 15 dB, under Rayleigh fading. The number of relays that optimizes the upper and lower bounds are shown for comparison.

APPENDICES

A. OPTIMIZATION PROBLEM

For completeness of explanation, in the appendix we solve optimization problem (16), which can be recast as follows:

$$C = \max_{\gamma_1, \gamma_2} \frac{1}{2} \sum_{i=1}^{2} \log_2 (1 + \gamma_i \lambda_i)$$

$$\text{s.t.} \quad \sum_{i=1}^{2} \gamma_i = 2P, \tag{A.1}$$

$$\gamma_i \geq \frac{\Pi_{i=1}^{2} (1 + \gamma_i \lambda_i) - 1}{|b_n|^2},$$

which is convex in both $\gamma_1 \in R^+$ and $\gamma_2 \in R^+$. The Lagrange dual function of the problem is

$$L(\gamma_1, \gamma_2, \mu, \nu) = \sum_{i=1}^{2} \log (1 + \gamma_i \lambda_i) - \mu \left(\sum_{i=1}^{2} \gamma_i - 2P \right)$$
$$+ \nu \left(\gamma_1 - \frac{\Pi_{i=1}^{2}(1 + \gamma_i \lambda_i) - 1}{|b_n|^2} \right), \tag{A.2}$$

where μ and ν are the Lagrange multipliers for first and second constraints, respectively. The three KKT conditions

(necessary and sufficient for optimality) of the dual problem are

$$\text{(i)} \quad \frac{\lambda_i}{1 + \gamma_i \lambda_i} - \mu + \nu \frac{d}{d\gamma_i} \left(\gamma_i - \frac{\Pi_{i=1}^{2}(1 + \gamma_i \lambda_i) - 1}{|b_n|^2} \right) = 0$$

$$\text{for } i \in \{1, 2\},$$

$$\text{(ii)} \quad \mu \left(\sum_{i=1}^{2} \gamma_i - 2P \right) = 0,$$

$$\text{(iii)} \quad \nu \left(\gamma_1 - \frac{\Pi_{i=1}^{2}(1 + \gamma_i \lambda_i) - 1}{|b_n|^2} \right) = 0. \tag{A.3}$$

Notice that the set $(\nu^*, \gamma_1^*, \gamma_2^*, \mu^*)$:

$$\nu^* = 0, \qquad \gamma_i^* = \left(\frac{1}{\mu^*} - \frac{1}{\lambda_i} \right)^+, \qquad \frac{1}{\mu^*} = P + \frac{1}{2} \sum_{i=1}^{2} \frac{1}{\lambda_i}, \tag{A.4}$$

satisfies KKT conditions hence yielding the optimum solution.[7] However, taking into account that optimal primal points must satisfy the two constraints in (A.1), and that

$$\left. \begin{array}{c} \sum_{i=1}^{2} \gamma_i = 2P \\[6pt] \gamma_1 \geq \dfrac{\Pi_{i=1}^{2}(1 + \gamma_i \lambda_i) - 1}{|b_n|^2} \end{array} \right\} \longrightarrow \gamma_1 \geq \gamma^c = \phi + \sqrt{\phi^2 + \frac{2P}{\lambda_1}} \in R^+ \tag{A.5}$$

with $\phi = (1/\mu^* - 1/\lambda_i) - |b_n|^2/2\lambda_1\lambda_2$. Then, the result in optimum power allocation is

$$\gamma_1^* = \max \left\{ \left(\frac{1}{\mu^*} - \frac{1}{\lambda_i} \right), \gamma^c \right\},$$

$$\gamma_2^* = 2P - \gamma_1^*, \tag{A.6}$$

$$\frac{1}{\mu^*} = P + \frac{1}{2} \sum_{i=1}^{2} \frac{1}{\lambda_i}.$$

B. CONCAVITY OF $\mathcal{C}_N$

In the appendix, we prove the concavity of capacity $\mathcal{C}_n$ (defined in (18) based on (9)) over random variables $|a|^2$, $\sum_{i=1}^{n} |c_i|^2$, and $|b_n|^2$. To do so, we first rewrite the function under study as a composition of functions:

$$\mathcal{C}_n = \mathcal{C}(\max(\Gamma_1(\mathbf{x}), \Gamma_2(\mathbf{x}))) + \mathcal{C}(\min(\Psi_1(\mathbf{x}), \Psi_2(\mathbf{x}))), \tag{B.7}$$

---

[7] Using standard notation, we define $(A)^+ = \max\{A, 0\}$.

where $\mathbf{x} = [|a|^2, \sum_{i=1}^n |c_i|^2, |b_n|^2]$ and

$$\Gamma_1(\mathbf{x}) = \left(\frac{1}{\mu_n} - \frac{1}{|a|^2}\right)|a|^2, \quad \Gamma_1 : \mathbf{R}^{3+} \longrightarrow \mathbf{R},$$

$$\Gamma_2(\mathbf{x}) = \gamma_n^c(\mathbf{x})|a|^2, \quad \Gamma_2 : \mathbf{R}^{3+} \longrightarrow \mathbf{R},$$

$$\Psi_1(\mathbf{x}) = \left(\frac{1}{\mu_n} - \frac{1}{|a|^2 + \sum_{i=1}^n |c_i|^2}\right)$$
$$\times \left(|a|^2 + \sum_{i=1}^n |c_i|^2\right), \quad \Psi_1 : \mathbf{R}^{3+} \longrightarrow \mathbf{R},$$

$$\Psi_2(\mathbf{x}) = (2P - \gamma_n^c(\mathbf{x}))\left(|a|^2 + \sum_{i=1}^n |c_i|^2\right), \quad \Psi_2 : \mathbf{R}^{3+} \longrightarrow \mathbf{R}.$$
$$\text{(B.8)}$$

First, we notice that pointwise maximum and pointwise minimum functions are nondecreasing functions with Hessian equal to zero. Next, computing the Hessian of $\Gamma_1(\mathbf{x})$ and $\Gamma_2(\mathbf{x})$ (respect to $\mathbf{x}$), it is shown that both are concave functions. Therefore, from [26, pages 86-87], we derive that $\max(\Gamma_1(\mathbf{x}), \Gamma_2(\mathbf{x}))$ is concave on $\mathbf{x}$. Accordingly, we may show that $\Psi_1(\mathbf{x})$ and $\Psi_2(\mathbf{x})$ are also concave functions, and so is $\min(\Psi_1(\mathbf{x}), \Psi_2(\mathbf{x}))$. Hence, considering that the sum of concave functions is always concave, and that $\mathcal{C}(x)$ is a concave nondecreasing function, we derive that $\mathcal{C}_n$ is concave on $\mathbf{x}$.

C. INTERSECTION OF CAPACITY CURVES

In this appendix, we analyze the intersection point κ_o of curves $f_1(\kappa) = \log_2(P \cdot \kappa N)$ and $f_2(\kappa) = \log_2(1 + 2P(Q(1 - \kappa)\kappa \cdot N)/(Q(1 - \kappa) + \kappa \cdot N))$ for a given number of relays N. To do so, we set $f_1(\kappa_o) = f_2(\kappa_o)$ to obtain[8]

$$Q(1 - \kappa_o) \approx \kappa_o \cdot N. \quad \text{(C.9)}$$

From approximation 3 in Section 5, equality above is equivalent to

$$\Pr\{|b|^2 \le \kappa_o \cdot N\} = 1 - \kappa_o \quad \text{(C.10)}$$

with $b \sim \mathcal{CN}(0, (d_o/d)^\alpha)$ and d the source-to-relay random distance. Furthermore, making use of the cdf in (23), we obtain

$$\kappa_o = 1 - \Pr\{|b|^2 \le \kappa_o \cdot N\}$$
$$= \int_0^{d^+} e^{-(x/d_o)^\alpha \kappa_o \cdot N} f_d(x) dx. \quad \text{(C.11)}$$

We can now apply Jensen's inequality for convex functions, in order to lower bound the integral as

$$\kappa_o \ge e^{-(E\{x\}/d_o)^\alpha \kappa_o \cdot N} \quad \text{(C.12)}$$

with $E\{x\} = \int_0^{d^+} x f_d(x) dx$. Equality is satisfied whenever the relays position are not random but deterministic, that is,

$f_d(x) = \delta(x - d_r)$. Next, from [23], we directly solve inequality (C.12) over κ_o as

$$\kappa_o(N) \ge \frac{W_0(\rho N)}{\rho N} \quad \text{(C.13)}$$

with $\rho = -(E\{x\}/d_o)^\alpha$ a fixed constant in $(0, 1)$, and $W_0(\cdot)$ the branch zero of the Lambert W function.

This solution is applicable for every possible random distribution of relays.

ACKNOWLEDGMENTS

The material of this paper was partially presented at the 39th Asilomar Conference on Signals, Systems and Computers, Pacific Grove, Calif, November 2005 and at the IEEE Wireless Communications and Networking Conference (WCNC), Las Vegas, Nev, March 2006. This work was partially supported by the Spanish Ministry of Science and Education grant TEC2005-08122-C03-02/TCM (ULTRARED) and TEC2006-10459/TCM (PERSEO), by the European Comission under project IST-2005-27402 (WIP) and by Generalitat de Catalunya under Grant SGR-2005-00690.

REFERENCES

[1] I. Telatar, "Capacity of multi-antenna Gaussian channels," *European Transactions on Telecommunications*, vol. 10, no. 6, pp. 585–595, 1999.

[2] S. M. Alamouti, "A simple transmit diversity technique for wireless communications," *IEEE Journal on Selected Areas in Communications*, vol. 16, no. 8, pp. 1451–1458, 1998.

[3] S. Vishwanath, N. Jindal, and A. Goldsmith, "Duality, achievable rates, and sum-rate capacity of Gaussian MIMO broadcast channels," *IEEE Transactions on Information Theory*, vol. 49, no. 10, pp. 2658–2668, 2003.

[4] D. Tse and P. Viswanath, *Fundamentals of Wireless Communications*, Cambridge University Press, Cambridge, UK, 1st edition, 2005.

[5] E. Zimmermann, P. Herhold, and G. Fettweis, "On the performance of cooperative relaying protocols in wireless networks," *European Transactions on Telecommunications*, vol. 16, no. 1, pp. 5–16, 2005.

[6] G. Kramer, M. Gastpar, and P. Gupta, "Cooperative strategies and capacity theorems for relay networks," *IEEE Transactions on Information Theory*, vol. 51, no. 9, pp. 3037–3063, 2005.

[7] D. Chen and J. N. Laneman, "The diversity-multiplexing tradeoff for the multiaccess relay channel," in *Proceedings of the 40th Annual Conference on Information Sciences and Systems*, pp. 1324–1328, Princeton, NJ, USA, March 2006.

[8] A. Sendonaris, E. Erkip, and B. Aazhang, "User cooperation diversity—part I: system description," *IEEE Transactions on Communications*, vol. 51, no. 11, pp. 1927–1938, 2003.

[9] A. Høst-Madsen and J. Zhang, "Capacity bounds and power allocation for wireless relay channels," *IEEE Transactions on Information Theory*, vol. 51, no. 6, pp. 2020–2040, 2005.

[10] T. Cover and A. El Gamal, "Capacity theorems for the relay channel," *IEEE Transactions on Information Theory*, vol. 25, no. 5, pp. 572–584, 1979.

[11] J. N. Laneman, "Cooperative diversity in wireless networks: algorithms and architectures," Ph.D. Dissertation, Massachusetts Institute of Technology, Cambridge, Mass, USA, 2002.

---

[8] Approximation (C.9) is obtained neglecting the effect of 1 within the logarithm in $f_2(\kappa)$, assuming sufficiently large transmitted power P.

[12] J. N. Laneman, D. Tse, and G. W. Wornell, "Cooperative diversity in wireless networks: efficient protocols and outage behavior," *IEEE Transactions on Information Theory*, vol. 50, no. 12, pp. 3062–3080, 2004.

[13] A. F. Dana, M. Sharif, R. Gowaikar, B. Hassibi, and M. Effros, "Is broadcast plus multiaccess optimal for Gaussian wireless networks?" in *Proceedings of the 37th Asilomar Conference on Signals, Systems, and Computers*, vol. 2, pp. 1748–1752, Pacific Grove, Calif, USA, November 2003.

[14] R. Nabar, H. Bölcskei, and F. W. Kneubühler, "Fading relay channels: performance limits and space-time signal design," *IEEE Journal on Selected Areas in Communications*, vol. 22, no. 6, pp. 1099–1109, 2004.

[15] A. del Coso and C. Ibars, "Achievable rate for Gaussian multiple relay channels with linear relaying functions," in *Proceedings of IEEE International Conference on Acoustics, Speech, and Signal Processing (ICASSP '07)*, vol. 3, pp. 505–508, Honolulu, Hawaii, USA, April 2007.

[16] A. El Gamal, M. Mohseni, and S. Zahedi, "Bounds on capacity and minimum energy-per-bit for AWGN relay channels," *IEEE Transactions on Information Theory*, vol. 52, no. 4, pp. 1545–1561, 2006.

[17] A. del Coso and C. Ibars, "Partial decoding for synchronous and asynchronous Gaussian multiple relay channels," in *Proceedings of the International Conference on Communications (ICC '07)*, pp. 713–718, Glasgow, Scotland, UK, June 2007.

[18] A. Høst-Madsen, "On the capacity of wireless relaying," in *Proceedings of the 56th IEEE Vehicular Technology Conference (VTC '02)*, vol. 3, pp. 1333–1337, Vancouver, BC, Canada, September 2002.

[19] A. El Gamal, "Capacity theorems for relay channels," in *Proceedings of MSRI Workshop on Mathematics of Relaying and Cooperation in Communication Networks*, Berkeley, Calif, USA, April 2006.

[20] J. N. Laneman and G. W. Wornell, "Distributed space-time-coded protocols for exploiting cooperative diversity in wireless networks," *IEEE Transactions on Information Theory*, vol. 49, no. 10, pp. 2415–2425, 2003.

[21] M. Dohler, *Virtual antenna arrays*, Ph.D. thesis, King's College London, London, UK, 2003.

[22] I. Maric and R. Yates, "Bandwidth and power allocation for cooperative strategies in Gaussian relay networks," in *Proceedings of the 38th Asilomar Conference on Signals, Systems and Computers*, vol. 2, pp. 1907–1911, Pacific Grove, Calif, USA, November 2004.

[23] R. M. Corless, G. H. Gonnet, D. E. G. Hare, D. J. Jeffrey, and D. E. Knuth, "On the Lambert W function," *Advances in Computational Mathematics*, vol. 5, no. 4, pp. 329–359, 1996.

[24] T. Cover and J. Thomas, *Elements of Information Theory*, Wiley Series in Telecommunications, Wiley-Interscience, New York, NY, USA, 1991.

[25] H. David, *Order Statistics*, John Wiley & Sons, New York, NY, USA, 2nd edition, 1981.

[26] S. Boyd and L. Vandenberghe, *Convex Optimization*, Cambridge University Press, Cambridge, UK, 1st edition, 2004.

Hindawi Publishing Corporation
EURASIP Journal on Wireless Communications and Networking
Volume 2007, Article ID 90401, 13 pages
doi:10.1155/2007/90401

Research Article

A Simplified Constant Modulus Algorithm for Blind Recovery of MIMO QAM and PSK Signals: A Criterion with Convergence Analysis

Aissa Ikhlef and Daniel Le Guennec

IETR/SUPELEC, Campus de Rennes, Avenue de la Boulaie, CS 47601, 35576 Cesson-Sévigné, France

Received 31 October 2006; Revised 18 June 2007; Accepted 3 September 2007

Recommended by Monica Navarro

The problem of blind recovery of QAM and PSK signals for multiple-input multiple-output (MIMO) communication systems is investigated. We propose a simplified version of the well-known constant modulus algorithm (CMA), named simplified CMA (SCMA). The SCMA cost function consists in projection of the MIMO equalizer outputs on one dimension (either real or imaginary part). A study of stationary points of SCMA reveals the absence of any undesirable local stationary points, which ensures a perfect recovery of all signals and a global convergence of the algorithm. Taking advantage of the phase ambiguity in the solution of the new cost function for QAM constellations, we propose a modified cross-correlation term. It is shown that the proposed algorithm presents a lower computational complexity compared to the constant modulus algorithm (CMA) without loss in performances. Some numerical simulations are provided to illustrate the effectiveness of the proposed algorithm.

1. INTRODUCTION

In the last decade, the interest in blind source separation (BSS) techniques has been important. The problem of blind recovery of multiple independent and identically distributed (i.i.d.) signals from their linear mixture in a multiple-input multiple-output (MIMO) system arises in many applications such as spatial division multiple access (SDMA), multiuser communications (such as CDMA for code division multiple access), and more recently Bell Labs layered space-time (BLAST) [1–3]. The aim of blind signals separation is to retrieve source signals without the use of a training sequence, which can be expensive or impossible in some practical situations. Another interesting class of blind methods is blind identification. Unlike blind source separation, the aim of blind identification is to find an estimate of the MIMO channel matrix [4–6]. Once this estimate has been obtained, the source signals can be efficiently recovered using MIMO detection methods, such as maximum likelihood (ML) [7] and BLAST [8] detection methods. The main difference between blind source separation and blind identification is that in the first case, the source signals are recovered directly from the observations, whereas in the second case, a MIMO detection algorithm

is needed, which may increase complexity (complexity depends on used methods). Note that, unlike BSS techniques, ML detector is nonlinear but optimum and suffers from high complexity. Sphere decoding [7] allows to reduce considerably ML detector complexity. On the other hand, the performance of MIMO detection methods depends strongly on the quality of the channel estimate which results from blind identification. In this paper, we consider the problem of blind source separation of MIMO instantaneous channel.

In literature, the constant modulus of many communication signals, such as PSK and 4-QAM signals, is a widely used property in blind source separation and blind equalization. The initial idea can be traced back to Sato [9], Godard [10], and Treichler et al. [11, 12]. The algorithms are known as CMAs. The first application after blind equalization was blind beamforming [13, 14] and more recently blind signals separation [15, 16]. In the case of constant modulus signals, CMA has proved reasonable performances and desired convergence requirements. On the other hand, the CMA yields a degraded performance for nonconstant modulus signals such as the quadrature amplitude modulation (QAM) signals, because the CMA projects all signal points onto a single modulus.

In order to improve the performance of the CMA for QAM signals, the so-called modified constant modulus algorithm (MCMA) [17], known as MMA for multimodulus algorithm, has been proposed [18–20]. This algorithm, instead of minimizing the dispersion of the magnitude of the equalizer output, minimizes the dispersion of the real and imaginary parts separately; hence the MMA cost function can be considered as a sum of two one-dimensional cost functions. The MMA provides much more flexibility than the CMA and is better suited to take advantage of the symbol statistics related to certain types of signal constellations, such as non-square and very dense constellations [18]. Please notice that both CMA and MMA are two-dimensional (i.e., employ both real and imaginary part of the equalizer outputs). Another class of algorithm has been proposed recently and named constant norm algorithm (CNA), whose CMA represents a particular case [21, 22].

In this paper, we propose a simplified version of the CMA cost function named simplified CMA (SCMA) and based only on one dimension (either real or imaginary part), as opposed to CMA. The major advantage of SCMA is its low complexity compared to that of CMA and MMA. Because, instead of using both real and imaginary parts as in CMA and MMA, only one dimension, the real or imaginary part, is considered in SCMA, which makes it very attractive for practical implementation especially when complexity issue arises such as in user's side. We will demonstrate that only the existing stationary points of the SCMA cost function correspond to a perfect recovery of all source signals except for the phase and permutation indeterminacy. We will show that the phase rotation is not the same for QAM, 4-PSK, and P-PSK ($P \geq 8$). Moreover, in order to reduce the complexity further, we will introduce a modified cross-correlation term by taking advantage of the phase ambiguity of the SCMA cost function for QAM constellations. An adaptive implementation by means of the stochastic gradient algorithm (SGA) will be described. A part of the results presented in this paper (QAM case with its convergence analysis) was previously reported in [23].

The remainder of the paper is organized as follows. In Section 2, the problem formulation and assumptions are introduced. In Section 3, we describe the SCMA criterion. The convergence analysis of the proposed cost function is carried out in Section 4. Section 5 introduces a modified cross-correlation constraint for QAM constellations. In Section 6, we present an adaptive implementation of the algorithm. Finally, Section 7 presents some numerical results.

2. PROBLEM FORMULATION

We consider a linear data model which takes the following form:

$$\mathbf{y}(n) = \mathbf{H}\mathbf{a}(n) + \mathbf{b}(n), \tag{1}$$

where $\mathbf{a}(n) = [a_1(n), \ldots, a_M(n)]^T$ is the ($M \times 1$) vector of the source signals, $\mathbf{H}$ is the ($N \times M$) MIMO linear memoryless channel, $\mathbf{y}(n) = [y_1(n), \ldots, y_N(n)]^T$ is the ($N \times 1$) vector of the received signals, and $\mathbf{b}(n) = [b_1(n), \ldots, b_N(n)]^T$ is the

($N \times 1$) noise vector. M and N represent the number of transmit and receive antennas, respectively.

In the case of the MIMO frequency selective channel (convolutive model), the system can be reduced to the model in (1) tanks to the linear prediction method presented in [5]. Afterwards, blind source separation methods can be applied. The following assumptions are considered:

(1) $\mathbf{H}$ has full column rank M,
(2) the noise is additive white Gaussian independent from the source signals,
(3) the source signals are independent and identically distributed (i.i.d), mutually independent $E[\mathbf{a}\mathbf{a}^H] = \sigma_a^2 \mathbf{I}_M$, and drawn from QAM or PSK constellations.

Please notice that these assumptions are not very restrictive and satisfied in BLAST scheme whose corresponding model is given in (1). Moreover, throughout this paper by QAM constellation we mean only square QAM constellation. In order to recover the source signals, the received signal $\mathbf{y}(n)$ is processed by an ($N \times M$) receiver matrix $\mathbf{W} = [\mathbf{w}_1, \ldots, \mathbf{w}_M]$. Then, the receiver output can be written as

$$\begin{aligned} \mathbf{z}(n) &= \mathbf{W}^T\mathbf{y}(n) = \mathbf{W}^T\mathbf{H}\mathbf{a}(n) + \mathbf{W}^T\mathbf{b}(n) \\ &= \mathbf{G}^T\mathbf{a}(n) + \tilde{\mathbf{b}}(n), \end{aligned} \tag{2}$$

where $\mathbf{z}(n) = [z_1(n), \ldots, z_M(n)]^T$ is the ($M \times 1$) vector of the receiver output, $\mathbf{G} = [\mathbf{g}_1, \ldots, \mathbf{g}_M] = \mathbf{H}^T\mathbf{W}$ is the ($M \times M$) global system matrix, and $\tilde{\mathbf{b}}(n)$ is the filtered noise at the receiver output.

The purpose of blind source separation is to find the matrix $\mathbf{W}$ such that $\mathbf{z}(n) = \hat{\mathbf{a}}(n)$ is an estimate of the source signals.

Please note that in blind signals separation, the best that can be done is to determine $\mathbf{W}$ up to a permutation and scalar multiple [3]. In other words, $\mathbf{W}$ is said to be a separation matrix if and only if

$$\mathbf{G}^T = \mathbf{W}^T\mathbf{H} = \mathbf{P}\Lambda, \tag{3}$$

where $\mathbf{P}$ is a permutation matrix and Λ a nonsingular diagonal matrix.

Throughout this paper, we use small and capital boldface letters to denote vectors and matrices, respectively. The symbols $(\cdot)^*$ and $(\cdot)^T$ denote the complex conjugate and transpose, respectively, $(\cdot)^H$ is the Hermitian transpose, and $\mathbf{I}_p$ is the ($p \times p$) identity matrix.

3. THE PROPOSED CRITERION

Unlike the CMA algorithm [10], whose aim consists in constraining the modulus of the equalizer outputs to be on a circle (projection onto a circle), we suggest to project the equalizer outputs onto one dimension (either real or imaginary part). To do so, we suggest to penalize the deviation of the square of the real (imaginary) part of the equalizer outputs from a constant.

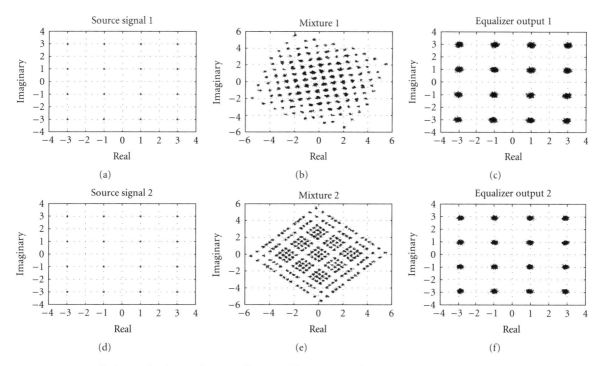

FIGURE 1: 16-QAM constellation. Left column: the constellations of the transmitted signals, middle column: the constellations of the received signals (mixtures), right column: the constellations of the recovered signals.

For the ℓth equalizer, we suggest to optimize the following criterion:

$$\min_{\mathbf{w}_\ell} \mathcal{J}(\mathbf{w}_\ell) = E\left[\left(z_{R,\ell}(n)^2 - R\right)^2\right], \quad \ell = 1, \ldots, M, \quad (4)$$

where $z_{R,\ell}(n)$ denotes the real part of the ℓth equalizer output $z_\ell(n) = \mathbf{w}_\ell^T \mathbf{y}(n)$ and R is the dispersion constant fixed by assuming a perfect equalization with respect to the zero forcing (ZF) solution, and is defined as

$$R = \frac{E\left[a_R(n)^4\right]}{E\left[a_R(n)^2\right]}, \quad (5)$$

where $a_R(n)$ denotes the real part of the source signal $a(n)$.

The term on the right side of the equality (4) prevents the deviation of the square of the real part of the equalizer outputs from a constant. The minimization of (4) allows the recovery of only one signal at each equalizer output (see proof in Section 4). But the algorithm minimization (4) does not ensure the recovery of all source signals because it may converge in order to recover the same source signal at many outputs. In order to avoid this problem, we suggest to use a cross-correlation term due to its computational simplicity. Then (4) becomes

$$\min_{\mathbf{w}_\ell} \mathcal{J}(\mathbf{w}_\ell) = E\left[\left(z_{R,\ell}(n)^2 - R\right)^2\right] + \alpha \sum_{i=1}^{\ell-1} \left|r_{\ell i}(n)\right|^2, \quad \ell = 1, \ldots, M, \quad (6)$$

where $\alpha \in \mathbb{R}^+$ is the mixing parameter and $r_{\ell i}(n) = E[z_\ell(n) z_i^*(n)]$ is the cross-correlation between the ℓth and

the ith equalizer outputs and prevents the extraction of the same signal at many outputs. Then the first term in (6) ensures the recovery of only one signal at each equalizer output and the cross-correlation term ensures that each equalizer output is different from the other ones; this results in the recovery of all source signals (see Section 4). In the following sections, we name (6) the cross-correlation simplified CMA (CC-SCMA) criterion. In (6) we could also use the imaginary part thanks to the symmetry of the QAM and PSK constellations. Since the analysis is the same for the imaginary part, throughout this paper, we only consider the real part.

4. CONVERGENCE ANALYSIS

Theorem 1. *Let M be i.i.d. and mutually independent signals $a_i(n), i = 1, \ldots, M$, which share the same statistical properties, are drawn from QAM or PSK constellations and are transmitted via an $(M \times N)$ MIMO linear memoryless channel and without the presence of noise. Provided that the weighting factor α is chosen to satisfy $\alpha \geq 2E[a_R^4]/\sigma_a^4 d^2$ (where $d = 1$ for QAM and P-PSK ($P \geq 8$) constellations and $d = \sqrt{2}$ for 4-PSK constellation), the algorithm in (6) will converge to a setting that corresponds, in the absence of any noise, to a perfect recovery of all transmitted signals, and the only stable minima are the Dirac-type vector taking the following form: $\mathbf{g}_\ell = [0, \ldots, 0, d_\ell e^{j\phi_\ell}, 0, \ldots, 0]^T$, where $\mathbf{g}_\ell$ is the ℓth column vector of $\mathbf{G}$, d_ℓ is the amplitude, and ϕ_ℓ is the phase rotation of the nonzero element. The pair (d_ℓ, ϕ_ℓ) is given by $\{1, \text{modulo } (\pi/2)\}, \{\sqrt{2}, \text{modulo } [(2k+1)\pi/4]\}$, and $\{1, \text{arbitrary in } [0, 2\pi]\}$ for QAM, 4-PSK, and P-PSK ($P \geq 8$) constellations, respectively.*

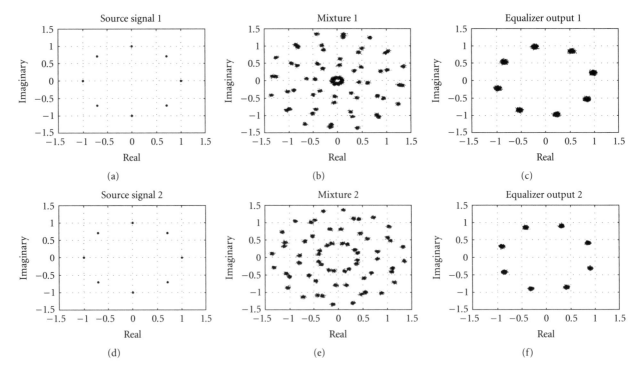

FIGURE 2: 8-PSK constellation. Left column: the constellations of the transmitted signals, middle column: the constellations of the received signals (mixtures), right column: the constellations of the recovered signals.

Proof. For simplicity, the analysis is restricted to noise-free case, that is,

$$\mathbf{z}(n) = \mathbf{G}^T \mathbf{a}(n). \tag{7}$$

Note that due to the assumed full column rank of $\mathbf{H}$, all results in the $\mathbf{G}$ domain will translate to the $\mathbf{W}$ domain as well. For convenience, the stationary points study will be carried out in the $\mathbf{G}$ domain [24].

Considering (7), the cross-correlation term in (6) can be simplified as

$$
\begin{aligned}
E[z_\ell(n)z_i^*(n)] &= E[\mathbf{w}_\ell^T \mathbf{H}\mathbf{a}(n)\mathbf{w}_i^H \mathbf{H}^* \mathbf{a}(n)^*] \\
&= \mathbf{g}_\ell^T E[\mathbf{a}(n)\mathbf{a}(n)^H]\mathbf{g}_i^* = \sigma_a^2 \mathbf{g}_\ell^T \mathbf{g}_i^* = \sigma_a^2 \mathbf{g}_i^H \mathbf{g}_\ell,
\end{aligned} \tag{8}
$$

where we use the fact that $\mathbf{w}_i^T \mathbf{H} = \mathbf{g}_i^T$.

Using (8) in (6), we get

$$
\begin{aligned}
\mathcal{J}(\mathbf{g}_\ell) = {}& E\left[\left(z_{R,\ell}(n)^2 - R\right)^2\right] \\
& + \alpha\sigma_a^4 \sum_{i=1}^{\ell-1} |\mathbf{g}_i^H \mathbf{g}_\ell|^2, \quad \ell = 1,\dots,M.
\end{aligned} \tag{9}
$$

From (9), we first notice that the adaptation of each $\mathbf{g}_\ell$ depends only on $\mathbf{g}_1,\dots,\mathbf{g}_{\ell-1}$. Then, we can begin by the first output, because $\mathbf{g}_1$ is optimized independently from all the other vectors $\mathbf{g}_2,\dots,\mathbf{g}_M$. Hence, for the first equalizer, $\mathbf{g}_1$, we have

$$\min_{\mathbf{g}_1} \mathcal{J}(\mathbf{g}_1) = E\left[\left(z_{R,1}(n)^2 - R\right)^2\right]. \tag{10}$$

By developing (10), we get (for notation convenience, in the following, we will omit the time index n)

$$\mathcal{J}(\mathbf{g}_1) = E[z_{R,1}^4] - 2\mathrm{RE}[z_{R,1}^2] + R^2. \tag{11}$$

Because the development is not the same for QAM, 4-PSK, and P-PSK ($P \geq 8$) constellations, we will enumerate the proof of each case separately. $\qquad\square$

4.1. QAM case

After a straightforward development of the terms in (11) with respect to statistical properties of QAM signals (see Appendix A), (11) can be written as

$$
\begin{aligned}
\mathcal{J}(\mathbf{g}_1) = {}& E[a_R^4]\left(\sum_{k=1}^{M} |g_{k1}|^2 - 1\right)^2 \\
& + \beta\left[\left(\sum_{k=1}^{M} |g_{k1}|^2\right)^2 - \sum_{k=1}^{M} |g_{k1}|^4\right] \\
& + 2\beta\sum_{k=1}^{M} g_{R,k1}^2 g_{I,k1}^2 - E[a_R^4] + R^2,
\end{aligned} \tag{12}
$$

where

$$
\begin{aligned}
\mathbf{g}_1 &= [g_{11},\dots,g_{M1}]^T, \\
g_{k1} &= g_{R,k1} + jg_{I,k1}, \\
\beta &= 3E^2[a_R^2] - E[a_R^4] = -\kappa_{a_R} > 0,
\end{aligned} \tag{13}
$$

and $\kappa_{a_R} = E[a_R^4] - 3E^2[a_R^2]$ represents the kurtosis of the real parts of the symbols. It is always negative in the case of PSK and QAM signals.

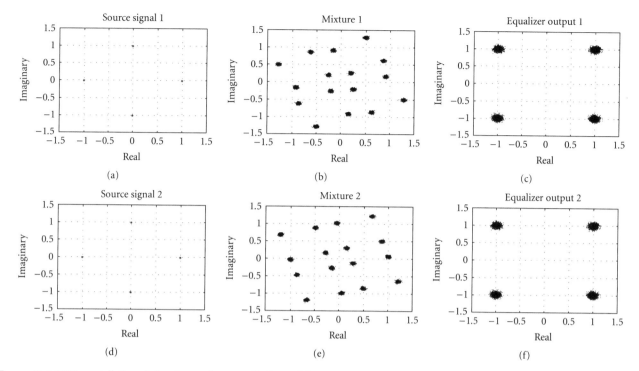

FIGURE 3: 4-PSK constellation. Left column: the constellations of the transmitted signals, middle column: the constellations of the received signals (mixtures), right column: the constellations of the recovered signals.

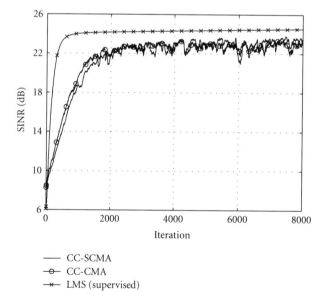

FIGURE 4: Performance comparison in term of SINR of the proposed algorithm (CC-SCMA) with CC-CMA and supervised LMS.

The minimum of (11) can be found easily by replacing the equalizer output in (12) by any of the transmitted signals, it is given by

$$J_{\min} = E\left[\left(a_R^2 - R\right)^2\right] = E[a_R^4] - 2\text{RE}[a_R^2] + R^2. \quad (14)$$

From (5), we have

$$\text{RE}[a_R^2] = E[a_R^4]. \quad (15)$$

Then

$$J_{\min} = -E[a_R^4] + R^2. \quad (16)$$

Comparing (12) and (16), we can write

$$\mathcal{J}(\mathbf{g_1}) = J_{\min} + \beta\left[\left(\sum_{k=1}^{M}|g_{k1}|^2\right)^2 - \sum_{k=1}^{M}|g_{k1}|^4\right] + E[a_R^4]\left(\sum_{k=1}^{M}|g_{k1}|^2 - 1\right)^2 + 2\beta\sum_{k=1}^{M}g_{R,k1}^2 g_{I,k1}^2. \quad (17)$$

Since $\beta > 0$ and that

$$\left(\sum_{k=1}^{M}|g_{k1}|^2\right)^2 \geq \sum_{k=1}^{M}|g_{k1}|^4, \quad (18)$$

we have

$$\beta\left[\left(\sum_{k=1}^{M}|g_{k1}|^2\right)^2 - \sum_{k=1}^{M}|g_{k1}|^4\right] \geq 0. \quad (19)$$

According to (19), $\mathcal{J}(\mathbf{g_1})$ is composed only of positive terms. Then, minimizing $\mathcal{J}(\mathbf{g_1})$ is equivalent to finding $\mathbf{g_1}$, which minimizes all terms simultaneously. One way to find the minimum of (17) is to look for a solution that cancels the gradients of each term separately. From (16), we know that

$J_{\min}$ is a constant ($\partial J_{\min}/\partial g_{\ell 1}^* = 0$), hence we only deal with the reminder terms. For that purpose, let us have

$$\mathcal{J}(\mathbf{g}_1) = J_{\min} + J_1(\mathbf{g}_1) + J_2(\mathbf{g}_1) + J_3(\mathbf{g}_1), \quad (20)$$

where

$$J_1(\mathbf{g}_1) = \beta\left[\left(\sum_{k=1}^{M}|g_{k1}|^2\right)^2 - \sum_{k=1}^{M}|g_{k1}|^4\right],$$

$$J_2(\mathbf{g}_1) = E[a_R^4]\left(\sum_{k=1}^{M}|g_{k1}|^2 - 1\right)^2, \quad (21)$$

$$J_3(\mathbf{g}_1) = 2\beta\sum_{k=1}^{M}g_{R,k1}^2 g_{I,k1}^2.$$

When we compute the derivatives of $J_1(\mathbf{g}_1)$, $J_2(\mathbf{g}_1)$, and $J_3(\mathbf{g}_1)$ with respect to $g_{\ell 1}^*$, we find

$$\frac{\partial J_1(\mathbf{g}_1)}{\partial g_{\ell 1}^*} = 2\beta g_{\ell 1}\left(\sum_{k=1}^{M}|g_{k1}|^2 - |g_{\ell 1}|^2\right) = 0$$

$$\Longrightarrow \sum_{k=1,k\neq\ell}^{M}|g_{k1}|^2 = 0, \quad (22)$$

$$\frac{\partial J_2(\mathbf{g}_1)}{\partial g_{\ell 1}^*} = 2E[a_R^4]g_{\ell 1}\left(\sum_{k=1}^{M}|g_{k1}|^2 - 1\right) = 0$$

$$\Longrightarrow \sum_{k=1}^{M}|g_{k1}|^2 = 1, \quad (23)$$

$$\frac{\partial J_3(\mathbf{g}_1)}{\partial g_{\ell 1}^*} = 2\beta(g_{R,\ell 1}g_{I,\ell 1}^2 + jg_{R,\ell 1}^2 g_{I,\ell 1}) = 0$$

$$\Longrightarrow g_{R,\ell 1} = 0 \quad \text{or} \quad g_{I,\ell 1} = 0 \quad \text{or} \quad g_{R,\ell 1} = g_{I,\ell 1} = 0. \quad (24)$$

Equation (22) implies that only one entry, $g_{\ell 1}$, of $\mathbf{g}_1$ is nonzero and the others are zeros. Equation (23) indicates that the modulus of this entry must be equal to one ($|g_{\ell 1}|^2 = 1$). Finally, from (24) either the real part or the imaginary part must be equal to zero. As a result of (23), the squared modulus of the nonzero part is equal to one, that is, either $g_{R,\ell 1}^2 = 1$ and $g_{I,\ell 1}^2 = 0$ or $g_{R,\ell 1}^2 = 0$ and $g_{I,\ell 1}^2 = 1$. Therefore, the solution $g_{\ell 1}$ is either a pure real or a pure imaginary with modulus equal to one, which corresponds to

$$g_{\ell 1} = e^{jm_1(\pi/2)}, \quad (25)$$

where m_1 is an arbitrary integer.

This solution shows that the minimization of $\mathcal{J}(\mathbf{g}_1)$ forces the equalizer output to form a constellation that corresponds to the source constellation with a modulo $\pi/2$ phase rotation.

From (22), (23), (24), and (25), we can conclude that the only stable minima for $\mathbf{g}_1$ take the following form: $\mathbf{g}_1 = [0,\ldots,0,e^{jm_1(\pi/2)},0,\ldots,0]^T$, that is, only one entry is nonzero, pure-real, or pure-imaginary with modulus equal to one, which can be at any of the M positions and all the other ones are zeros. This solution corresponds to the recov-

ery of only one source signal and cancels the others. For the second equalizer $\mathbf{g}_2$, from (9), we have

$$\mathcal{J}(\mathbf{g}_2) = E\left[(z_{R,2}(n)^2 - R)^2\right] + \alpha\sigma_a^4|\mathbf{g}_1^H \mathbf{g}_2|^2, \quad (26)$$

this means that the adaptation of $\mathbf{g}_2$ depends on $\mathbf{g}_1$.

We examine the convergence of $\mathbf{g}_2$ once $\mathbf{g}_1$ has converged to one signal, because the adaptation of $\mathbf{g}_1$ is realized independently from the other $\mathbf{g}_i$. For the sake of simplicity, and without loss of generality, we consider that $\mathbf{g}_1$ has converged to the first signal, that is,

$$\mathbf{g}_1 = [de^{j\varphi} \mid 0,\ldots,0]^T, \quad d = 1, \varphi = m_1\frac{\pi}{2}, \quad (27)$$

then

$$|\mathbf{g}_1^H \mathbf{g}_2|^2 = d^2|g_{12}|^2. \quad (28)$$

Using this and the result in Appendix A, $\mathcal{J}(\mathbf{g}_2)$ in (26) can be expressed as follows:

$$\mathcal{J}(\mathbf{g}_2) = E[a_R^4]\left(\sum_{k=1}^{M}|g_{k2}|^2 - 1\right)^2$$

$$+ \beta\left[\left(\sum_{k=1}^{M}|g_{k2}|^2\right)^2 - \sum_{k=1}^{M}|g_{k2}|^4\right] + 2\beta\sum_{k=1}^{M}g_{R,k2}^2 g_{I,k2}^2$$

$$- E[a_R^4] + R^2 + \alpha\sigma_a^4 d^2|g_{12}|^2. \quad (29)$$

If we differentiate (29) directly, with respect to g_{12}^*, and then cancel the operation result, we get

$$\frac{\partial \mathcal{J}(\mathbf{g}_2)}{\partial g_{12}^*} = 2E[a_R^4]g_{12}\left(\sum_{k=1}^{M}|g_{k2}|^2 - 1\right) + 2\beta g_{12}\sum_{k=2}^{M}|g_{k2}|^2$$

$$+ 2\beta(g_{R,12}g_{I,12}^2 + jg_{R,12}^2 g_{I,12}) + \alpha\sigma_a^4 d^2 g_{12} = 0. \quad (30)$$

By canceling both real and imaginary parts of (30), we have

$$g_{R,12} = 0 \quad \text{or} \quad \chi + 2\beta g_{I,12}^2 = 2E[a_R^4] - \alpha d^2\sigma_a^4,$$
$$g_{I,12} = 0 \quad \text{or} \quad \chi + 2\beta g_{R,12}^2 = 2E[a_R^4] - \alpha d^2\sigma_a^4, \quad (31)$$

where $\chi = 2E[a_R^4]\sum_{k=1}^{M}|g_{k2}|^2 + 2\beta\sum_{k=2}^{M}|g_{k2}|^2 \geq 0$.

However, since $\chi + 2\beta g_{I,12}^2 \geq 0$ and $\chi + 2\beta g_{R,12}^2 \geq 0$, the theorem's condition $2E[a_R^4] - \alpha\sigma_a^4 d^2 \leq 0$ requires that $g_{R,12} = 0$ and $g_{I,12} = 0$, that is, $g_{12} = 0$.

Hence, $\mathbf{g}_2$ will take the form

$$\mathbf{g}_2 = [0 \mid \bar{\mathbf{g}}_2^T]^T, \quad (32)$$

which results in

$$\mathbf{g}_1^H \mathbf{g}_2 = 0. \quad (33)$$

Therefore, (26) is reduced to

$$\min_{\bar{\mathbf{g}}_2} \mathcal{J}(\bar{\mathbf{g}}_2) = E\left[(z_{R,2}^2(n) - R)^2\right], \quad (34)$$

where the second equalizer output $z_2 = \mathbf{g}_2^T \mathbf{a} = \bar{\mathbf{g}}_2^T \bar{\mathbf{a}}$, with $\bar{\mathbf{a}} = [a_2, \ldots, a_M]^T$.

Equation (34) has the same form as (10). Hence the analysis is exactly the same as described previously. Consequently, the stationary points of (34) will take the form $\bar{\mathbf{g}}_2 = [0, \ldots, 0, e^{jm_2(\pi/2)}, 0, \ldots, 0]^T$, which corresponds to $\mathbf{g}_2 = [0 \mid 0, \ldots, 0, e^{jm_2(\pi/2)}, 0, \ldots, 0]^T$. Hence $\mathbf{g}_2$ will recover perfectly a different signal than the one already recovered by $\mathbf{g}_1$. Without loss of generality, again, we assume that the single nonzero element of $\mathbf{g}_2$ is in its second position, that is, $\mathbf{g}_2 = [0, e^{jm_2(\pi/2)}, 0, \ldots, 0]^T$.

If we continue in the same manner for each $\mathbf{g}_i$, we can see that each $\mathbf{g}_i$ converges to a setting, in which zeros have the positions of the already recovered signals and its remaining entries contain only one nonzero element; this corresponds to the recovery of a different signal, and this process continues until all signals have been recovered.

On the basis of this analysis, we can conclude that the minimization of the suggested cost function in the case of QAM signals ensures a perfect recovery of all source signals and that the recovered signals correspond to the source signals with a possible permutation and a modulo $\pi/2$ phase rotation.

4.2. P-PSK case $(P \geq 8)$

On the basis of the results in Appendix B, we have

$$\mathcal{J}(\mathbf{g}_1) = E[a_R^4]\left(\sum_{k=1}^{M} |g_{k1}|^2 - 1\right)^2$$
$$+ \beta\left[\left(\sum_{k=1}^{M} |g_{k1}|^2\right)^2 - \sum_{k=1}^{M} |g_{k1}|^4\right] - E[a_R^4] + R^2. \tag{35}$$

And from (16), $J_{\min} = -E[a_R^4] + R^2$ is a constant. If we cancel the derivatives of the first and second terms on the right side of (35), we obtain

$$\sum_{k=1}^{M} |g_{k1}|^2 = 1, \tag{36}$$

$$\sum_{k=1, k \neq \ell}^{M} |g_{k1}|^2 = 0. \tag{37}$$

Therefore, (36) and (37) dictate that the solution must take the form

$$\mathbf{g}_1 = [0, \ldots, 0, e^{j\phi_\ell}, 0, \ldots, 0]^T, \tag{38}$$

where $\phi_\ell \in [0, 2\pi]$ is an arbitrary phase in the ℓth position of $\mathbf{g}_1$ which can be at any of the M possible positions.

The solution $\mathbf{g}_1$ has only one nonzero entry with a modulus equal to one, and all the other ones are zeros. This solution corresponds to the recovery of only one source signal and cancels the other ones. With regard to the other vectors, the analysis is exactly the same as the one in the case of QAM signals.

Then, we can say that the minimization of the SCMA criterion, in the case of P-PSK ($P \geq 8$) signals, ensures the recovery of all signals except for an arbitrary phase rotation for each recovered signal.

4.3. 4-PSK case

On the basis of the results found in Appendix C, $J(\mathbf{g}_1)$ can be written as

$$\mathcal{J}(\mathbf{g}_1) = E^2[a_R^2]\left[3\left(\sum_{k=1}^{M} |g_{k1}|^2\right)^2 - \sum_{k=1}^{M} |g_{k1}|^4 \right.$$
$$\left. - 4\sum_{k=1}^{M} |g_{k1}|^2 - 4\sum_{k=1}^{M} g_{R,k1}^2 g_{I,k1}^2\right] + R^2. \tag{39}$$

In order to find the stationary points of (39), we cancel its derivative

$$\frac{\partial \mathcal{J}(\mathbf{g}_1)}{\partial g_{\ell 1}^*} = E^2[a_R^2]\left[6g_{\ell 1}\sum_{k=1}^{M} |g_{k1}|^2 - 2g_{\ell 1}|g_{\ell 1}|^2 - 4g_{\ell 1} \right.$$
$$\left. - 4(g_{R,\ell 1}g_{I,\ell 1}^2 + jg_{R,\ell 1}^2 g_{I,\ell 1})\right] = 0. \tag{40}$$

By canceling both real and imaginary parts of (40), we have

$$3g_{R,\ell 1}\sum_{k=1}^{M} |g_{k1}|^2 - g_{R,\ell 1}|g_{\ell 1}|^2 - 2g_{R,\ell 1} - 2g_{R,\ell 1}g_{I,\ell 1}^2 = 0,$$

$$3g_{I,\ell 1}\sum_{k=1}^{M} |g_{k1}|^2 - g_{I,\ell 1}|g_{\ell 1}|^2 - 2g_{I,\ell 1} - 2g_{I,\ell 1}g_{R,\ell 1}^2 = 0. \tag{41}$$

According to (41),

$$g_{R,\ell 1} = g_{I,\ell 1}. \tag{42}$$

Then (41) can be reduced to

$$6\sum_{k=1, k \neq \ell}^{M} g_{R,k1}^2 + 2g_{R,\ell 1}^2 - 2 = 0. \tag{43}$$

Thus

$$g_{R,\ell 1}^2 = -3\sum_{k=1, k \neq \ell}^{M} g_{R,k1}^2 + 1. \tag{44}$$

Finally, we find

$$g_{R,\ell 1}^2 = -3(p-1)g_{R,\ell 1}^2 + 1, \tag{45}$$

where $1 \leq p \leq M$ is the number of nonzero elements in $\mathbf{g}_1$, which gives

$$g_{R,\ell 1}^2 = g_{I,\ell 1}^2 = \begin{cases} \dfrac{1}{(3p-2)}, & \text{if } \ell \in F_p, \\ 0, & \text{otherwise,} \end{cases} \quad \forall p = 1, \ldots, M, \tag{46}$$

where F_p is any p-element subset of $\{1, \ldots, M\}$.

Now, we study separately the stationary points for each value of p.

(i) $p = 1$: in this case, $\mathbf{g}_1$ has only one non zero entry, with

$$g_{R,\ell 1} = \pm 1, \qquad g_{I,\ell 1} = \pm 1, \tag{47}$$

that is,

$$g_{\ell 1} = c_\ell e^{j\phi_\ell}, \tag{48}$$

where $c_\ell = \sqrt{2}$ and $\phi_\ell = (2q + 1)(\pi/4)$, with q is an arbitrary integer. Therefore, $\mathbf{g}_1 = [0, \ldots, 0, c_\ell e^{j\phi_\ell}, 0, \ldots, 0]^T$ is the global minimum.

(ii) $p \geq 2$: in this case, the solutions have at least two nonzero elements in some positions of $\mathbf{g}_1$. All nonzero elements have the same squared amplitude of $2/(3p - 2)$.

Let us consider the following perturbation:

$$\mathbf{g}_1' = \mathbf{g}_1 + \mathbf{e}, \tag{49}$$

where $\mathbf{e} = [e_1, \ldots, e_M]^T$ is an $(M \times 1)$ vector whose norm $\|\mathbf{e}\|^2 = \mathbf{e}^H \mathbf{e}$ can be made arbitrarily small and is chosen so that its nonzero elements are only in positions where the corresponding elements of $\mathbf{g}_1$ are nonzero:

$$e_\ell \neq 0 \Longleftrightarrow \ell \in F_p. \tag{50}$$

Let this perturbation be orthogonal with $\mathbf{g}_1$, that is, $\mathbf{e}^H \mathbf{g}_1 = 0$. Then, we have

$$\sum_{\ell \in F_p} |g_{\ell 1}'|^2 = \sum_{\ell \in F_p} |g_{\ell 1}|^2 + \sum_{\ell \in F_p} |e_\ell|^2. \tag{51}$$

We now define, as ε_ℓ, the difference between the squared magnitudes of $g_{\ell 1}'$ and $g_{\ell 1}$, that is,

$$|g_{\ell 1}'|^2 = |g_{\ell 1}|^2 + \varepsilon_\ell, \quad \varepsilon_\ell \in R, \varepsilon_\ell \neq 0 \Longleftrightarrow \ell \in F_p, \tag{52}$$

where

$$\sum_{\ell \in F_p} \varepsilon_\ell = \sum_{\ell \in F_p} |e_\ell|^2. \tag{53}$$

We assume that

$$(g_{R,\ell 1}')^2 = g_{R,\ell 1}^2 + \frac{\varepsilon_\ell}{2}, \qquad (g_{I,\ell 1}')^2 = g_{I,\ell 1}^2 + \frac{\varepsilon_\ell}{2}. \tag{54}$$

By evaluating $\mathcal{J}(\mathbf{g}_1')$, we find

$$\begin{aligned}
\mathcal{J}(\mathbf{g}_1') = E^2[a_R^2]\Bigg[&3\left(\sum_{\ell \in F_p} |g_{\ell 1}|^2 + \varepsilon_\ell\right)^2 - \sum_{\ell \in F_p}\left(|g_{\ell 1}|^2 + \varepsilon_\ell\right)^2 \\
&- 4\sum_{\ell \in F_p}\left(|g_{\ell 1}|^2 + \varepsilon_\ell\right) - 4\sum_{\ell \in F_p}\left(g_{R,\ell 1}^2 + \frac{\varepsilon_\ell}{2}\right) \\
&\times \left(g_{I,\ell 1}^2 + \frac{\varepsilon_\ell}{2}\right)\Bigg] + R^2,
\end{aligned}$$

$$\begin{aligned}
\mathcal{J}(\mathbf{g}_1') = E^2[a_R^2]\Bigg[&3\left(\sum_{\ell \in F_p} |g_{\ell 1}|^2\right)^2 - \sum_{\ell \in F_p} |g_{\ell 1}|^4 \\
&- 4\sum_{\ell \in F_p} |g_{\ell 1}|^2 - 4\sum_{\ell \in F_p} g_{R,\ell 1}^2 g_{I,\ell 1}^2\Bigg] + R^2 \\
+ E^2[a_R^2]\Bigg[&6\sum_{\ell \in F_p} |g_{\ell 1}|^2 \sum_{\ell \in F_p} \varepsilon_\ell - 4\sum_{\ell \in F_p} |g_{\ell 1}|^2 \varepsilon_\ell \\
&- 4\sum_{\ell \in F_p} \varepsilon_\ell + 3\left(\sum_{\ell \in F_p} \varepsilon_\ell\right)^2 - 2\sum_{\ell \in F_p} \varepsilon_\ell^2\Bigg].
\end{aligned} \tag{55}$$

Using (46) in (55), and after some simplifications, we get

$$\mathcal{J}(\mathbf{g}_1') = \mathcal{J}(\mathbf{g}_1) + E^2[a_R^2]\left[3\left(\sum_{\ell \in F_p} \varepsilon_\ell\right)^2 - 2\sum_{\ell \in F_p} \varepsilon_\ell^2\right]. \tag{56}$$

It exists $\varepsilon_\ell \in \mathbb{R}, (\ell \in F_p)$ so that

$$3\left(\sum_{\ell \in F_p} \varepsilon_\ell\right)^2 - 2\sum_{\ell \in F_p} \varepsilon_\ell^2 < 0. \tag{57}$$

Then, $\exists \varepsilon_\ell \in \mathbb{R}, (\ell \in F_p)$ so that

$$\mathcal{J}(\mathbf{g}_1') < \mathcal{J}(\mathbf{g}_1). \tag{58}$$

Hence, $\mathcal{J}(\mathbf{g}_1)$ cannot be a local minimum.

Now we consider another perturbation which takes the form

$$g_{\ell 1}' = \begin{cases} \sqrt{1 + \xi}\, g_{\ell 1} & \text{if } \ell \in F_p, \\ 0 & \text{otherwise}, \end{cases} \tag{59}$$

where ξ is a small positive constant.

By evaluating $\mathcal{J}(\mathbf{g}_1')$, we obtain

$$\begin{aligned}
\mathcal{J}(\mathbf{g}_1') = E^2[a_R^2]\Bigg[&3\left(\sum_{\ell \in F_p}(1 + \xi)|g_{\ell 1}|^2\right)^2 \\
&- \sum_{\ell \in F_p}(1 + \xi)^2|g_{\ell 1}|^4 - 4\sum_{\ell \in F_p}(1 + \xi)|g_{\ell 1}|^2 \\
&- 4\sum_{\ell \in F_p}(1 + \xi)^2 g_{R,\ell 1}^2 g_{I,\ell 1}^2\Bigg] + R^2.
\end{aligned} \tag{60}$$

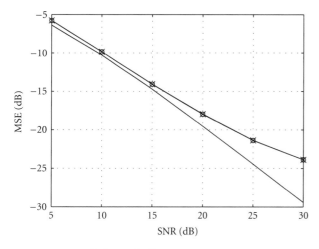

FIGURE 5: *MSE* versus *SNR* of CC-SCMA, MCC-SCMA, and supervised MMSE.

Using (36) and after some simplifications, we get

$$\mathcal{J}(\mathbf{g}_1') = \mathcal{J}(\mathbf{g}_1) + \frac{4\xi^2 p}{3p - 2} E^2[a_R^2].$$ (61)

Therefore, we always have

$$\mathcal{J}(\mathbf{g}_1') > \mathcal{J}(\mathbf{g}_1), \quad \forall p \in N^{+*}.$$ (62)

Hence, $\mathbf{g}_1$ cannot be a local maximum.

Then, on the basis of (58) and (62), $\mathbf{g}_1$ is a saddle point for $p \geq 2$.

Therefore, the only stable minima correspond to $p = 1$.

We conclude that the only stable minima take the form $\mathbf{g}_1 = [0,\ldots,0,c_\ell e^{j\phi_\ell},0,\ldots,0]^T$, which ensure the extraction of only one source signal and cancel the other ones. For the remainder of the analysis, we proceed exactly as we did for QAM signals.

Finally, in order to conclude this section, we can say that the minimization of the cost function in (6) ensures the recovery of all source signals in the case of source signals drawn from QAM or PSK constellations.

5. MODIFIED CROSS-CORRELATION TERM

In the previous section, we have seen that, in the case of QAM constellation, the signals are recovered with modulo $\pi/2$ phase rotation. By taking advantage of this result, we suggest to use, instead of cross-correlation term in (6), the following term:

$$\sum_{m=1}^{\ell-1} E^2[z_{R,\ell}(n)z_{R,m}(n)] + E^2[z_{R,\ell}(n)z_{I,m}(n)], \quad \ell = 1,\ldots,M.$$ (63)

Using (63) in (6), instead of the classical cross-correlation term, the criterion becomes

$$\mathcal{J}_\ell(n) = E[(z_{R,\ell}^2(n) - R)^2]$$
$$+ \alpha \sum_{m=1}^{\ell-1} (E^2[z_{R,\ell}(n)z_{R,m}(n)] + E^2[z_{R,\ell}(n)z_{I,m}(n)]),$$
$$\ell = 1,\ldots,M.$$ (64)

Please note that in (64) the multiplications in cross-correlation terms are not complexes, as opposed to (6) which reduces complexity.

The cost function in (64) is named modified cross-correlation SCMA (MCC-SCMA).

Remark 1. The new cross-correlation term could be also used by the MMA algorithm, because it recovers QAM signals with modulo $\pi/2$ phase rotation.

In the following section, the complexity of the modified cross-correlation term will be discussed and compared with the classical cross-correlation term.

6. IMPLEMENTATION AND COMPUTATIONAL COMPLEXITY

6.1. Implementation

In order to implement (6) and (64), we suggest to use the classical stochastic gradient algorithm (SGA) [25]. The general form of the SGA is given by

$$\mathbf{W}(n + 1) = \mathbf{W}(n) - \frac{1}{2}\mu\nabla_{\mathbf{W}}(\mathcal{J}),$$ (65)

where $\nabla_{\mathbf{W}}(\mathcal{J})$ is the gradient of $\mathcal{J}$ with respect to $\mathbf{W}$.

6.1.1. For CC-SCMA

The equalizer update equation at the nth iteration is written as

$$\mathbf{w}_\ell(n + 1) = \mathbf{w}_\ell(n) - \mu e_\ell(n)\mathbf{y}^*(n), \quad \ell = 1,\ldots,M,$$ (66)

where the constant which arises from the differentiation of (6) is absorbed within the step size μ. $e_\ell(n)$ is the instantaneous error $e_\ell(n)$ for the ℓth equalizer given by

$$e_\ell(n) = (z_{R,\ell}^2(n) - R)z_{R,\ell}(n) + \frac{\alpha}{2} \sum_{m=1}^{\ell-1} \hat{r}_{\ell m}(n)z_m(n),$$ (67)

where the scalar quantity $\hat{r}_{\ell m}$ represents the estimate of $r_{\ell m}$, it can be recursively computed as [25]

$$\hat{r}_{\ell m}(n + 1) = \lambda\hat{r}_{\ell m}(n) + (1 - \lambda)z_\ell(n)z_m^*(n),$$ (68)

where $\lambda \in [0, 1]$ is a parameter that controls the length of the effective data window in the estimation.

Please note that since $E[\hat{r}_{\ell m}(n)] = E[z_\ell(n)z_m^*(n)]$, then the estimator $\hat{r}_{\ell m}(n)$ is unbiased.

TABLE 1: Comparison of the algorithms complexity against weight update.

Algorithm	Multiplications	Additions
CC-CMA	$2M(3M + 4N + 1) - 2$	$4M(M + 2N - 1)$
CC-SCMA	$6M(M + N) - 2$	$M(4M + 6N - 5) + 1$
MCC-SCMA	$4M(M + N) - 1$	$2M(M + 2N - 1)$

6.1.2. For MCC-SCMA

We have exactly the same equation as (66), but the instantaneous error signal of the ℓth equalizer is given by

$$
\begin{aligned}
e_\ell(n) = {} & (z_{R,\ell}^2(n) - R)z_{R,\ell}(n) \\
& + \frac{\alpha}{2}\sum_{m=1}^{\ell-1}\left[\hat{r}_{RR,\ell m}(n)z_{R,m}(n) + \hat{r}_{RI,\ell m}(n)z_{I,m}(n)\right],
\end{aligned}
\tag{69}
$$

where

$$
\begin{aligned}
\hat{r}_{RR,\ell m}(n + 1) &= \lambda\hat{r}_{RR,\ell m}(n) + (1 - \lambda)z_{R,\ell}(n)z_{R,m}^*(n), \\
\hat{r}_{RI,\ell m}(n + 1) &= \lambda\hat{r}_{RI,\ell m}(n) + (1 - \lambda)z_{R,\ell}(n)z_{I,m}^*(n).
\end{aligned}
\tag{70}
$$

6.2. Complexity

We consider the computational complexity of (66) for one iteration and for all equalizer outputs. With

(i) for CC-SCMA,

$$
e_\ell(n) = (z_{R,\ell}(n)^2 - R)z_{R,\ell}(n) + \frac{\alpha}{2}\sum_{m=1}^{\ell-1}\hat{r}_{\ell m}(n)z_m(n);
\tag{71}
$$

(ii) for MCC-SCMA (modified cross-correlation SCMA),

$$
\begin{aligned}
e_\ell(n) = {} & (z_{R,\ell}(n)^2 - R)z_{R,\ell}(n) \\
& + \frac{\alpha}{2}\sum_{m=1}^{\ell-1}\left[\hat{r}_{RR,\ell m}(n)z_{R,m}(n) + \hat{r}_{RI,\ell m}(n)z_{I,m}(n)\right];
\end{aligned}
\tag{72}
$$

(iii) for CC-CMA (cross-correlation CMA)

$$
e_\ell(n) = (|z_\ell(n)|^2 - R)z_\ell(n) + \frac{\alpha}{2}\sum_{m=1}^{\ell-1}\hat{r}_{\ell m}(n)z_m(n).
\tag{73}
$$

According to Table 1, the CC-SCMA presents a low complexity compared to that of CC-CMA. On a more interesting note, the results in Table 1 show that the use of modified cross-correlation term reduce significantly the complexity. Please note that the number of operations is per iteration.

7. NUMERICAL RESULTS

Some numerical results are now presented in order to confirm the theoretical analysis derived in the previously sec-

tions. For that purpose, we use the signal to interference and noise ratio (SINR) defined as

$$
\begin{aligned}
\mathrm{SINR}_k &= \frac{|g_{kk}|^2}{\sum_{\ell,\ell \neq k}|g_{\ell k}|^2 + \mathbf{w}_k^T\mathbf{R_b}\mathbf{w}_k^*}, \\
\mathrm{SINR} &= \frac{1}{M}\sum_{k=1}^{M}\mathrm{SINR}_k,
\end{aligned}
\tag{74}
$$

where SINR_k is the signal-to-interference and noise ratio at the kth output. $g_{ij} = \mathbf{h}_i^T\mathbf{w}_j$, where $\mathbf{w}_j$ and $\mathbf{h}_i$ are the jth and ith column vector of matrices $\mathbf{W}$ and $\mathbf{H}$, respectively. $\mathbf{R_b} = E[\mathbf{bb}^H] = \sigma_b^2\mathbf{I}_N$ is the noise covariance matrix. The source signals are assumed to be of unit variance.

The SINR is estimated via the average of 1000 independent trials. Each estimation is based on the following model. The system inputs are independent, uniformly distributed and drawn from 16-QAM, 4-PSK, and 8-PSK constellations. We considered M transmit and N receive spatially decorrelated antennas. The channel matrix $\mathbf{H}$ is modeled by an $(N \times M)$ matrix with independent and identically distributed (i.i.d.), complex, zero-mean, Gaussian entries. We considered $\alpha = 1$ (this value satisfy the theorem condition) and $\lambda = 0.97$. The variance of noise is determined according to the desired Signal-to-Noise Ratio (SNR).

Figures 1, 2, and 3 show the constellations of the source signals, the received signals, and the receiver outputs (after convergence) using the proposed algorithm for 16-QAM, 8-PSK, and 4-PSK constellations, respectively. We have considered that SNR = 30 dB, $M = 2$, $N = 2$, and that $\mu = 5 \times 10^{-3}$. Please note that the constellations on Figures 1, 2, and 3 are given before the phase ambiguity is removed (this ambiguity can be solved easily by using differential decoding).

In Figure 1, we see that the algorithm recovers the 16-QAM source signals, but up to a modulo $\pi/2$ phase rotation which may be different for each output. Figure 2 shows that the 8-PSK signals are recovered with an arbitrary phase rotation. In Figure 3, the 4-PSK signals are recovered with a $(2k + 1)(\pi/4)$ phase rotation and an amplitude of $\sqrt{2}$. Theses results are in accordance with the theoretical analysis given in Section 4.

In order to compare the performances of CC-SCMA and CC-CMA, the same implementation is considered for both algorithms (see Section 6). We have considered $M = 2$, $N = 3$, SNR = 25 dB, and the step sizes were chosen so that the algorithms have sensibly the same steady-state performances. We have also used the supervised least-mean square algorithm (LMS) as a reference.

Figure 4 represents the SINR performance plots for the proposed approach and the CC-CMA algorithm. We observe that the speed of convergence of the proposed approach is very close to that of the CC-CMA. Hence, it represents a good compromise between performance and complexity.

Figure 5 represents the mean-square error (MSE) versus SNR. In order to verify the effectiveness of the modified cross-correlation term, we have considered $M = 2$, $N = 3$, 16-QAM, and $\mu = 0.02$ for both algorithms. In this figure, the supervised minimum mean square (MMSE) receiver serves as reference. We observe that CC-SCMA and MCC-SCMA have almost the same behavior. So, in the case of QAM signals, it is preferable to use a modified cross-correlation term because of its low complexity and of its similar performances compared to the one of the classical cross-correlation term.

8. CONCLUSION

In this paper, we have proposed a new globally convergent algorithm for the multiple-input multiple-output (MIMO) adaptive blind separation of QAM and PSK signals. The criterion is based on one dimension (either real or imaginary) and consists in penalizing the deviation of the real (or the imaginary) part from a constant. It was demonstrated that the proposed approach is globally convergent to a setting that recovers perfectly, in the absence of noise, all the source signals. A modification for the cross-correlation constraint in the case of QAM constellation has been suggested. Our algorithm has shown a low computational complexity compared to that of CMA, especially when the modified cross-correlation constraint is used, which makes it attractive for implementation in practical applications. Simulation results have shown that the suggested algorithm has a good performance despite its lower complexity.

APPENDICES

A. QAM CASE

From (11),

$$\mathcal{J}(\mathbf{g}_1) = E[z_{R,1}^4] - 2\mathrm{RE}[z_{R,1}^2] + R^2. \tag{A.1}$$

We have

$$z_1(n) = \mathbf{g}_1^T \mathbf{a}(n), \tag{A.2}$$

where

$$\begin{aligned}
\mathbf{g}_1 &= [g_{11}, \ldots, g_{M1}]^T, \quad g_{k1} = g_{R,k1} + j g_{I,k1}, \\
\mathbf{a} &= [a_1, \ldots, a_M]^T, \quad a_k = a_{R,k} + j a_{I,k}.
\end{aligned} \tag{A.3}$$

Then

$$z_{R,1}(n) = \sum_{k=1}^{M} (g_{R,k1} a_{R,k} - g_{I,k1} a_{I,k}). \tag{A.4}$$

For i.i.d. and mutually independent source signals that drawn from square QAM constellation, we have

$$\begin{aligned}
& E[a_R] = E[a_I] = 0, \\
& E[a_R^m] = E[a_I^m], \quad \forall m, \\
& E[a_{R,k}^m a_{I,\ell}^n] = E[a_{R,k}^m] E[a_{I,\ell}^n], \quad \forall k, \ell, m, n, \\
& E[a_{R,k}^m a_{R,\ell}^n] = E[a_{I,k}^m a_{I,\ell}^n] = \begin{cases} E[a_R^{m+n}], & \text{if } k = \ell, \\ E[a_R^m] E[a_R^n], & \text{otherwise.} \end{cases}
\end{aligned} \tag{A.5}$$

We have

$$E[z_{R,1}^2] = E\left\{ \left[\sum_{k=1}^{M} (g_{R,k1} a_{R,k} - g_{I,k1} a_{I,k}) \right]^2 \right\}$$

$$= \sum_{k=1}^{M} \sum_{\ell=1}^{M} E[(g_{R,k1} a_{R,k} - g_{I,k1} a_{I,k})(g_{R,\ell 1} a_{R,\ell} - g_{I,\ell 1} a_{I,\ell})]$$

$$= \sum_{k=1}^{M} \sum_{\ell=1}^{M} (g_{R,k1} g_{R,\ell 1} E[a_{R,k} a_{R,\ell}] - g_{R,k1} g_{I,\ell 1} E[a_{R,k} a_{I,\ell}]$$

$$- g_{I,k1} g_{R,\ell 1} E[a_{I,k} a_{R,\ell}]) + g_{I,k1} g_{I,\ell 1} E[a_{I,k} a_{I,\ell}]. \tag{A.6}$$

Using (A.5), we obtain

$$E[z_{R,1}^2] = E[a_R^2] \sum_{k=1}^{M} |g_{k1}|^2. \tag{A.7}$$

Similarly

$$E[z_{R,1}^4] = E\left\{ \left[\sum_{k=1}^{M} (g_{R,k1} a_{R,k} - g_{I,k1} a_{I,k}) \right]^4 \right\}$$

$$= \sum_{k=1}^{M} \sum_{\ell=1}^{M} \sum_{m=1}^{M} \sum_{n=1}^{M} E[(g_{R,k1} a_{R,k} - g_{I,k1} a_{I,k})$$
$$\times (g_{R,\ell 1} a_{R,\ell} - g_{I,\ell 1} a_{I,\ell})$$
$$\times (g_{R,m1} a_{R,m} - g_{I,m1} a_{I,m})$$
$$\times (g_{R,n1} a_{R,n} - g_{I,n1} a_{I,n})]. \tag{A.8}$$

Developing (A.8) and using (A.5), we have the following three cases:

(i) $k = \ell = m = n$:

$$E[z_{R,1}^4] = E[a_R^4] \sum_{k=1}^{M} (g_{R,k1}^4 + g_{I,k1}^4) + 6E^2[a_R^2] \sum_{k=1}^{M} (g_{R,k1}^2 g_{I,k1}^2); \tag{A.9}$$

(ii) $k = \ell \neq m = n$:

$$E[z_{R,1}^4] = 3E^2[a_R^2] \sum_{k=1}^{M} \sum_{\ell=1, \ell \neq k}^{M} |g_{k1}|^2 |g_{\ell 1}|^2; \tag{A.10}$$

(iii) otherwise: (A.8) is equal to zero.

From (A.9) and (A.10), (A.8) becomes

$$E[z_{R,1}^4] = E[a_R^4] \sum_{k=1}^{M} (g_{R,k1}^4 + g_{I,k1}^4) + 6E^2[a_R^2] \sum_{k=1}^{M} (g_{R,k1}^2 g_{I,k1}^2)$$

$$+ 3E^2[a_R^2] \sum_{k=1}^{M} \sum_{\ell=1, \ell \neq k}^{M} |g_{k1}|^2 |g_{\ell 1}|^2. \tag{A.11}$$

Then

$$E[z_{R,1}^4] = E[a_R^4]\sum_{k=1}^M |g_{k1}|^4 - 2E[a_R^4]\sum_{k=1}^M g_{R,k1}^2 g_{I,k1}^2$$

$$+ 6E^2[a_R^2]\sum_{k=1}^M g_{R,k1}^2 g_{I,k1}^2 \qquad (A.12)$$

$$+ 3E^2[a_R^2]\sum_{k=1}^M \sum_{\ell=1,\ell\neq k}^M |g_{k1}|^2 |g_{\ell 1}|^2.$$

On the other hand, we have

$$\left[\left(\sum_{k=1}^M |g_{k1}|^2\right)^2 - \sum_{k=1}^M |g_{k1}|^4\right] = \sum_{k=1}^M \sum_{\ell=1,\ell\neq k}^M |g_{k1}|^2 |g_{\ell 1}|^2. \qquad (A.13)$$

Substituting (A.13) into (A.12), we get

$$E[z_{R,1}^4] = E[a_R^4]\sum_{k=1}^M |g_{k1}|^4$$

$$+ 3E^2[a_R^2]\left[\left(\sum_{k=1}^M |g_{k1}|^2\right)^2 - \sum_{k=1}^M |g_{k1}|^4\right] + 2\beta\sum_{k=1}^M g_{R,k1}^2 g_{I,k1}^2, \qquad (A.14)$$

where $\beta = 3E^2[a_R^2] - E[a_R^4]$.

Using (A.6) and (A.14) in (A.1), we obtain

$$\mathcal{J}(\mathbf{g}_1)$$

$$= E[a_R^4]\sum_{k=1}^M |g_{k1}|^4 + 3E^2[a_R^2]\left[\left(\sum_{k=1}^M |g_{k1}|^2\right)^2 - \sum_{k=1}^M |g_{k1}|^4\right]$$

$$+ 2\beta\sum_{k=1}^M g_{R,k1}^2 g_{I,k1}^2 - 2E[a_R^2]\sum_{k=1}^M |g_{k1}|^2 + R^2$$

$$= E[a_R^4]\left(\sum_{k=1}^M |g_{k1}|^2\right)^2 - E[a_R^4]\left(\sum_{k=1}^M |g_{k1}|^2\right)^2$$

$$+ E[a_R^4]\sum_{k=1}^M |g_{k1}|^4 + 3E^2[a_R^2]\left[\left(\sum_{k=1}^M |g_{k1}|^2\right)^2 - \sum_{k=1}^M |g_{k1}|^4\right]$$

$$+ 2\beta\sum_{k=1}^M g_{R,k1}^2 g_{I,k1}^2 - 2E[a_R^2]\sum_{k=1}^M |g_{k1}|^2 + R^2 + E[a_R^4] - E[a_R^4]. \qquad (A.15)$$

Rearranging terms, we get

$$\mathcal{J}(\mathbf{g}_1) = E[a_R^4]\left(\sum_{k=1}^M |g_{k1}|^2 - 1\right)^2$$

$$+ \beta\left[\left(\sum_{k=1}^M |g_{k1}|^2\right)^2 - \sum_{k=1}^M |g_{k1}|^4\right] + 2\beta\sum_{k=1}^M g_{R,k1}^2 g_{I,k1}^2$$

$$- E[a_R^4] + R^2. \qquad (A.16)$$

Finally, we get (12).

B. P-PSK CASE ($P \geq 8$)

In the case of P-PSK ($P \geq 8$), (A.5) hold and moreover we have

$$E[a_{R,k}^2 a_{I,\ell}^2] = \begin{cases} \dfrac{1}{3}E[a_R^4] & \text{if } k = \ell, \\ E^2[a_R^2] & \text{otherwise.} \end{cases} \qquad (B.17)$$

Using (A.5) in (A.6),

$$E[z_{R,1}^2] = E[a_R^2]\sum_{k=1}^M |g_{k1}|^2. \qquad (B.18)$$

Using (A.5), and (B.17) in (A.8), we find

(i) $k = \ell = m = n$:

$$E[z_{R,1}^4] = E[a_R^4]\sum_{k=1}^M |g_{k1}|^4; \qquad (B.19)$$

(ii) $k = \ell \neq m = n$:

$$E[z_{R,1}^4] = 3E^2[a_R^2]\sum_{k=1}^M \sum_{\ell=1,\ell\neq k}^M |g_{k1}|^2 |g_{\ell 1}|^2; \qquad (B.20)$$

(iii) otherwise: (A.8) is equal to zero.
Then

$$E[z_{R,1}^4] = E[a_R^4]\sum_{k=1}^M (g_{R,k1}^4 + g_{I,k1}^4)$$

$$+ 3E^2[a_R^2]\sum_{k=1}^M \sum_{\ell=1,\ell\neq k}^M |g_{k1}|^2 |g_{\ell 1}|^2. \qquad (B.21)$$

By developing the above equation as in the QAM case, we get

$$\mathcal{J}(\mathbf{g}_1) = E[a_R^4]\left(\sum_{k=1}^M |g_{k1}|^2 - 1\right)^2 + \beta\left[\left(\sum_{k=1}^M |g_{k1}|^2\right)^2 - \sum_{k=1}^M |g_{k1}|^4\right]$$

$$- E[a_R^4] + R^2. \qquad (B.22)$$

C. 4-PSK CASE

For 4-PSK signals,

$$E[a_{R,k}^2 a_{I,\ell}^2] = \begin{cases} 0 & \text{if } k = \ell, \\ E^2[a_R^2] & \text{otherwise.} \end{cases} \qquad (C.23)$$

Considering (A.5), and (C.23), and proceeding in the same way as in the case of P-PSK signals ($P \geq 8$), we can easily find

$$\mathcal{J}(\mathbf{g}_1) = E^2[a_R^2]\left[3\left(\sum_{k=1}^M |g_{k1}|^2\right)^2 - \sum_{k=1}^M |g_{k1}|^4 - 4\sum_{k=1}^M |g_{k1}|^2 - 4\sum_{k=1}^M g_{R,k1}^2 g_{I,k1}^2\right] + R^2. \qquad (C.24)$$

ACKNOWLEDGMENT

The authors are grateful to the anonymous referees for their constructive critics and valuable comments.

REFERENCES

[1] G. J. Foschini, "Layered space-time architecture for wireless communication in a fading environment when using multi-element antennas," *Bell Labs Technical Journal*, vol. 1, no. 2, pp. 41–59, 1996.

[2] A. Mansour, A. K. Barros, and N. Ohnishi, "Blind separation of sources: methods, assumptions and applications," *IEICE Transactions on Fundamentals of Electronics, Communications and Computer Sciences*, vol. E83-A, no. 8, pp. 1498–1512, 2000.

[3] A. Cichocki and S.-I. Amari, *Adaptive Blind Signal and Image Processing: Learning Algorithms and Applications*, John Wiley & Sons, New York, NY, USA, 2003.

[4] K. Abed-Meraim, J.-F. Cardoso, A. Y. Gorokhov, P. Loubaton, and E. Moulines, "On subspace methods for blind identification of single-input multiple-output FIR systems," *IEEE Transactions on Signal Processing*, vol. 45, no. 1, pp. 42–55, 1997.

[5] A. Gorokhov and P. Loubaton, "Blind identification of MIMO-FIR systems: a generalized linear prediction approach," *Signal Processing*, vol. 73, no. 1-2, pp. 105–124, 1999.

[6] J. K. Tugnait, "Identification and deconvolution of multichannel linear non-Gaussian processes using higher order statistics and inverse filter criteria," *IEEE Transactions on Signal Processing*, vol. 45, no. 3, pp. 658–672, 1997.

[7] B. Hassibi and H. Vikalo, "On the sphere-decoding algorithm—II: generalizations, second-order statistics, and applications to communications," *IEEE Transactions on Signal Processing*, vol. 53, no. 8, pp. 2819–2834, 2005.

[8] B. Hassibi, "An efficient square-root algorithm for BLAST," in *Proceedings of the IEEE International Conference on Acoustics, Speech, and Signal Processing (ICASSP '00)*, vol. 2, pp. 737–740, Istanbul, Turkey, June 2000.

[9] Y. Sato, "A method of self-recovering equalization for multilevel amplitude-modulation systems," *IEEE Transactions on Communications*, vol. 23, no. 6, pp. 679–682, 1975.

[10] D. Godard, "Self-recovering equalization and carrier tracking in two-dimensional data communication systems," *IEEE Transactions on Communications*, vol. 28, no. 11, pp. 1867–1875, 1980.

[11] J. Treichler and B. Agee, "A new approach to multipath correction of constant modulus signals," *IEEE Transactions on Acoustics, Speech, and Signal Processing*, vol. 31, no. 2, pp. 459–472, 1983.

[12] M. G. Larimore and J. Treichler, "Convergence behavior of the constant modulus algorithm," in *Proceedings of the IEEE International Conference on Acoustics, Speech, and Signal Processing (ICASSP '83)*, vol. 1, pp. 13–16, Boston, Mass, USA, April 1983.

[13] J. Treichler and M. G. Larimore, "New processing techniques based on the constant modulus adaptive algorithm," *IEEE Transactions on Acoustics, Speech, and Signal Processing*, vol. 33, no. 2, pp. 420–431, 1985.

[14] R. Gooch and J. Lundell, "The CM array: an adaptive beamformer for constant modulus signals," in *Proceedings of the IEEE International Conference on Acoustics, Speech, and Signal Processing (ICASSP '86)*, vol. 11, pp. 2523–2526, Tokyo, Japan, April 1986.

[15] L. Castedo, C. J. Escudero, and A. Dapena, "A blind signal separation method for multiuser communications," *IEEE Transactions on Signal Processing*, vol. 45, no. 5, pp. 1343–1348, 1997.

[16] C. B. Papadias and A. J. Paulraj, "A constant modulus algorithm for multiuser signal separation in presence of delay spread using antenna arrays," *IEEE Signal Processing Letters*, vol. 4, no. 6, pp. 178–181, 1997.

[17] K. N. Oh and Y. O. Chin, "Modified constant modulus algorithm: blind equalization and carrier phase recovery algorithm," in *Proceedings of the IEEE International Conference on Communications (ICC '95)*, vol. 1, pp. 498–502, Seattle, Wash, USA, June 1995.

[18] J. Yang, J.-J. Werner, and G. A. Dumont, "The multimodulus blind equalization and its generalized algorithms," *IEEE Journal on Selected Areas in Communications*, vol. 20, no. 5, pp. 997–1015, 2002.

[19] L. M. Garth, J. Yang, and J.-J. Werner, "Blind equalization algorithms for dual-mode CAP-QAM reception," *IEEE Transactions on Communications*, vol. 49, no. 3, pp. 455–466, 2001.

[20] P. Sansrimahachai, D. B. Ward, and A. G. Constantinides, "Blind source separation for BLAST," in *Proceedings of the 14th International Conference on Digital Signal Processing (DSP '02)*, vol. 1, pp. 139–142, Santorini, Greece, July 2002.

[21] A. Goupil and J. Palicot, "Constant norm algorithms class," in *proceedings of the 11th European Signal Processing Conference (EUSIPCO '02)*, vol. 1, pp. 641–644, Toulouse, France, September 2002.

[22] A. Ikhlef, D. Le Guennec, and J. Palicot, "Constant norm algorithms for MIMO communication systems," in *Proceedings of the 13th European Signal Processing Conference (EUSIPCO '05)*, Antalya, Turkey, September 2005.

[23] A. Ikhlef and D. Le Guennec, "Blind recovery of MIMO QAM signals : a criterion with its convergence analysis," in *Proceedings of the 14th European Signal Processing Conference (EUSIPCO '06)*, Florence, Italy, September 2006.

[24] C. B. Papadias, "Globally convergent blind source separation based on a multiuser kurtosis maximization criterion," *IEEE Transactions on Signal Processing*, vol. 48, no. 12, pp. 3508–3519, 2000.

[25] S. Haykin, *Adaptive Filter Theory*, Prentice-Hall, Upper Saddle River, NJ, USA, 4th edition, 2002.

Hindawi Publishing Corporation
EURASIP Journal on Wireless Communications and Networking
Volume 2007, Article ID 57175, 10 pages
doi:10.1155/2007/57175

Research Article

Employing Coordinated Transmit and Receive Beamforming in Clustering Double-Directional Radio Channel

Chen Sun,[1] Makoto Taromaru,[1] and Takashi Ohira[2]

[1] *ATR Wave Engineering Laboratories, 2-2-2 Hikaridai, Keihanna Science City, Kyoto 619-0288, Japan*
[2] *Department of Information and Computer Sciences, Toyohashi University of Technology, 1-1 Hibariga-oka, Toyohashi 441-8580, Japan*

Received 31 October 2006; Accepted 1 August 2007

Recommended by Robert W. Heath Jr.

A novel beamforming (BF) system that employs two switched beam antennas (SBAs) at both ends of the wireless link in an indoor double-directional radio channel (DDRC) is proposed. The distributed directivity gain (DDG) and beam pattern correlation in DDRC are calculated. The channel capacity of the BF system is obtained from an analytical model. Using the channel capacity and outage capacity as performance measures, we show that the DDG of the BF system directly increases the average signal-to-noise ratio (SNR) of the wireless link, thus achieving a direct increase of the ergodic channel capacity. By jointly switching between different pairs of transmit (Tx) and receive (Rx) directional beam patterns towards different wave clusters, the system provides diversity gain to combat against multipath fading, thus reducing the outage probability of the random channel capacity. Furthermore, the performance of the BF system is compared with that of a multiple-input multiple-output (MIMO) system that is set up using linear antenna arrays. Results show that in a low-SNR environment, the BF system outperforms the MIMO system in the same clustering DDRC.

1. INTRODUCTION

Along with the evolution of wireless technologies, broadband Internet access and multimedia services are expected to be available for commercial mobile subscribers. This enthusiasm has created a need for high-data-rate wireless transmission systems. Study on adaptive antenna systems has demonstrated their potential in increasing the spectrum efficiency of a wireless radio channel. The beamforming (BF) ability of the adaptive antennas increases the transmission range, reduces delay spread, and suppresses interference [1–4]. In a multipath rich environment, they provide diversity gain to counteract multipath fading [5, 6]. Installing antenna arrays at both transmitter (Tx) and receiver (Rx) sides builds up a multiple-input multiple-output (MIMO) system [7]. The MIMO system increases the data rate of wireless transmission by sending multiple data streams though multiple Tx and Rx antennas over fading channels [8].

The potential increase in the spectrum efficiency of the MIMO systems has been evaluated under various wireless propagation channel models. Recent investigation on wireless channel model has introduced the concept of double-directional radio channel (DDRC) that incorporates directional information at both Tx and Rx sides of a wireless link [9]. The techniques of extracting directional information at both ends of the wireless link are presented in [10, 11]. In [11] Wallace and Jensen extend the Saleh-Valenzuela's clustering model [12] and the Spencer's model [13] to include both angle-of-arrival (AOA) information and angle-of-departure (AOD) information to describe an indoor DDRC with clustering phenomenon, which is here after referred to as *clustering DDRC*. Incorporating spatial information at both ends of the link, the statistical model provides a better description of a wireless propagation channel, thus allowing a better evaluation of the system performance in a practical situation than other models. The channel capacity that can be achieved by a MIMO system in a clustering DDRC model is evaluated in [11]. The predicted capacities match perfectly with the measured data. This justifies the clustering DDRC model being a better description of a practical wireless propagation channel independently of specific antenna characteristics.

However, the predicted capacity of a MIMO system in a clustering DDRC model is lower than that predicted under an ideal statistical channel model, which assumes statistically independent fading at antenna array elements [7]. The reason is that the potential capacity of MIMO systems relies on the richness of scattering waves. The clustering phenomenon of impinging waves in the Wallace's model [11] leads to an increase of fading correlation among antenna array elements, thus resulting in a lower-channel capacity.

In addition to the capacity loss of a MIMO system in a clustering channel, the implementation of a MIMO system poses practical problems, such as the power consumption limitation and the size constraints of commercial mobile terminals. A MIMO receiver that is equipped with an antenna array consists of multiple RF channels connected to individual array elements so that spatial processing (such as BF, diversity combining) is carried out at digital stage. The power consumption and fabrication cost of the MIMO system increase with the number of array elements.

To circumvent the afore-mentioned limitations of the MIMO system and to increase the channel capacity of wireless communication in a clustering DDRC, we propose a BF system that employs two switched beam antennas (SBAs) with predefined directional beam patterns at both ends of the wireless link. Knowing that the main energy of the signals is transported spatially through wave clusters [14], the directional beam patterns at both Tx and Rx sides are directed towards the wave clusters. Employing the directional beam pattern in the presence of distributed waves, the distributed directivity gains (DDGs) at both Tx and Rx antennas directly increase the average SNR of the wireless link. By switching between different pairs of Tx and Rx directional beam patterns, the system provides a diversity freedom to further improve the performance.

A potential application of the proposed technique is an ad hoc network. It is a self-configuring network of mobile terminals without a wireless backbone, such as access points [3]. Commercial applications of adaptive antenna technology at these battery-powered mobile terminals (e.g., laptops, PDAs) require adaptive antennas that have low-power consumption and low-fabrication cost. The parasitic array antennas have shown their potential in fulfilling the goal [3]. Therefore, in this paper, the Tx and Rx SBAs are realized by two electronically steerable parasitic array radiator (ESPAR) antennas [15–17]. We do not employ a multibeam antennas as their structures consist of a number of individual RF channel that lead to undesirable high-power consumption and fabrication cost [3].

The rest of the paper is organized as follows. In Section 2, we describe the clustering DDRC model, the SBA and the BF system. In Section 3, we calculate the DDG at both ends of the wireless link. The correlation of different wireless links that are set up with different pairs of transmit and receive beam patterns is obtained. Based on these results, an analytical model of the BF system is built. In Section 4, we calculate the channel capacity of the BF system based on the analytical model and compare the performance of the BF system with a MIMO system in the same clustering DDRC model. Finally, in Section 5, we conclude the paper.

2. SYSTEM MODEL

In order to present the system model of the proposed BF system, we review the clustering DDRC model first. After that, the ESPAR antennas that operate as SBAs in the proposed system are described. Finally, the BF system that employs two ESPAR antennas at both ends of the wireless link in a clustering DDRC is given.

2.1. Clustering DDRC

In [12] Saleh and Valenzuela describe the clustering phenomenon of impinging waves in an indoor multipath channel, that is, the result of scattering and reflection on small objects, rough surface, and so forth, along the paths of wave propagation In [13]Spencer et al. extend Saleh's model according to the measurement results to include spatial information of the clusters and waves within each cluster. The Spencer's model is further extended in [11] to describe a clustering DDRC, which takes into account the spatial information at both Tx and Rx sides.

The spatial channel response (SCR) of a clustering DDRC is written as [11]

$$
h(\phi^r, \phi^t) = \frac{1}{\sqrt{LK}} \sum_{l=0}^{L-1} \sum_{k=0}^{K-1} \alpha_{l,k} \\
\times \delta(\phi^r - \Phi_l^r - \omega_{l,k}^r) \delta(\phi^t - \Phi_l^t - \omega_{l,k}^t),
\tag{1}
$$

where superscripts $(\cdot)^r$ and $(\cdot)^t$ denote, respectively, the Rx and Tx sides of a wireless link. Let ϕ^r and ϕ^t represent the AOA and AOD, respectively. The nominal AOA and AOD of the lth cluster are denoted by Φ_l^r and Φ_l^t, which are uniformly distributed over the azimuthal plane [13]. Here, L is the total number of wave clusters. In the following discussion, we call the waves within each cluster as *rays* and let $\omega_{l,k}^r$ and $\omega_{l,k}^t$ denote, respectively, the AOA and AOD of the kth ray within the lth cluster. Within each wave cluster, there are totally K rays. Furthermore, $\omega_{l,k}^r$ and $\omega_{l,k}^t$ follow a Laplacian distribution centered at Φ_l^r and Φ_l^t, respectively, as

$$
f_\omega(\omega_{l,k} \mid \Phi_l) = \frac{1}{\sqrt{2}\sigma} e^{-|\sqrt{2}(\omega_{l,k}-\Phi_l)/\sigma|},
\tag{2}
$$

where σ is the angular spread of rays within each cluster. In (1), the amplitude of $\alpha_{l,k}$ associated with the kth ray in the lth cluster is assumed to follow a Rayleigh distribution [11, 13] with the mean power following a double-exponential distribution. That is,

$$
E[\alpha_{l,k}^2] = E[\alpha_{0,0}^2] e^{-\Gamma_l/\Gamma} e^{-\Gamma_{l,k}/\gamma},
\tag{3}
$$

where $E[\cdot]$ is the expectation operator. $\alpha_{0,0}^2$ is the power decay of the first ray of the first cluster with $E[\alpha_{0,0}^2] = 1$. Γ and γ are the constant decay time of cluster and rays, respectively. Γ_l is the delay of the first arrival ray of the lth cluster and $\Gamma_0 = 0$ nanosecond. It is exponentially distributed and conditioned on Γ_{l-1} as

$$
f_\Gamma(\Gamma_l \mid \Gamma_{l-1}) = \Lambda e^{-\Lambda(\Gamma_l - \Gamma_{l-1})}, \quad 0 \le \Gamma_{l-1} < \Gamma_l < \infty.
\tag{4}
$$

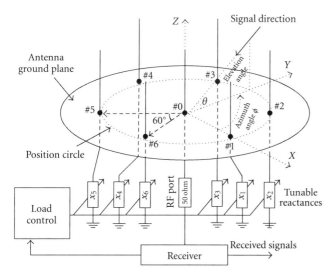

FIGURE 1: Structure of a 7-element ESPAR antenna. The elements are monopoles with interelement spacing of a quarter wavelength.

The delay of rays $\tau_{l,k}$ within each cluster is also exponentially distributed and conditioned on the previous delay $\tau_{l,k-1}$ as

$$f_\tau(\tau_{l,k} \mid \tau_{l,k-1}) = \lambda e^{-\lambda(\tau_{l,k} - \tau_{l,k-1})}, \quad 0 \le \tau_{l,k-1} < \tau_{l,k} < \infty. \tag{5}$$

Here, Λ and λ are the constant arrival rates of clusters and rays, respectively. The statistics of rays and clusters are assumed being independent [11, 13], that is,

$$f_{\tau,\omega}(\tau_{l,k}, \omega_{l,k} \mid \tau_{l,k-1}) = f_\tau(\tau_{l,k} \mid \tau_{l,k-1}) f_\omega(\omega_{l,k}). \tag{6}$$

In this study, we set $\Gamma = 30$ nanoseconds, $\gamma = 20$ nanoseconds, $1/\Lambda = 30$ nanoseconds, $1/\lambda = 5$ nanoseconds, and $\sigma = 20°$ to represent an indoor or urban microscopic environment as in [12].

2.2. ESPAR antenna

In this paper, two parasitic array antennas named as ESPAR [15, 16] are employed at both ends of the wireless link to produce Tx and Rx directional beam patterns. In this subsection, we briefly explain the operation principle of an ESPAR antenna.

Figure 1 shows the structure of a seven-element ESPAR antenna. There, one active central monopole is surrounded by parasitic elements on a circle of radius of a quarter wavelength on the circular grounded base plate. The central monopole is connected to an RF receiver and each parasitic monopole is loaded with a tunable reactance. Let $s(t)$ be the far-field impinging wave from direction ϕ, the output signal at the RF port is written as

$$y(t) = \mathbf{w}_{\text{ESP}}^T \boldsymbol{\alpha}(\phi) s(t) + n(t), \tag{7}$$

where $\boldsymbol{\alpha}(\phi)$ is the 7-by-1 dimensional steering vector defined based upon the array geometry of the ESPAR antenna. It is written as

$$\boldsymbol{\alpha}(\phi) = [\, \text{A1} \;\; \text{A2} \;\; \text{A3} \;\; \text{A4} \;\; \text{A5} \;\; \text{A6} \;\; \text{A7} \,]^T, \tag{8}$$

where A1 $= 1$, A2 $= e^{j(\pi/2)\cos(\phi)}$, A3 $= e^{j(\pi/2)\cos(\phi - \pi/3)}$, A4 $= e^{j(\pi/2)\cos(\phi - 2\pi/3)}$, A5 $= e^{j(\pi/2)\cos(\phi - \pi)}$, A6 $= e^{j(\pi/2)\cos(\phi - 4\pi/3)}$, A7 $= e^{j(\pi/2)\cos(\phi - 5\pi/3)}$, In (7), $n(t)$ denotes an additive white Gaussian noise (AWGN) component. Superscript $(\cdot)^T$ denotes transpose. $\mathbf{w}_{\text{ESP}}$ is written as

$$\mathbf{w}_{\text{ESP}} = z_0 (\mathbf{Y}^{-1} + \mathbf{X})^{-1} \mathbf{u}_1, \tag{9}$$

where $\mathbf{Y}$ is a mutual admittance matrix of array elements. The values of entities of $\mathbf{Y}$ are calculated using a method of moment (MoM) Numerical Electromagnetic Code (NEC) simulator [18] and are given in [16]. Matrix $\mathbf{X}$ is a diagonal matrix given by $\text{diag}[z_0, jx_1, \ldots, jx_6]$, where $x_i(i = 1, 2, \ldots, 6)$ [Ω] is the reactance loaded at the array parasitic elements as shown in Figure 1. The characteristic impedance at the central RF port is $z_0 = 50$ [Ω]. In (9), $\mathbf{u}_1$ is defined as $\mathbf{u}_1 = [\, 1 \;\; 0 \;\; 0 \;\; 0 \;\; 0 \;\; 0 \;\; 0 \,]^T$. From (7), we can see that $\mathbf{w}_{\text{ESP}}$ is equivalent to an array BF weight vector. Here, we write the normalized $\mathbf{w}_{\text{ESP}}$ as

$$\mathbf{w} = \frac{\mathbf{w}_{\text{ESP}}}{\sqrt{\mathbf{w}_{\text{ESP}}^H \mathbf{w}_{\text{ESP}}}}. \tag{10}$$

The normalized azimuthal directional beam pattern for a given weight vector $\mathbf{w}$ is written as

$$g(\phi \mid \mathbf{w}) = \frac{|\mathbf{w}^T \boldsymbol{\alpha}(\phi)|^2}{(1/2\pi) \int_0^{2\pi} \mathbf{w}^T \boldsymbol{\alpha}(\phi) \boldsymbol{\alpha}^H(\phi) \mathbf{w}^* \, d\phi}, \tag{11}$$

where superscript $(\cdot)^*$ is the complex-conjugate. In this study we have neglected the effect of load mismatch. For simplicity, we only consider the impinging waves that are copolarized with the array elements in the azimuthal plane. Since $\mathbf{w}$ is dependent on the tunable reactances loaded at those parasitic elements, changing the values of reactance, the beam adjusts the beam pattern of an ESPAR antenna. This model describes the beam pattern control based on tuning the loaded reactances at the parasitic elements. It enables us to examine the system performance at different setting of reactance values. The validity of this model has been proved by experiments and simulation based on (7)–(11) in [16].

In this paper, we use the ESPAR to generate a few predefined directional beam patterns and employ the antenna as an SBA. When the reactance values are set to $(-9\,000\,000)$ [Ω], the ESPAR antenna forms a directional beam pattern pointing to $0°$ in the azimuthal plane. The 3D beam pattern that is produced using the NEC simulator is shown in Figure 2. The azimuthal plane of this simulated beam pattern using an NEC simulator is compared with the calculated beam pattern with MATLAB using (11) and the measured beam pattern in Figure 3. Since the three beam patterns match closely, we use (11) to describe a beam pattern that is generated by an ESPAR antenna in the following analysis. By shifting the reactance values, the antenna produces six consecutive beam patterns as shown in Figure 4. The reactance values for those six directional beam patterns are listed in Table 1. In the following analysis, we use $\mathbf{w}_m, m = 1, 2, \ldots, 6$, to denote the ESPAR weight vector that is described in (10) for the six directional beam patterns pointing to $(m-1) \times 60°$ in the azimuthal plane. $g(\phi \mid \mathbf{w}_m)$ is the corresponding normalized directional beam pattern as described in (11).

TABLE 1: Reactance values for different directional beam patterns.

Equivalent weight	Beam azimuthal direction	x_1	x_2	x_3	x_4	x_5	x_6
w_1	0°	−90	0	0	0	0	0
w_2	60°	0	−90	0	0	0	0
w_3	120°	0	0	−90	0	0	0
w_4	180°	0	0	0	−90	0	0
w_5	240°	0	0	0	0	−90	0
w_6	300°	0	0	0	0	0	−90

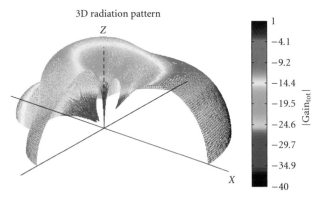

FIGURE 2: Three-dimensional beam pattern of a 7-element ESPAR antenna that is calculated by NEC simulator. X = [−9 000 000] Ω. The directional beam pattern is pointing to 0° in the azimuthal plane.

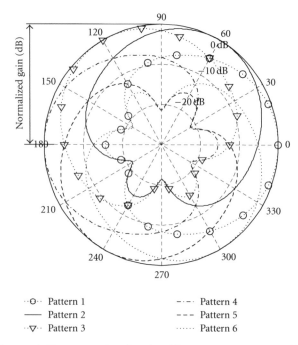

··O·· Pattern 1   —·—· Pattern 4
—— Pattern 2   ——— Pattern 5
··▽·· Pattern 3   ········ Pattern 6

FIGURE 4: Six consecutive directional beam patterns generated by shifting the reactance values.

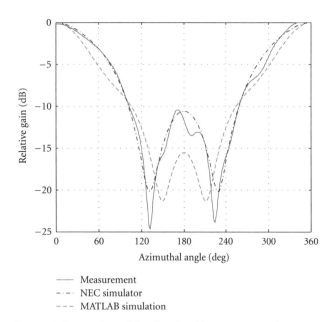

—— Measurement
—·—· NEC simulator
——— MATLAB simulation

FIGURE 3: Comparison of the directional beam patterns of measurement and simulation using MATLAB and NEC simulator.

2.3. BF system

Two ESPAR antennas are employed at both ends of the wireless link. When both the Tx and Rx directional beam patterns are steered towards one of the existing clusters, the link is set up. This prerequisites a cluster identification process.

In [9, 13, 19], wave clusters are identified in spatial-temporal domain through visual observation. In this paper, we assume that the clusters' spatial information is known at both Tx and Rx sides, and the directional beam patterns at both ends of the link can be steered towards a common wave cluster. As shown in Figure 5, there are two clusters of waves at both the Tx and Rx sides. Both Tx and Rx can choose one out of six predefined directional beam patterns pointing to a common wave cluster.

3. PERFORMANCE EVALUATION

Given the above model, the question that arises is how much capacity this system can provide. In order to calculate the channel capacity, in this section, we examine the directivity gain and the beam pattern correlation in a clustering DDRC model. The channel capacity will be given in Section 4.

3.1. DDG

Firstly, we examine the directivity gain of the beam pattern for the system model. Analyzing the directivity gain in

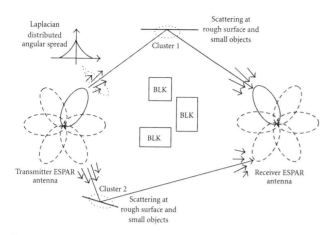

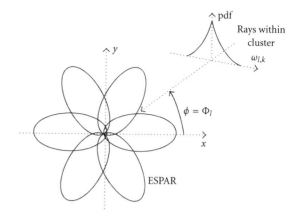

Figure 5: System model of employing directional antennas at both ends of the wireless link in a clustering DDRC.

Figure 6: The impinging waves from a cluster with nominal direction Φ_l on directional beam patterns of the ESPAR antenna.

the presence of angular distribution of the impinging waves invokes calculating the DDG [20]. Since we only consider waves in the azimuthal plane, the DDG can be given as

$$D(\Phi, \sigma, \mathbf{w}) = \eta_{\text{ant}} 2\pi \int_0^{2\pi} f_\omega(\phi \mid \Phi) g(\phi \mid \mathbf{w}) d\phi, \quad (12)$$

where $f_\omega(\phi \mid \Phi)$ is the probability density function (pdf) of the angular distribution of impinging waves and $g(\phi \mid \mathbf{w})$ is given in (11). Here, η_{ant} is the antenna efficiency that is assumed to be unity. For simplicity, we omit η_{ant} in the following analysis.

Figure 6 shows the situation when there is one cluster centered at Φ_l. The AOAs of rays within the cluster follow Laplacian distribution as expressed in (2). As stated previously, the ESPAR antenna can produce six consecutive directional beam patterns by shifting the reactance values. Let us now calculate the DDG for the directional beam pattern that points to $0°$ in the azimuthal plane, that is, we use $\mathbf{w}_1$ in (11), while the nominal direction of the cluster is moving from $0°$ to $180°$. In this case, we can calculate the DDG as

$$D(\Phi, \sigma, \mathbf{w}_1) = \frac{\sqrt{2}\pi \int_0^{2\pi} |\mathbf{w}_1^T \boldsymbol{\alpha}(\phi)|^2 e^{-|\sqrt{2}(\phi-\Phi)/\sigma|} d\phi}{\sigma \int_0^{2\pi} \mathbf{w}_1^T \boldsymbol{\alpha}(\phi) \boldsymbol{\alpha}^H(\phi) \mathbf{w}_1^* d\phi}. \quad (13)$$

The DDG corresponding to different angular spreads is also shown in Figure 7. For the Laplacian distributed scenarios, using a directional beam pattern pointing to a wave cluster provides a significant DDG. However, the DDG degrades as the center of the cluster moves away from the direction of the beam pattern's maximum gain. Please note that when the waves are uniformly distributed over the azimuthal plane, using a directional beam pattern does not provide any DDG.

The DDG calculated in Figure 7 considers only one cluster of impinging waves. For multiple clusters (e.g., two clusters), we can write the angular distribution according to (2) as

$$f_\omega(\phi \mid \Phi_1, \Phi_2) = \frac{1}{2\sqrt{2}\sigma} (e^{-|\sqrt{2}(\phi-\Phi_1)/\sigma_1|} + e^{-|\sqrt{2}(\phi-\Phi_2)/\sigma_2|}). \quad (14)$$

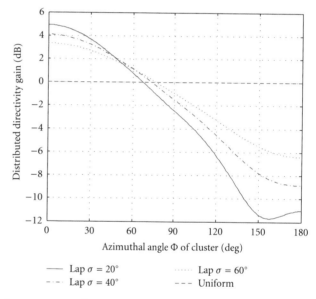

Figure 7: DDG for the directional beam pattern by the ESPAR antennas in different spatially distributed radio channels.

We also follow the assumption in [11, 13] that the angular spreads of all clusters are the same, that is, $\sigma_1 = \sigma_2$. Inserting (14) into (12), we obtain the DDG when there are two clusters as

$$D(\Phi_1, \Phi_2, \sigma, \mathbf{w}_1) = \frac{1}{2}D(\Phi_1, \sigma, \mathbf{w}_1) + \frac{1}{2}D(\Phi_2, \sigma, \mathbf{w}_1). \quad (15)$$

3.2.   Link gain

By now, we have calculated the DDG at one (Tx or Rx) side of the wireless link. Because both the Tx and Rx employ directional beam patterns, the joint directivity gain incorporates the directivity gains at both sides of the link. This is in accordance with the concept of DDRC. Given the SRC of a clustering DDRC in (1) and the beam pattern in (11), the spatial channel model of the BF system is given by

$$h_{\mathrm{BF}} = \sqrt{g(\phi^r \mid \mathbf{w}_m^r)} * h(\phi^r, \phi^t) * \sqrt{g(\phi^t \mid \mathbf{w}_m^t)}$$

$$= \frac{1}{\sqrt{LK}} \sum_{l=0}^{L-1} \sum_{k=0}^{K-1} \alpha_{l,k} \sqrt{g(\Phi_l^r + \omega_{l,k}^r \mid \mathbf{w}_m^r)} \sqrt{g(\Phi_l^t + \omega_{l,k}^t \mid \mathbf{w}_m^t)},$$

$$(16)$$

where $\mathbf{w}_m^r$ and $\mathbf{w}_m^t$ are the BF weight vectors at the Rx side and Tx sides, respectively. Taking the expectation of $|h_{\mathrm{BF}}|^2$ in (16), we obtain the link gain as

$$G = E[h_{\mathrm{BF}} h_{\mathrm{BF}}^H] = \frac{1}{LK} \sum_{l=0}^{L-1} \sum_{k=0}^{K-1} E[\alpha_{l,k} \alpha_{l,k}^*]$$

$$\times E[g(\Phi_l^r + \omega_{l,k}^r \mid \mathbf{w}_m^r)] E[g(\Phi_l^t + \omega_{l,k}^t \mid \mathbf{w}_m^t)].$$

$$(17)$$

For a given cluster realization, we take the expectation over rays. To simplify the model, we assume that $\alpha_{l,k}$ of each ray follows a complex normal distribution $CN(0, |\alpha_{l,0}|^2)$ with zero mean and a variance that is equal to the mean power decay of the first arrival ray within the lth cluster. Thus, we have

$$\frac{1}{K} \sum_{k=0}^{K-1} E(\alpha_{l,k} \alpha_{l,k}^*) = e^{-\Gamma_l/\Gamma}, \quad \text{for } l = 0, 1, \ldots, L-1. \quad (18)$$

Furthermore, according to (4), we assume that $\Gamma_l = l - 1/\Lambda$ for $l = 0, 1, \ldots, L-1$. Therefore, given a channel cluster realization, that is, the locations of clusters, the link gain is given by

$$G = \frac{1}{LK} \sum_{l=0}^{L-1} E(\alpha_{l,k} \alpha_{l,k}^*) \int_0^{2\pi} f_\omega(\phi \mid \Phi_l^r) g(\phi \mid \mathbf{w}_m^r) d\phi$$

$$\times \int_0^{2\pi} f_\omega(\phi \mid \Phi_l^t) g(\phi \mid \mathbf{w}_m^t) d\phi$$

$$= \frac{1}{L} [\ D(\Phi_0^r, \sigma, \mathbf{w}_m^r) \ \ldots \ D(\Phi_{L-1}^r, \sigma, \mathbf{w}_m^r) \ ] \quad (19)$$

$$\times \text{diag}\Big[1 \ e^{-1/\Lambda\Gamma} \ \ldots \ e^{-(L-1)/\Lambda\Gamma} \Big]$$

$$\times [\ D(\Phi_0^t, \sigma, \mathbf{w}_m^t) \ \ldots \ D(\Phi_{L-1}^t, \sigma, \mathbf{w}_m^t) \ ]^T.$$

The physical meanings of (19) are that, firstly, each route through a pair of Tx and Rx directional beam patterns pointing to a common cluster is associated with a mean power decay that increases exponentially with the cluster's delay time. Therefore, directional beam patterns should be steered towards early arrival clusters for a high-average link gain. Secondly, the mean link gain of the BF system decreases as the number of clusters increases. The channel is closer to a multipath-rich environment where the diversity scheme is more advantageous over the BF system. Finally, there exists extra diversity freedom by switching among different pairs of Tx and Rx beams to select different clusters as transmission routes in that the fading statistics of the rays within different clusters are assumed to be independent according to [11, 13]. The correlation between different routes is induced due to beam pattern overlapping.

Now we consider the case where there exist two clusters. For the simplicity of analysis, we arbitrarily set the positions of both clusters at both Tx and Rx sides as $\Phi_0^r = \Phi_0^t = 0°$ and $\Phi_1^r = \Phi_1^t = 120°$. By jointly pointing directional beam patterns towards $0°$ using $\mathbf{w}_1$ at both sides of the link, the link provides a joint gain of

$$G_1 = \frac{1}{2} \{ D(\Phi_0^r, \sigma, \mathbf{w}_1^r) D(\Phi_0^t, \sigma, \mathbf{w}_1^t)$$

$$+ e^{-1/\Lambda\Gamma} D(\Phi_1^r, \sigma, \mathbf{w}_1^r) D(\Phi_1^t, \sigma, \mathbf{w}_1^t) \}. \quad (20)$$

When both Tx and Rx point towards $120°$ using $\mathbf{w}_3$ the link gain is

$$G_2 = \frac{1}{2} \{ D(\Phi_0^r, \sigma, \mathbf{w}_3^r) D(\Phi_0^t, \sigma, \mathbf{w}_3^t)$$

$$+ e^{-1/\Lambda\Gamma} D(\Phi_1^r, \sigma, \mathbf{w}_3^r) D(\Phi_1^t, \sigma, \mathbf{w}_3^t) \}. \quad (21)$$

Therefore, the link gain difference between these two branches can be obtained as

$$\eta = $$
$$\frac{D(\Phi_0^r, \sigma, \mathbf{w}_1^r) D(\Phi_0^t, \sigma, \mathbf{w}_1^t) + e^{-1/\Lambda\Gamma} D(\Phi_1^r, \sigma, \mathbf{w}_1^r) D(\Phi_1^t, \sigma, \mathbf{w}_1^t)}{D(\Phi_0^r, \sigma, \mathbf{w}_3^r) D(\Phi_0^t, \sigma, \mathbf{w}_3^t) + e^{-1/\Lambda\Gamma} D(\Phi_1^r, \sigma, \mathbf{w}_3^r) D(\Phi_1^t, \sigma, \mathbf{w}_3^t)}. \quad (22)$$

The weight vectors $\mathbf{w}_3^r$ and $\mathbf{w}_3^t$ produce directional beam patterns at both ends of the link towards $120°$, whereas $\mathbf{w}_1^r$ and $\mathbf{w}_1^t$ produce directional beam pattern towards $0°$. Therefore, $D(\Phi_0^r, \sigma | \mathbf{w}_3^r)$, $D(\Phi_0^t, \sigma | \mathbf{w}_3^t)$, $D(\Phi_1^r, \sigma | \mathbf{w}_1^r)$, and $D(\Phi_1^t, \sigma | \mathbf{w}_1^t)$ in (22) represent the DDGs when the beam patterns have a $120°$ misalignment. From Figure 7, we know that the DDG values with such misalignment are relatively small. Furthermore, we assume that the beam pattern for $0°$ and $120°$ are same. As stated in the previous section, we assume $\Gamma = 30$ nanoseconds, $1/\Lambda = 30$ nanosecond, $\sigma = 20°$. Thus, (22) is simplified to

$$\eta \approx \frac{D(\Phi_0^r, \sigma, \mathbf{w}_1^r) D(\Phi_0^t, \sigma, \mathbf{w}_1^t)}{e^{-1/\Lambda\Gamma} D(\Phi_1^r, \sigma, \mathbf{w}_3^r) D(\Phi_1^t, \sigma, \mathbf{w}_3^t)} \approx e = 4.34 \,\text{dB}. \quad (23)$$

Therefore, by switching directional beam patterns jointly at both sides of the channel, we set up two wireless links with a 4.34 dB difference of link gains. We approximate the gain of each wireless link as

$$G_1 \approx D(\Phi_0^r, \sigma, \mathbf{w}_1^r) D(\Phi_0^t, \sigma, \mathbf{w}_1^t),$$
$$G_2 \approx e^{-1/\Lambda\Gamma} D(\Phi_1^r, \sigma, \mathbf{w}_3^r) D(\Phi_1^t, \sigma, \mathbf{w}_3^t). \quad (24)$$

From Figure 7, we know that $G_1 = 2 \times 4.9 = 9.8\,\text{dB}$ and $G_2 = 5.5\,\text{dB}$ for the channel with $\sigma = 20°$. Please note that the link gain calculated in our model is normalized such that when omnidirectional antennas are installed at both ends of the link, the link gain G is 0 dB.

3.3. Channel correlation

As explained previously, by jointly switching different Tx and Rx beam patterns, the two wireless links can be exploited to achieve certain diversity advantage. The correlation between the two wireless links is induced due to beam pattern overlapping. In this subsection, we examine the beam pattern correlation.

We assume that there are two clusters in the clustering DDRC as shown in Figure 5. Furthermore, each cluster's center is assumed to be aligned with the direction of a beam pattern. Therefore, there is no beam pattern direction misalignment. Given the distribution of rays $f_\omega(\phi|\Phi_1, \Phi_2)$ in (14), the correlation for the two directional beam patterns $g(\phi|\mathbf{w}_m)$ and $g(\phi|\mathbf{w}_n)$ can be expressed as

$$
\rho_{m,n} = \frac{|\int_0^{2\pi} f_\omega(\phi|\Phi_1, \Phi_2)\mathbf{w}_m^T\boldsymbol{\alpha}(\phi)\boldsymbol{\alpha}^H(\phi)\mathbf{w}_n^* d\phi\,|}{\sqrt{\int_0^{2\pi} f_\omega(\phi|\Phi_1, \Phi_2)g(\phi|\mathbf{w}_m)d\phi}}
$$
$$
\times \frac{|\int_0^{2\pi} f_\omega(\phi|\Phi_1, \Phi_2)\mathbf{w}_m^T\boldsymbol{\alpha}(\phi)\boldsymbol{\alpha}^H(\phi)\mathbf{w}_n^* d\phi\,|}{\sqrt{\int_0^{2\pi} f_\omega(\phi|\Phi_1, \Phi_2)g(\phi|\mathbf{w}_n)d\phi}}, \tag{25}
$$

where $f(\phi)$ is given in (14). The correlation $\rho_{m,n}$ is shown in Figure 8 for different values of σ. The markers in the figure indicate the correlation for six discrete beam patterns, whereas the dashed lines show the correlation for continuously rotated beam patterns. We notice that the correlation decreases as the angular spread of rays increases. The back lobe of the beam patterns as shown in Figure 4 induces the high correlation for two beam patterns pointing the opposite directions (180° angular separation).

Knowing the correlation of any given two beam patterns at each side of the wireless link, the complex correlation of two link channels constructed by jointly pointing directional beam patterns towards the same clusters is given as $\rho = \rho_{m,n}^r\rho_{m,n}^t$. For the channel realization with $\sigma = 20°$, $\Phi_0^r = \Phi_0^t = 0°$, and $\Phi_1^r = \Phi_1^t = 120°$, we obtain the channel correlation as $\rho = 0.8 \times 0.8 = 0.64$.

Using the DDG and beam pattern correlation, we set up an analytical model of the BF system. In the following section, we examine the channel capacity of the BF system based on this analytical model.

4. NUMERICAL RESULT

In Section 3, we have obtained the link gains and the correlation of two wireless links. In this section, we examine the link performance of the BF system and compare it with a MIMO system. To simplify the performance evaluation process and make the system independent of particular modulation schemes and coding techniques, we use the Shannon's theoretical channel capacity and the outage rate to evaluate the link performance and investigate the benefits from beam pattern diversity gain.

4.1. Outage probability relative to output SNR

In the case of two clusters, after the Tx and Rx have identified the clusters, two link channels can be established by jointly directing beam patterns at both Tx and Rx sides towards the common cluster. Switching between these two channels pro-

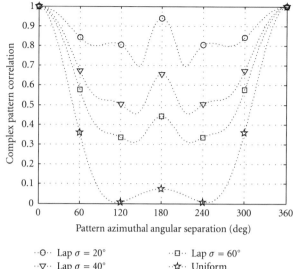

FIGURE 8: Complex correlation of beam patterns in the presence of a given angular distribution property of the channel.

vides a diversity gain. In this paper, we denote the link from Tx to Rx through the cluster at $\Phi_0^r = \Phi_0^t = 0°$ as channel no.1, and denote that through the cluster at $\Phi_1^r = \Phi_1^t = 120°$ as channel no.2. In the previous subsections, we have calculated that their complex envelop correlation is 0.8×0.8, and the difference of link gains is 4.34 dB. Without selection combining (SC), each link experiences a Rayleigh fading process. The outage rate of the output SNR relative to the input SNR at each channel is given by

$$
\Pr(r < x) = F_r(x) = 1 - \exp\left(-\frac{x}{G_1}\right), \quad \text{for } i = 1, \text{ or } 2, \tag{26}
$$

where r is the instantaneous link gain.

The outage rate with the SC of these two channels is obtained by calculating

$$
\Pr(r < x) = F_r(x) = 1 - \exp\left(-\frac{x}{G_1}\right)Q(b, a|\rho|)
$$
$$
- \exp\left(-\frac{x}{G_2}\right)(1 - Q(b|\rho|, a)). \tag{27}
$$

$Q(a, b)$ is the Marcum's Q-function:

$$
Q(a, b) = \int_b^\infty \exp\left\{-1/2(a^2 + x^2)\right\}I_0(ax)x\,dx, \tag{28}
$$

and I_0 is the modified Bessel function:

$$
a = \sqrt{\frac{2x}{G_1(1 - |\rho|^2)}},
$$
$$
b = \sqrt{\frac{2x}{G_2(1 - |\rho|^2)}}. \tag{29}
$$

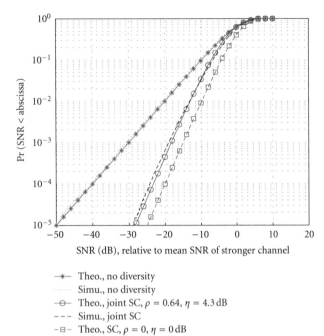

FIGURE 9: The outage rates of different channels using directional antennas at both ends of the wireless links.

The outage rates of relative output SNR from the channels without and with SC technique are shown in Figure 9, respectively. Given the DDGs for the two channels and their correlation in the presence of two wave clusters, we calculate the theoretical outage rate with the SC of two channels using (26) and (27). The curve without a diversity approach shows the typical Rayleigh fading. The curve with the square markers shows the situation when ideal SC of two channels with equal SNRs and independent fading processes is employed. Compared to the curve without diversity, we can see that even the two channels have a 4.3 dB difference of link gains and a correlation of $\rho = 0.64$, significant diversity gain is still achieved by jointly switching between different pairs of Tx and Rx directional beam patterns.

Simulation is also carried out for the same channel realization. Totally, 1000 ray realizations are created. In addition, for each ray realization, 1000 fading channels are randomly generated. At each instant channel realization, the Tx and Rx ESPAR antennas jointly switch to a common cluster that gives the maximum instantaneous link gain. The output link gain after SC is used to compute the outage rate. The simulation results match closely with the theoretical results and show significant performance. This verifies our analytical model. In the following analysis, we will use this analytical model to examine the performance of the BF system and compare it with a MIMO system.

Please note that the difference between the theoretical and simulated results is due to (a) the simplification of distribution of amplitude of rays within each cluster as a normal distribution with the variance equal to the mean power decay of the first arrival ray in (18): (b) the approximation of the difference between the link gains of two channels in (23).

4.2. Channel capacity

Given an instantaneous link gain r and the average SNR, the channel capacity of the BF system is given by

$$C_{\mathrm{BF}} = \log(1 + r\mathrm{SNR}). \tag{30}$$

Using the outage probabilities for the BF system without and with diversity given in (26) and (27), respectively, we obtain the outage probably of C_{BF} as

$$\Pr(C_{\mathrm{BF}} < C_{\mathrm{out}}) = F_{\mathrm{BF}}(C_{\mathrm{out}}) = \Pr\left(r < \frac{2^{C_{\mathrm{out}}} - 1}{\mathrm{SNR}}\right)$$
$$= F_r\left(\frac{2^{C_{\mathrm{out}}} - 1}{\mathrm{SNR}}\right). \tag{31}$$

In order to compare the performance of the BF system with that of a MIMO system, we also numerically obtain the channel capacity of a MIMO system that is set up with linear antenna arrays of half-wavelength interelement spacing at both ends of the wireless link. The MIMO channel capacity is dependent on the array orientation with respect to the position of wave clusters [21]. To compare the BF system and the MIMO system at a given cluster realization, we assume that at each side of the link, the directions of the two clusters with respect to the end-fire direction of the linear array are 0° and 120°, respectively. All the elements of the linear arrays have the same omnidirectional beam pattern. Thus, the DDG of each element is 0 dB. The steering vector of the linear array that consists of M array elements reads

$$\boldsymbol{\alpha}(\phi) = \begin{bmatrix} 1 & e^{-j\pi\cos(\phi)} & \dots & e^{-j\pi(M-1)\cos(\phi)} \end{bmatrix}^T. \tag{32}$$

We employ Wallace's method [11] to obtain the MIMO channel model in a clustering DDRC. In the following analysis, we refer to this MIMO channel model as Wallace's MIMO model. Given the SRC of a clustering DDRC in (1), the spatial channel model for the MIMO system in the clustering DDRC is given by

$$\mathbf{H} = \boldsymbol{\alpha}(\phi^r) * h(\phi^r, \phi^t) * \boldsymbol{\alpha}^T(\phi^t), \tag{33}$$

with each matrix entry being normalized as $E[|\mathbf{H}(i,j)|^2] = 1$. In this study, we assume that the linear arrays at both ends of the link have two elements, that is, $M = 2$. For the MIMO system in an ideal statistical model [7], the entries of $\mathbf{H}$ are given as independent and identically distributed (iid) complex Gaussian random variables (RV), that is, $\mathbf{H}(i,j) \sim CN(0,1)$. We here after refer to this MIMO channel model as Foschini's MIMO channel. The channel capacity of a 2×2 MIMO system is given by

$$C_{\mathrm{MIMO}} = \log_2 \det\left(\mathbf{I} + \frac{\mathrm{SNR}}{2}\mathbf{H}\mathbf{H}^H\right), \tag{34}$$

where superscript $(\cdot)^H$ denotes conjugate transpose, and $\mathbf{I}$ is an identity matrix. Furthermore, $\det(\cdot)$ denotes matrix determinant.

The channel capacities for both BF and MIMO systems when SNR = 10 dB are shown in Figure 10. The line "omni-omni, gain = 0 dB" refers to the situation when one omnidirectional antenna (DDG is 0 dB) is installed at each side of

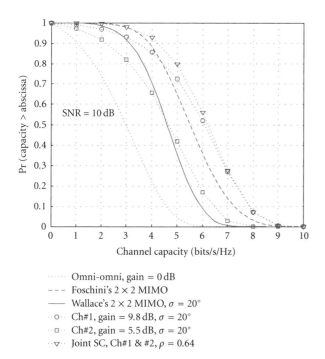

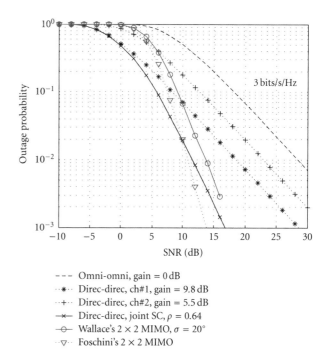

Figure 10: Shannon's theoretical channel capacities of both BF and MIMO systems in different scenarios.

Figure 11: Outage rates of channel capacity at 3 bits/sec/Hz for both BF and MIMO systems in different scenarios.

the link. The dashed line shows the theoretical channel capacity of the Foschini's 2×2 MIMO system as in [7]. This channel capacity is predicted based on the assumption of rich multipath. The solid line in Figure 10 shows the channel capacity of the Wallace's 2×2 MIMO system in a clustering DDRC model [11]. As expected, in a clustering channel, the channel capacity of a MIMO system is lower than that predicted in ideal statistical model.

When directional beam patterns are jointly employed towards the first arrival cluster, the Tx and Rx directional antennas jointly provide a DDG of $G_1 = 9.8$ dB to the wireless channel, which significantly increases the channel capacity. It is interesting to note that the ergodic capacity achieved by the BF system is even higher than that of the Foschini's 2×2 MIMO channel [7]. As shown in Figure 10, SC of these two links channels does not provide significant gain in terms of ergodic capacity. This is consistent with the conclusion in [22] that the influence of the diversity gain on the ergodic channel capacity is negligible. However, as will be shown in the following analysis, the diversity advantage of SC does significantly reduce the outage probability, thus improving the stability of wireless transmission.

Figure 11 shows the outage rate for $C_{out} = 3$ bits/sec/Hz at different SNR situations. Compared with the curve of a scalar Rayleigh fading channel "omni-omni, gain = 0 dB," that is, one onmidirectional antenna is installed at each end of the wireless link, the outage curves for both channel #1 and channel #2 are moved to the left by the joint DDG. This is due to the fact that the DDG obtained from using directional antennas directly increase the average SNR of the wireless link. When SC of the two channels is employed, the outage probability is further reduced by the diversity gain. Thus,

the required SNR to achieve the same outage rate at a given channel capacity is significantly reduced.

In the same figure, the outage probabilities at $C_{out} = 3$ bits/s/Hz of the 2×2 MIMO system in both ideal statistical channel and clustering DDRC are also plotted. It is interesting to note that in a low-SNR environment, the BF system with SC outperforms the Wallace's MIMO system in terms of the outage probability. This shows that in a low-SNR environment, the BF system provides a better wireless link in terms of stability than that of a MIMO system in the same clustering environment.

5. CONCLUSION AND DISCUSSION

In this paper, we proposed a BF system that directs both Tx and Rx directional beam patterns towards a common wave cluster. In a clustering DDRC, we obtained the DDG and beam pattern correlation for the BF system. Based on these results, we built an analytical model through which the channel capacity of the BF system was obtained. The close match between the simulation and analytical results justifies our model. Using the outage probability of the theoretical channel capacity, we have shown that the DDG of the BF directly increases the mean SNR of the wireless link, thus achieving a direct increase of the ergodic channel capacity. By jointly switching between different pairs of Tx and Rx directional beam patterns towards different wave clusters, the system also provides diversity gain to combat against multipath fading, thus reducing the outage probability of the channel capacity.

By comparing the proposed BF system with a MIMO system, we have shown that for a given cluster realization in a clustering DDRC, the BF system provides a higher-ergodic channel capacity. This is because of the significant DDG that is achieved by employing directional antennas in a clustering channel. Furthermore, in a low-SNR environment, the BF system that employs SC outperforms the MIMO system in terms of outage probability of the channel capacity. Thus, the BF system provides a wireless link that is of higher stability than that of the MIMO system.

From the viewpoint of practical implementations of adaptive antennas, if the beamforming network (BFN) is implemented at the RF stage, we can employ a directional antenna at a transceiver that has only one RF channel. This simplifies the system structure and reduces the DC power consumption. As an implementation example, the ESPAR antennas are employed in this study. The MIMO systems, on the other hand, require independent RF channels, increasing system complexity and cost. Since performance gains similar to MIMO can be obtained with the less expensive BF system for realistic environments with cluster-type scattering, the results suggest that BF is a good candidate for advanced indoor wireless communications networks.

REFERENCES

[1] K. Sheikh, D. Gesbert, D. Gore, and A. Paulraj, "Smart antennas for broadband wireless access networks," *IEEE Communications Magazine*, vol. 37, no. 11, pp. 100–105, 1999.

[2] J. Litva and T. K.-Y. Lo, *Digital Beamforming in Wireless Communications*, Artech House, Boston, Mass, USA, 1996.

[3] C. Sun and N. C. Karmakar, "Adaptive array antennas," in *Encyclopedia of RF and Microwave Engineering*, K. Chang, Ed., p. 5832, John Wiley & Sons, New York, NY, USA, 2005.

[4] C. Passerini, M. Missiroli, G. Riva, and M. Frullone, "Adaptive antenna arrays for reducing the delay spread in indoor radio channels," *Electronics Letters*, vol. 32, no. 4, pp. 280–281, 1996.

[5] D. G. Brennan, "Linear diversity combining techniques," *Proceedings of the IEEE*, vol. 91, no. 2, pp. 331–356, 2003.

[6] M. Schwartz, W. R. Bennett, and S. Stein, *Communication Systems and Techniques*, McGraw-Hill, New York, NY, USA, 1966.

[7] G. J. Foschini and M. J. Gans, "On limits of wireless communications in a fading environment when using multiple antennas," *Wireless Personal Communications*, vol. 6, no. 3, pp. 311–335, 1998.

[8] A. F. Naguib, N. Seshádri, and A. R. Calderbank, "Increasing data rate over wireless channels," *IEEE Signal Processing Magazine*, vol. 17, no. 3, pp. 76–92, 2000.

[9] M. Steinbauer, A. F. Molisch, and E. Bonek, "The double-directional radio channel," *IEEE Antennas & Propagation Magazine*, vol. 43, no. 4, pp. 51–63, 2001.

[10] C. Sun and N. C. Karmakar, "Direction of arrival estimation with a novel single-port smart antenna," *EURASIP Journal on Applied Signal Processing*, vol. 2004, no. 9, pp. 1364–1375, 2004.

[11] J. W. Wallace and M. A. Jensen, "Modeling the indoor MIMO wireless channel," *IEEE Transactions on Antennas and Propagation*, vol. 50, no. 5, pp. 591–599, 2002.

[12] A. Saleh and R. Valenzuela, "A statistical model for indoor multipath propagation," *IEEE Journal on Selected Areas in Communications*, vol. 5, no. 2, pp. 128–137, 1987.

[13] Q. H. Spencer, B. D. Jeffs, M. A. Jensen, and A. L. Swindlehurst, "Modeling the statistical time and angle of arrival characteristics of an indoor multipath channel," *IEEE Journal on Selected Areas in Communications*, vol. 18, no. 3, pp. 347–360, 2000.

[14] M. A. Jensen and J. W. Wallace, "A review of antennas and propagation for MIMO wireless communications," *IEEE Transactions on Antennas and Propagation*, vol. 52, no. 11, pp. 2810–2824, 2004.

[15] T. Ohira and J. Cheng, "Analog smart antennas," in *Adaptive Antenna Arrays: Trends and Applications*, S. Chandran, Ed., pp. 184–204, Springer, New York, NY, USA, 2004.

[16] C. Sun, A. Hirata, T. Ohira, and N. C. Karmakar, "Fast beamforming of electronically steerable parasitic array radiator antennas: theory and experiment," *IEEE Transactions on Antennas and Propagation*, vol. 52, no. 7, pp. 1819–1832, 2004.

[17] H. Kawakami and T. Ohira, "Electrically steerable passive array radiator (ESPAR) antennas," *IEEE Antennas & Propagation Magazine*, vol. 47, no. 2, pp. 43–50, 2005.

[18] http://www.qsl.net/wb6tpu/swindex.html.

[19] C.-C. Chong, C.-M. Tan, D. I. Laurenson, S. McLaughlin, M. A. Beach, and A. R. Nix, "A new statistical wideband spatio-temporal channel model for 5-GHz band WLAN systems," *IEEE Journal on Selected Areas in Communications*, vol. 21, no. 2, pp. 139–150, 2003.

[20] R. Vaugh and J. B. Andersen, *Channels, Propagation and Antennas for Mobile Communications*, IEE Electromagnetic Waves Series, Institution of Electrical Engineers, Glasgow, UK, 2003.

[21] X. Li and Z.-P. Nie, "Effect of array orientation on performance of MIMO wireless channels," *IEEE Antennas and Wireless Propagation Letters*, vol. 3, no. 1, pp. 368–371, 2004.

[22] C. G. Günther, "Comment on "estimate of channel capacity in Rayleigh fading environment"," *IEEE Transactions on Vehicular Technology*, vol. 45, no. 2, pp. 401–403, 1996.

Hindawi Publishing Corporation
EURASIP Journal on Wireless Communications and Networking
Volume 2007, Article ID 25757, 12 pages
doi:10.1155/2007/25757

Research Article

Inter- and Intrasite Correlations of Large-Scale Parameters from Macrocellular Measurements at 1800 MHz

Niklas Jaldén, Per Zetterberg, Björn Ottersten, and Laura Garcia

ACCES Linnaeus Center, KTH Signal Processing Lab, Royal Institute of Technology, 100 44 Stockholm, Sweden

Received 15 November 2006; Accepted 31 July 2007

Recommended by A. Alexiou

The inter- and intrasite correlation properties of shadow fading and power-weighted angle spread at both the mobile station and the base station are studied utilizing narrowband multisite MIMO measurements in the 1800 MHz band. The influence of the distance between two base stations on the correlation is studied in an urban environment. Measurements have been conducted for two different situations: widely separated as well as closely located base stations. Novel results regarding the correlation of the power-weighted angle spread between base station sites with different separations are presented. Furthermore, the measurements and analysis presented herein confirm the autocorrelation and cross-correlation properties of the shadow fading and the angle spread that have been observed in previous studies.

1. INTRODUCTION

As the demand for higher data rates increases faster than the available spectrum, more efficient spectrum utilization methods are required. Multiple antennas at both the receiver and the transmitter, so-called multiple input multiple output (MIMO) systems, is one technique to achieve high spectral efficiency [1, 2]. Since multiantenna communication systems exploit the spatial characteristics of the propagation environment, accurate channel models incorporating spatial parameters are required to conduct realistic performance evaluations. Since future systems may reuse frequency channels within the same cell to increase system capacity, the characterization of the communication channel, including correlation properties of spatial parameters, becomes more critical. Several measurement campaigns have been conducted to develop accurate propagation models for the design, analysis, and simulation of MIMO wireless systems [3–9]. Most of these studies are based on measurements of a single MIMO link (one mobile and one base station). Thus, these measurements may not capture all necessary aspects required for multiuser MIMO systems. From the measurement data collected, several parameters describing the channel characteristics can be extracted. This work primarily focusses on extracting some key parameters that capture the most essential characteristics of the environment, and that later can be used

to generate realistic synthetic channels with the purpose of link level simulations. To evaluate system performance with several base stations (BS) and mobile stations (MS), it has generally been assumed that all parameters describing the channels are independent from one link (single BS to single MS) to another [3, 10]. However, correlation between the channel parameters of different links may certainly exist, for example, when one BS communicates with two MSs that are located in the same vicinity, or vice versa. In this case, the radio signals propagate over very similar environments and hence, parameters such as shadow fading and/or spread in angle of arrival should be very similar. This has also been experimentally observed in some work where the autocorrelation of the so-called large scale (LS) is studied. These LS parameters, such as shadow fading, delay spread, and angle spread, are shown to have autocorrelation that decreases exponentially with a decorrelation distance of some tenths of meters [11, 12]. High correlation of these parameters is expected if the MS moves within a small physical area. We believe that this may also be the case for multiple BSs that are closely positioned. The assumption that the channel parameters for different links are completely independent may result in over/under estimation of the performance of the multiuser systems. Previous studies [13–15] have investigated the shadow fading correlation between two separate base station sites and found substantial correlation for closely located base stations. However, the intersite correlation of angle spreads has not been studied previously. Herein, multisite MIMO measurements have been conducted to address this issue. We investigate the existence of correlation between LS parameters on separate links using data collected in two extensive narrow-band measurement campaigns. The intra- and intersite correlations of the shadow fading and the power-weighted angle spread at the base and mobile stations are investigated. The analysis provides unique correlation results for base- and mobile-station angle spreads as well as log-normal (shadow) fading.

The paper is structured as follows: in Section 2 we give a short introduction to the concept of large-scale parameters and in Section 3 some relevant previous research is summarized. The two measurement campaigns are presented in Section 4. In Section 5, we state the assumptions on the channel model while Section 6 describes the estimation procedure. The results are presented in Section 7 and conclusions are drawn in Section 8.

2. INTRODUCTION TO LARGE-SCALE PARAMETERS

The wireless channel is very complex and consists of time varying multipath propagation and scattering. We consider channel modeling that aims at characterizing the radio media for relevant scenarios. One approach is to conduct measurements and "condense" the information of typical channels into a parameterized model that captures the essential statistics of the channel, and later create synthetic data with the same properties for evaluating link and system-level performance, and so on. Large-scale parameters are based on this concept. The term large-scale parameters was used [3] for a collection of quantities that can be used to describe the characteristics of a MIMO channel. This collection of parameters are termed large scale because they are assumed to be constant over "large" areas of several wavelengths. Further, these parameters are assumed to depend on the local environment of the transmitter and receiver. Some of the possible LS parameters are listed below:

   (i) shadow fading,

  (ii) angle of arrival (AoA) angle spread,

 (iii) angle of departure (AoD) angle spread,

 (iv) AoA elevation spread,

  (v) AoD elevation spread,

 (vi) cross polarization ratio,

(vii) delay spread.

This paper investigates only the shadow fading and the angle spread parameters. Shadow fading describes the variation in the received power around some local mean, which depends on the distance between the transmitter and receiver; see Section 6.1. The power-weighted angle spread describes the size of the sector or area from which the majority of the power is received. The angle spread parameter will be different for the transmitter (Tx) and receiver (Rx) sides of the link, since it largely depends on the amount of local scat-

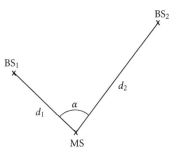

FIGURE 1: Model of the cross-correlation as a function of the relative distance and angle separation, also proposed in [16].

tering; see further in Section 5. A description of the other LS parameters may be found in [3].

3. PREVIOUS WORK

An early paper by Graziano [13] investigates the correlation of shadow fading in an urban macrocellular environment between one MS and two BSs. The correlation is found to be approximately 0.7-0.8 for small angles ($\alpha < 10°$), where α is defined as displayed in Figure 1. Later, Weitzen argued in [14] that the correlation for the shadow fading can be much less than 0.7 even for small angles, in disagreement with the results presented by Graziano. This was illustrated by analyzing measurement data collected in the downtown Boston area using one custom made MS and several pairs of BSs from an existing personal communication system. These results are reasonable since in most current systems the BS sites are widely spread over an area. If the angle α separating the two BSs is small, the relative distance is large, and a small relative distance corresponds to a large angle separation. Thus, a more appropriate model for the correlation of the shadow fading parameter is to assume that it is a function of the relative distance $d = \log_{10}(d_1/d_2)$ between the two BSs and the angle α separating them as proposed in [16]. The distances d_1 and d_2 are defined as in Figure 1. Further studies on the correlation of shadow fading between several sites can be found in, for example, [15, 17–19].

The angular spread parameter has been less studied. In [12], the autocorrelation of the angle spread at a single base station is studied and found to be well modeled by an exponential decay, and the angle spread is further found to be negatively correlated with shadow fading. However, to the authors' knowledge, the intersite correlation of the angle spread at the MS or BS has not been studied previously. Herein, we extend the analysis performed on the 2004 data in [20]. We also investigate data collected in 2005 and find substantial correlation between the shadow fading but less between the angular spreads. The low correlation of the spatial parameters may be important for future propagation modeling. The angle spread at the mobile station is studied and a distribution proposed. Further, we find that the correlation between the base station and mobile station angular spreads (of the same link) is significant for elevated base stations but virtually zero for base stations just above rooftop.

4. MEASUREMENT CAMPAIGNS

Two multiple-site MIMO measurement campaigns have been conducted by KTH in the Stockholm area using custom built multiple antenna transmitters and receivers. These measurements were carried out in the summer of 2004 and the autumn of 2005 and will in the following be refereed to as the 2004 and 2005 campaigns.

Because of measurement equipment shortcomings, the measured MIMO channels have unknown phase rotations. This is due to small unknown frequency offsets. In the 2004 campaign, these phase rotations are introduced at the mobile side and therefore the relation between the measured channel and the true channel is given by

$$\mathbf{H}_{\text{measured, 2004}} = \Lambda_f \mathbf{H}_{\text{true}}, \tag{1}$$

where $\Lambda_f = \text{diag}(\exp(j2\pi f_1 t), \dots, \exp(j2\pi f_n t))$ and $f_1, \dots, f_n$ are unknown. Similarly, the campaign of 2005 has unknown phase rotations at the base station side[1] resulting in the following relation:

$$\mathbf{H}_{\text{measured, 2005}} = \mathbf{H}_{\text{true}} \Lambda_f. \tag{2}$$

The frequencies changed up 5 Hz per second. However, the estimators that will be used are designed with these shortcomings in mind.

4.1. Measurement hardware

The hardware used for these measurements is the same as the hardware described in [21, 22]. The transmitter continuously sends a unique tone on each antenna in the 1800 MHz band. The tones are separated 1 kHz from each other. The receiver downconverts the signal to an intermediate frequency of 10 kHz, samples and stores the data on a disk. This data is later postprocessed to extract the channel matrices. The system bandwidth is 9.6 kHz, which allows narrow-band channel measurements with high sensitivity. The offline and narrow-band features simplify the system operation, since neither real-time constrains nor broadband equalization is required. For a thorough explanation of the radio frequency hardware, [23] may be consulted.

4.2. Antennas

In both measurements campaigns, Huber-Suhner dual-polarized planar antennas with slanted linear polarization ($\pm 45°$), SPA 1800/85/8/0/DS, were used at both the transmitter and the receiver. However, only one of the polarizations ($+45°$) was actually used in these measurements. The antennas were mounted in different structures on the mobile and base stations as described below. For more information on the antenna radiation patterns and so on, see [24].

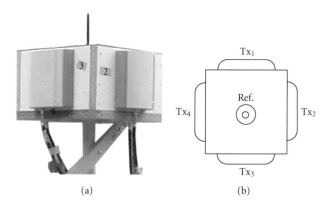

(a) (b)

FIGURE 2: Mobile station box antenna.

4.2.1. Base satation array

At the base station, the antenna elements were mounted on a metal plane to form a uniform linear array with 0.56 wavelength (λ) spacing. In the 2004 campaign, an array of four by four elements was used at the BS. However, the "columns" were combined using 4 : 1 combiners to produce four elements with higher vertical gain. The base stations in the 2005 campaign were only equipped with 2 elements.

4.2.2. Mobile station array

At the mobile side, the four antenna elements were mounted on separate sides of a wooden box as illustrated in Figure 2. This structure is similar to the uniform linear array using four elements. A wooden box is used so that the antenna radiation patterns are unaffected by the structure.

4.3. 2004 campaign

In this campaign uplink measurements were made using one 4-element box-antenna transmitter at the MS, see Figure 2, and three 4-element uniform linear arrays (ULA), with an antenna spacing of 0.56λ, at the receiving BSs. The BSs covered 3 sectors on two different sites. Site 1, Kårhuset-A, had one sector while site 2, Vanadis, had two sectors, B and C, separated some 20 meters and with boresights offset 120-degrees in angle. We define a sector by the area seen from the BS boresight $\pm 60°$. The environment where the measurements where conducted can be characterized as typical European urban with mostly six to eight storey stone buildings and occasional higher buildings and church towers. Figure 3 shows the location of the base station sites and the route covered by the MS. The BS sectors are displayed by the dashed lines in the figure, and the arrow indicates the antenna pointing direction. Sector A is thus the area seen between the dashed lines to the west of site Kårhuset. Sector B and sector C are the areas southeast and northeast of site Vanadis, respectively. A more complete description of the transmitter hardware and measurement conditions can be found in [25].

---

[1] In the 2004 campaign, the phase rotations are due to drifting and unlocked local oscillators in the four mobile transmitters, while in the 2005 campaign they are due to drifting sample-rates in the D/A and A/D converters.

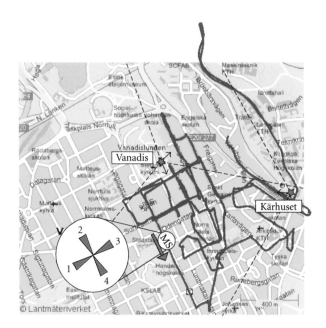

FIGURE 3: Measurement geography and travelled route for 2004 campaign.

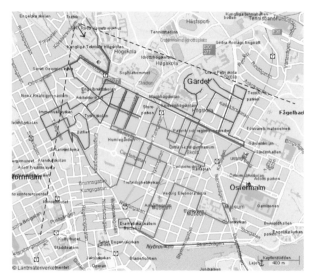

FIGURE 4: Measurement map and travelled route for the 2005 campaign.

4.4. 2005 campaign

In contrast to the previous campaign, the 2005 campaign collected data in the downlink. Two BSs with two antennas each were employed (the same type of antenna elements as in the 2004 campaign was used), each transmitting, simultaneously, one continuous tone separated 1 kHz in the 1800 MHz band. The two base stations were located on the same roof separated 50 meters, with identical boresight and therefore covering almost the same sector. The characteristics of the environment in the measured area are the same as 2004. The routes were different but with some small overlap. The MS was equipped with the 4-element box antenna as was used

in 2004, see Figure 2, to get a closer comparison between the two campaigns. In Figure 4, we see the location of the two BSs (in the upper left corner) and the measured trajectory which covered a distance of about 10 km. The arrow in the figure indicates the pointing direction of the base station antennas. The campaign measurements were conducted during two days, and the difference in color of the MS routes depicts which area was measured which day. The setups were identical on these two days.

5. PRELIMINARIES

Assume we have a system with M Tx antennas at the base station and K Rx antennas at the mobile station. Let $h_{k,m}(t)$ denote the narrow-band MIMO channel between the kth receiver antenna and the mth transmitter antenna. The narrow-band MIMO channel matrix is then defined as

$$\mathbf{H}(t) = \begin{pmatrix} h_{1,1}(t) & h_{1,2}(t) & \dots & h_{1,M}(t) \\ h_{2,1}(t) & \ddots & & \vdots \\ \vdots & & \ddots & \vdots \\ h_{K,1}(t) & \dots & \dots & h_{K,M}(t) \end{pmatrix}. \quad (3)$$

The channel is assumed to be composed of N propagation rays. The nth ray has angle of departure θ_k, angle of arrival α_k, gain g_k, and Doppler frequency f_k. The steering vector[2] of the transmitter given by $\mathbf{a}^{\mathrm{Tx}}(\theta_k)$ and that of the receiver is $\mathbf{a}^{\mathrm{Rx}}(\alpha_k)$. Thus, the channel is given by

$$\mathbf{H} = \sum_{k=1}^{N} g_k e^{j2\pi f_k t} \mathbf{a}^{\mathrm{Rx}}(\alpha_k) \left(\mathbf{a}^{\mathrm{Tx}}(\theta_k) \right)^H. \quad (4)$$

The ray parameters $(\theta_k, \alpha_k, g_k, \text{and } f_k)$ are assumed to be slowly varying and approximately constant for a distance of 30λ. Below, we define the shadow fading and the base station and the mobile station angle spread.

5.1. Shadow fading

The measured channel matrices are normalized so that they are independent of the transmitted power. The received power, P_{Rx}, at the MS is defined as

$$P_{\mathrm{Rx}} = E|\mathbf{H}|^2 P_{\mathrm{Tx}} = \sum_{k=1}^{N} |g_k|^2 |\mathbf{a}^{\mathrm{BS}}(\theta_k)|^2 |\mathbf{a}^{\mathrm{MS}},(\alpha_k)|^2 P_{\mathrm{Tx}}, \quad (5)$$

where P_{Tx} is the transmit power. The ratio of the received and the transmitted powers is commonly assumed to be related as [26]

$$\frac{P_{\mathrm{Rx}}}{P_{\mathrm{Tx}}} = \frac{K}{R^n} S_{\mathrm{SF}}, \quad (6)$$

---

[2] The steering vector $\mathbf{a}(\theta)$ can be seen a complex-valued vector of length equal to the number of antenna elements in the array. The absolute value of the kth element is the square root of the antenna gain of that element and the phase shift of the element relative to some common reference point. That is $a_k(\theta) = \sqrt{\tilde{a}_k(\theta)} e^{j\phi_k}$.

where K is a constant, proportional to the squared norms of the steering vectors that depend on the gain at the receiver and transmitter antennas as well as the carrier frequency, base station height, and so on. The distance separating the transmitter and receiver is denoted by R. The variable S_{SF} describes the slow variation in power, usually termed shadow fading, and is due to obstacles and obstruction in the propagation path. Expressing (6) in decibels (dB) and rearranging the terms in the path loss, which describe the difference between transmitted and received powers, we have

$$L = 10\log_{10}(P_{Tx}) - 10\log_{10}(P_{Rx})$$
$$= n10\log_{10}(R) - 10\log_{10}(K) - 10\log_{10}(S_{SF}), \quad (7)$$

where the logarithm is taken with base ten. Thus, the path loss is assumed to be linearly decreasing with log-distance separating the transmitter and receiver when measured in dB.

5.2. Base station power-weighted angle spread

The power-weighted angle spread at the base station, $\sigma^2_{AS,BS}$, is defined as

$$\sigma^2_{AS,BS} = \sum_{k=1}^{N} p_k (\theta_k - \bar{\theta})^2, \quad (8)$$

where $p_k = |g_k|^2$ is the power of the kth ray and the mean angle $\bar{\theta}$ is given by

$$\bar{\theta} = \sum_{k=1}^{N} p_k \bar{\theta}_k. \quad (9)$$

5.3. Mobile station power-weighted angle spread

The power-weighted angle spread at the mobile station, $\sigma^2_{AS,MS}$, is defined as

$$\sigma^2_{AS,MS} = \min_{\bar{\alpha}} \left\{ \frac{1}{\sum_{k=1}^{N} p_k} \sum_{k=1}^{N} p_k \left(\widetilde{\text{mod}}(\alpha_k - \bar{\alpha}) \right)^2 \right\}, \quad (10)$$

where $\widetilde{\text{mod}}$ is short for modulo and defined as

$$\widetilde{\text{mod}}(\alpha) = \begin{cases} \alpha + 180, & \text{when } \alpha < -180, \\ \alpha, & \text{when } |\alpha| < 180, \\ \alpha - 180, & \text{when } \alpha > 180. \end{cases} \quad (11)$$

The definition of the MS angle spread is equivalent to the circular spread definition in [10, Annex A]. In the following, the power-weighted angle spread will be refereed to as the angle spread.

6. PARAMETER ESTIMATION PROCEDURES

In the measurement equipment, the receiver samples the channel on all Rx antennas simultaneously at a rate which provides approximately 35 channel realizations per wavelength. The first step of estimating the LS parameters is

TABLE 1: Number of measured 30λ segments from each measuremaent campaign, and number of segments in each **BS** sector.

All data 2004	S_A	S_B	S_C	All data 2005
2089	1742	1636	453	1637

to segment the data into blocks of length 30λ. This corresponds to approximately a $5m$ trajectory, during which the ray-parameters are assumed to be constant, [12], and therefore the LS parameters are assumed to be constant as well. Then smaller data sets for each BS are constructed such that they only contain samples within the given BS's sector and blocks outside the BS's sector of coverage are discarded; see definition in Section 4.3. Table 1 shows the total number of measured 30λ segments from the campaigns as well as the number of segments within each BS sector.

6.1. Estimation of shadow fading

The fast fading due to multipath scattering varies with a distance on the order of a wavelength [26]. Thus, the first step to estimate the shadow fading is to remove the fast fading component. This is done by averaging the received power over the entire 30λ-segment and over all Tx and Rx antennas. The path loss component is estimated by calculating the least squares fit to the average received powers from all 30λ-segments against log-distance. The shadow fading, which is the variation around a local mean, is then estimated by subtracting the distant dependent path loss component from the average received power for each local area. This estimation method for the shadow fading is the same as in, for example, [12].

6.2. Estimation of the base station power-weighted angle spread

Although advanced techniques have been developed for estimating the power-weighted angle spread, [27–29], a simple estimation procedure will be used here. Previously reported estimation procedures use information from several antenna elements where both amplitude and phase information is available. In [25], the angle spread for the 2004 data set is estimated using a precalculated look-up table generated using the gain from a beam steered towards the angle of arrival. However, as explained in Section 4.4, the BSs used in 2005 were only equipped with two antenna elements with unknown frequency offsets, and thus a beam-forming approach, or more complex estimation methods, are not applicable. Therefore, we have devised another method to obtain reasonable estimates of the angle spread applicable to both our measurement campaigns. We cannot measure the angle of departure distribution itself, thus we will only consider its second-order moment, that is, the angle of departure spread. This method is similar to the previous one [25] in that a look-up table is used for determining the angle spreads. However here the cross-correlation between the signal envelopes is used instead of the beam-forming gain.

The look-up table, which contains the correlation coefficient as a function of the angle spread and the angle of departure, has been precalculated by generating data from a model with a Laplacian (power-weighted) AoD distribution, since this distribution has been found to have a very good fit to measurement data; see, for example, [30]. The details of the look-up table generation is described in Appendix A. Note that our method is similar to the method used in [31], where the correlation coefficient is studied as a function of the angle of arrival and the antenna separation. To estimate the angle spread with this approach, only the correlation coefficient between the envelopes of the received signals at the BS and the angle to the MS is calculated, where the latter is derived using the GPS information supplied by the measurements.

For the 2005 measurements, which were conducted with two antenna elements at the BS and four antennas at the MS, the cross-correlation between the signal envelopes at the BS is averaged over all four mobile antennas as

$$c_{1,2} = \sum_{k=1}^{4} \frac{\mathrm{E}\{(|H_{k,1}| - m_{k,1})(|H_{k,2}| - m_{k,2})\}}{\sigma_{k,1}\sigma_{k,2}}, \quad (12)$$

where

$$\begin{aligned} m_{k,1} &= \mathrm{E}\{|H_{k,1}|\}, \\ m_{k,2} &= \mathrm{E}\{|H_{k,2}|\}, \\ \sigma_{k,1}^2 &= \mathrm{E}\{(|H_{k,1}| - m_{k,1})^2\}, \\ \sigma_{k,2}^2 &= \mathrm{E}\{(|H_{k,2}| - m_{k,2})^2\}. \end{aligned} \quad (13)$$

For the 2004 measurements, where also the BS had 4 antennas, the average correlation coefficient over the three antenna pairs is used.

The performance of the estimation method presented above has been assessed by generating data from the SCM model, [10], then calculating the true angle spread (which is possible on the simulated data since all rays are known) and the estimated angle spread using the method described above. The results of this comparison are shown in Figure 5. From the estimates in the figure, it is readily seen that the angle spread estimate is reasonably unbiased, with a standard deviation of 0.1 log-degrees.

6.3. Estimation of the mobile station power-weighted angle spread

At the mobile station, an estimate of the power-weighted angle spread is extracted from the power levels of the four MS antennas. Accurate estimate cannot be expected, however, the MS angle spread is usually very large due to rich scattering at ground level in this environment and reasonable estimates can still be obtained as will be seen.

A first attempt is to use a four-ray model where the AoAs of the four rays are identical to the boresights of the four MS antennas, that is, $\alpha_n = 90°(n - 2.5)$. The powers of the four rays $p_1, \ldots, p_4$ are obtained from the powers of the four antennas, that is, the Euclidean norm of the rows of the channel matrices $\mathbf{H}$. These estimates are obtained by averaging the fast fading over the 30λ segments. From the powers the angle

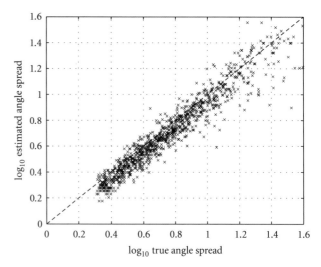

FIGURE 5: Performance of the angle spread estimator on SCM generated data.

spread is calculated using the circular model defined in (11) resulting in

$$\begin{aligned} &\hat{\sigma}_{\mathrm{AS,MS\text{-}fe}}^2 \\ &= \min_{\bar{\alpha}}\left\{\frac{1}{\sum_{n=1}^{4} p_n}\sum_{n=1}^{4} p_n\left(\widetilde{\mathrm{mod}}(90(n - 2.5) - \bar{\alpha})\right)^2\right\}, \end{aligned} \quad (14)$$

where $(\cdot)_{\mathrm{fe}}$ is short for first-estimate. As explained in [10, Annex A], the angle spread should be invariant to the orientation of the antenna, hence, knowledge of the moving direction of the MS is not required. The performance of the estimate is first evaluated by simulating a large number of widely different cases, using the SCM model, and estimating the spread based on four directional antennas as proposed here. The result is shown in Figure 6. The details of the simulation are described in Appendix B.

The results show that the angle spread is often overestimated using the proposed method. However, as indicated by Figure 6, a better second estimate $(\cdot)_{\mathrm{se}}$ is obtained by the following compensation:

$$\hat{\sigma}_{\mathrm{AS,MS\text{-}se}}^2 = \frac{(\hat{\sigma}_{\mathrm{AS,MS\text{-}fe}}^2 - 30)100}{70}. \quad (15)$$

The performance of this updated estimator is shown in Figure 7. The second estimate is reasonable when $\hat{\sigma}_{\mathrm{AS,MS\text{-}se}}^2 > 33$. When $\hat{\sigma}_{\mathrm{AS,MS\text{-}se}}^2 < 33$, the true angle spread may be anywhere from zero and $\hat{\sigma}_{\mathrm{AS,MS\text{-}se}}^2$. For small angle spreads, problems occur since all rays may fall within the bandwidth of a single-antenna. The estimated angle spread from our measurements at the MS is usually larger than 33°, thus this drawback in the estimation method has little impact on the final result. From the estimates in Figure 7, it is readily seen that the angle spread estimate is unbiased, with a standard deviation of 6 degrees.

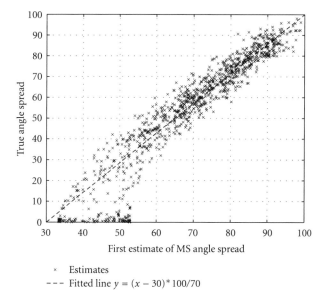

FIGURE 6: Performance of the first mobile station angle spread estimate.

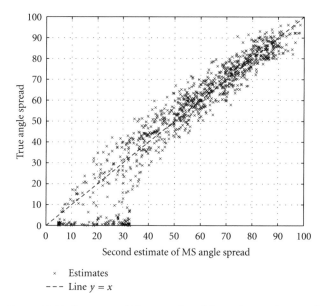

FIGURE 7: Performance of the second mobile station angle spread estimate.

7. RESULTS

In this section, the results of the analysis are presented in three parts. First the statistical information of the parameters is shown followed by their autocorrelation and cross-correlation properties.

7.1. Statistical properties

The first- and the second-order statistics of the LS parameters are estimated and shown in Table 3. The standard deviation of the shadow fading is given in dB while the angle spread at the BS is given in logarithmic degrees. Further, the MS an-

TABLE 2: Parmeters α and β for the beta best fit distribution to the angle spread at the mobile.

	2004:A	2004:B	2004:C	2005:1	2005:2
α	8.69	5.74	4.22	6.85	7.07
β	2.85	2.36	2.44	2.72	2.77

gle spread is given in degrees. The mean value of the shadow fading component is not tabulated since it is zero by definition. As seen from the histograms in Figure 8, which shows the statistics of the LS parameters for site 2004:B, the shadow fading and log-angle spread can be well modeled with a normal distribution. This agrees with observations reported in [12, 26]. The angle spread at the mobile on the other hand is better modeled by a scaled beta distribution, defined as

$$f(x, \alpha, \beta) = \frac{1}{B(\alpha, \beta)} \left(\frac{x}{\eta}\right)^{\alpha-1} \left(1 - \frac{x}{\eta}\right)^{\beta-1}, \qquad (16)$$

where $\eta = 360/\sqrt{12}$ is a normalization constant, equal to the maximum possible angle spread. The best fit shape parameters α and β for each of the measurement sets are tabulated in Table 2. The parameter $B(\alpha, \beta)$ is a constant which depends on α and β such that $\int_0^\eta f(x, \alpha, \beta) dx = 1$. The distributions of the parameters from all the other measured sites are similar, with statistics given in Table 3. From the table it is seen that the angle spread clearly depends on the height of the BS. The highest elevated BS, 2004:B, has the lowest angle spread and correspondingly, the BS at rooftop level, 2004:A, has the largest angle spread. The mean angle spreads at the base station are quite similar to the typical urban sites in [12] (0.74–0.95) and to those of the SCM urban macromodel (0.81–1.18) [10]. Furthermore, the standard deviations of the angle spread and the shadow fading found here, see Table 3, are somewhat smaller than those of [12]. One explanation for this could be that the measured propagation environments in 2004 and 2005 are more uniform than those measured in [12].

7.2. LS autocorrelation

The rate of change of the LS parameters is investigated by estimating the autocorrelation as a function of distance travelled by the MS. The autocorrelation functions for the large-scale parameters are shown in Figures 9 and 10, where the correlation coefficient between two variables is calculated as explained in Appendix C. Note that the autocorrelation functions can be well approximated by an exponential function with decorrelation distances as seen in Table 4. The decorrelation distance is defined as the distance for which the correlation has decreased to e^{-1}. Furthermore, it can be noted that these distances are very similar for the 2004 and the 2005 measurements, which is reasonable since the environments are similar. The exponential model has been proposed before, see [12], for the shadow fading and angle spread at the BS. The results shown herein indicate that this is a good model for the angle spread at the MS as well.

TABLE 3: Inter-BS correlation for measurement campaign 2004 site A.

	2004:A	2004:B	2004:C	2005:1	2005:2
std[SF]	5.6 dB	5.2 dB	5.4 dB	4.9 dB	4.9 dB
$E[\hat{\sigma}_{AS,BS}]$	1.2 ld	0.91 ld	0.85 ld	0.96 ld	0.87 ld
$std[\hat{\sigma}_{AS,BS}]$	0.25 ld	0.2 ld	0.23 ld	0.19 ld	0.17 ld
$E[\hat{\sigma}_{AS,MS}]$	75.1 deg	70.6 deg	65.9 deg	71.6 deg	72.2 deg
$std[\hat{\sigma}_{AS,MS}]$	15.7 deg	18.7 deg	19.2 deg	16.1 deg	16.9 deg

TABLE 4: Average decorrelation distane in maters for the estimated large-scale parameters.

	SF	$\hat{\sigma}_{AS,BS}$	$\hat{\sigma}_{AS,MS}$
$d_{decorr}(m)$	113	88	32

TABLE 5: Intra-BS correlation of LS parameters for measurement campaign 2004 site A.

	2004:A		
	SF	$\hat{\sigma}_{AS,MS}$	$\hat{\sigma}_{AS,BS}$
SF	1.00	−0.37	−0.46
$\hat{\sigma}_{AS,MS}$	−0.37	1.00	0.10
$\hat{\sigma}_{AS,BS}$	−0.46	0.10	1.00

7.3. Intrasite correlation

The intrasite correlation coefficients between different large-scale parameters at the same site are calculated for the two separate measurement campaigns. In Tables 5 and 6, the correlation coefficients for the two base stations, sectors A and B, from 2004 are shown, respectively. The last sector (C) is not shown since it is very similar to B and these parameters are based on a much smaller set of data, see Table 1. In Table 7, the same results are shown for the 2005 measurements. Since sites (2005:1 and 2005:2) show similar results and are from similar environments, the average correlation of the two is shown. It follows from mathematics that these tables are symmetrical, and in fact they only contain three significant values. The reason for showing nine values, instead of three, is to ease comparison with the intersite correlation coefficients shown in Tables 8–10. As seen from the tables, the angle spread is negatively correlated with shadow fading as was earlier found in for example [3, 12]. The cross-correlation coefficient between the shadow fading and base station angle spread is quite close to that of [12], that is −0.5 to −0.7. For the two cases where the BS is at rooftop level, Kårhuset, 2004:A, and the 2005 sites, there is no correlation between the angle spreads at the MS and the BS. However, for Vanadis, 2004:B, there is a positive correlation of 0.44. A possible explanation is that the BS is elevated some 10 meters over average rooftop height. Thus, no nearby scatterers exist and the objects that influence the angle spread at the BS are the same as the objects that influence the angle spread at the MS. A BS at rooftop on the other hand may have some nearby scatterers that will affect the angle of arrival and spread. In Figure 11, this is explained graphically. The stars are some of

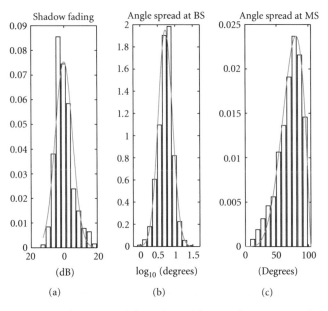

FIGURE 8: Histograms of the estimated large-scale parameters for site 2004:B.

the scatterers and the dark section of the circles depicts the area from which the main part of the signal power comes, that is the angle spread. In the left half of the picture, we see an elevated BS, without close scatterers, and therefore a large MS angle spread results in a large BS spread. In the right half of Figure 11, a BS at rooftop is depicted, with nearby scatterers, and we see how a small angle spread at the MS can result in a large BS angle spread (or the other way around).

7.4. Inter-BS correlation

The correlation coefficients between large-scale parameters at two separate sites are calculated for the data collected from both measurement campaigns. Only the data points which are common to both base station sectors, $S_i \cap S_j$, are used for this evaluation, that is, points that are within the ±60° beamwidth of both sites. As seen in section 4, describing the measurement campaigns, there is no overlap between site 2004:B and 2004:C if one considers ±60° sectors. For this specific case, the sector is defined as the area within ±70° of the BS's boresight, thus resulting in a 20° sector overlap. The results of this analysis are displayed in Tables 8, 9, and 10 for 2004:A-B, B-C, and 2005:1-2, respectively. As earlier shown in [20], the average correlation between the two sites 2004:A

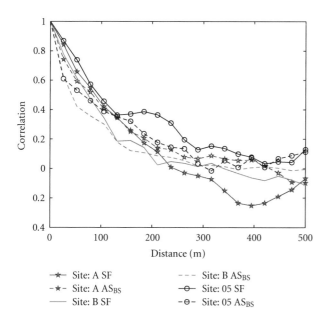

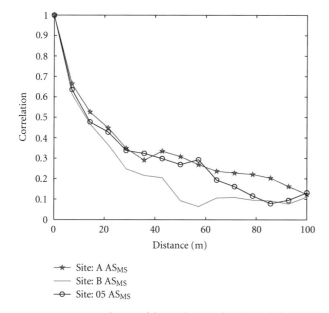

FIGURE 9: Autocorrelation of the shadow fading and the angle spread at the base station for both measurements.

FIGURE 10: Autocorrelation of the angle spread at the mobile station for both measurements.

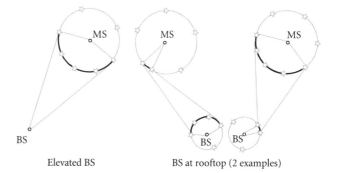

FIGURE 11: Model of correlation between angle spread at base station and mobile station.

TABLE 6: Intra-BS correlation of LS parameters for measurement campaign 2004 site B.

	2004:B		
	SF	$\hat{\sigma}_{AS,MS}$	$\hat{\sigma}_{AS,BS}$
SF	1.00	−0.54	−0.69
$\hat{\sigma}_{AS,MS}$	−0.54	1.00	0.44
$\hat{\sigma}_{AS,BS}$	−0.69	0.44	1.00

TABLE 7: Intra-BS correlation of LS parameters for measurement campaign 2005.

	2005		
	SF	$\hat{\sigma}_{AS,MS}$	$\hat{\sigma}_{AS,BS}$
SF	1.00	−0.25	−0.59
$\hat{\sigma}_{AS,MS}$	−0.25	1.00	0.11
$\hat{\sigma}_{AS,BS}$	−0.59	0.11	1.00

TABLE 8: Inter-BS correlation of all studied LS parameters between site A and site B from 2004 measurements.

		2004:A		
		SF	$\hat{\sigma}_{AS,MS}$	$\hat{\sigma}_{AS,BS}$
	SF	−0.14	0.08	−0.06
2004:B	$\hat{\sigma}_{AS,MS}$	−0.07	−0.05	0.03
	$\hat{\sigma}_{AS,BS}$	−0.04	−0.09	0.07

TABLE 9: Inter-BS correlation of all studied LS parameters between site B and site C from 2004 measurements.

		2004:B		
		SF	$\hat{\sigma}_{AS,MS}$	$\hat{\sigma}_{AS,BS}$
	SF	0.83	−0.23	−0.52
2004:C	$\hat{\sigma}_{AS,MS}$	−0.19	0.53	0.18
	$\hat{\sigma}_{AS,BS}$	−0.54	0.22	0.31

and 2004:B is close to zero. This is not surprising since the angular separation is quite large and the environments at the two separate sites are different. The correlations between sectors B and C of 2004 are similar as between sectors 1 and 2 of 2005. In both cases, the two BSs are on the same roof, and separated 20 and 50 meters for 2004 and 2005, respectively. As can be seen, these tables (Tables 8–10) are not symmetric. Thus the correlation of, for example, the shadow fading at BS 2005:1 and the angle spread at BS 2005:2 is not the same as the correlation of the shadow fading of BS 2005:2 and the an-

gle spread of BS 2005:1 ($\langle SF^{2005:1}, \hat{\sigma}_{AS,BS}^{2005:2} \rangle \neq \langle SF^{2005:2}, \hat{\sigma}_{AS,BS}^{2005:1} \rangle$), and so on. This is not surprising.

In Figure 12, the correlation coefficient is plotted against the angle separating the two base stations with the mobile in the vertex. The large variation of the curve is due to a lack of data. This may be surprising in the light of the quite

TABLE 10: Inter-BS correlation of all studied LS parameters between site B1 and B2 from 2005 measurements.

| | | 2005:1 | | |
		SF	$\hat{\sigma}_{AS,MS}$	$\hat{\sigma}_{AS,BS}$
	SF	0.85	−0.06	−0.45
2005:2	$\hat{\sigma}_{AS,MS}$	−0.05	0.46	0.04
	$\hat{\sigma}_{AS,BS}$	−0.27	0.18	0.33

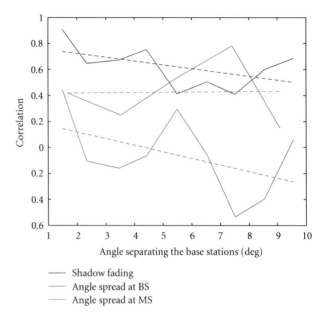

FIGURE 13: Intersite correlation of the large-scale parameters as a function of the angle separating the base stations for the 2005 measurements.

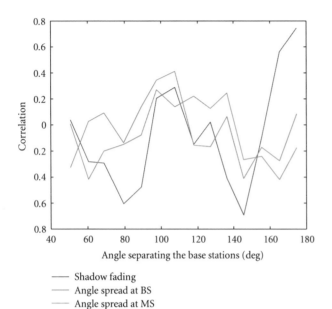

FIGURE 12: Intersite correlation of the large-scale parameters as a function of the angle separating the base stations for the 2004 measurements.

long measurement routes. However, due to the long decorrelation distances of the LS parameters (∼100 m), the number of independent observations is small. The high correlation for large angles of about 180° is mainly due to a very small data set available for this separation. Furthermore, this area of measurements is open with a few large buildings in the vicinity and thus the received power to both BSs is high.

If, on the other hand, the cross-correlation of the large-scale parameters between the two base stations from the 2005 measurements is studied, it is found that the correlation is substantial, see Table 10. Also, note that the correlation in angle spread is much smaller than the shadow fading. If the correlation is plotted as a function of the angle, separating the BSs as in Figure 13, a slight tendency of a more rapid drop in the correlation of angle spread than that of the shadow fading for increasing angles is seen. The intersite correlation results shown in Figures 12 and 13 are calculated disregarding the relative distance, see Figure 1. However, for the 2005 campaign this distance $d \approx 0$ is due to the location of the base stations.

The intersite correlation of the angle spread was calculated in the same way as the shadow fading. Only the measurement locations common to two sectors were used for these measurements. The angle spread is shown to have

smaller correlation than the shadow fading even for small angular separations. This indicates that it may be less important to include this correlation in future wireless channel models. It should be highlighted that the correlations shown in Table 10 are for angles $\alpha < 10°$ and a relative distance $|d = \log(d_1/d_2)| \approx 0$.

8. CONCLUSION

We studied the correlation properties of the three large-scale parameters shadow fading, base station power-weighted angle spread, and mobile station power-weighted angle spread. Two limiting cases were considered, namely when the base stations are widely separated, ∼900 m, and when they are closely positioned, some 20–50 meters apart.

The results in [12] on the distribution and autocorrelation of shadow fading and base station angle spread were confirmed although the standard deviations of the angular spread and shadow fading were slightly smaller in our measurements. The high interbase station shadow fading correlation, when base stations are close, as observed in [13], was also confirmed in this analysis. Our results also show that angular spread correlation exists at both the base station and the mobile station if the base station separation is small. However, the correlation in angular spread is significantly smaller than the correlation of the shadow fading. Thus it is less important to model this effect. For widely separated base stations, our results show that the base station and mobile station angular spreads as well as the shadow fading are uncorrelated.

The angle spread at the mobile was analyzed and a scaled beta distribution was shown to fit the measurements well. Further, we have also found that the base station and mobile station angular spreads are correlated for elevated base stations but uncorrelated for base station just above rooftop. Correlation can be expected if the scatters are only located close to the mobile station, which is the case for macrocellular environments, as illustrated in Figure 11.

In the future, it will be of interest to assess also the region in between the two limiting cases studied herein. Note that the limiting case of distances of 20–50 meters has a practical interest. For instance, the sectors of three-sector sites are sometimes not colocated but placed on different edges of a roof. The two base stations may also belong to different operators and the properties studied here could then be important when studying adjacent carrier interference.

APPENDICES

A. GENERATION OF ANGLE SPREAD LOOK-UP TABLE

The Laplacian angle of departure distribution is given by

$$P_A(\theta) = Ce^{(-|\theta-\theta_0|)/(\sigma_{\text{AoD}})}, \qquad (A.1)$$

where θ_0 is the nominal direction of the mobile and σ_{AoD} is angle-of-departure spread. The variable C is a constant such that $\int_{-\pi}^{\pi} P_A(\theta)d\theta = 1$. When generating data, the channel covariance matrix is first estimated as

$$\mathbf{R} = \int_{\theta=-180°}^{180°} P_A(\theta)\mathbf{a}(\theta)\mathbf{a}^*(\theta)d\theta, \qquad (A.2)$$

where $\mathbf{a}(\theta)$ is the array steering vector which is given by

$$\mathbf{a}(\theta) = p(\theta)\big[1, \exp\big(-j2\pi d_{\text{spacing}}\sin(\theta)\big)\big]^T, \qquad (A.3)$$

$p(\theta)$ is the (amplitude) antenna element diagrams of the array, and d_{spacing} is the spacing between the antenna elements given in wavelengths. In our case the element diagrams are approximated by

$$p^2(\theta) = \max(10^{1.4}\cos^2(\theta), 10^{-0.2}), \qquad (A.4)$$

and the antenna element spacing is 0.56 wavelength. The procedure for calculating the look-up table is then (1) fix angle spread and nominal angle of arrival, (2) calculate the covariance matrix $\mathbf{R}$ and it is eigendecomposition, (3) generate data from the model and calculate the envelope correlation.

The choice of Laplacian (power-weighted) AoD distribution over others, such as the Gaussian one, does only affect the estimation results marginally due to the short antenna spacing distance. This is further explained in [31].

B. EVALUATION OF THE MOBILE STATION ANGLE SPREAD ESTIMATOR

To test the estimator of the (power-weighted) RMS angle spread at the mobile-station side, some propagation channels were generated. Each channel had random number of clusters which was equally distributed between 1 and 10. The AoA of each cluster is uniformly distributed between 0° and 360°. The powers of the clusters are log-normally distributed with a standard deviation of 8 dB. Each cluster is modeled

with between 1 and 100 rays (all with equal power) which are uniformly distributed within the cluster width. The cluster widths are uniformly distributed between 0 and 10 degrees. One-thousand propagation (completely independent) channels are drawn from this model. The powers of the four antennas are calculated based on the powers of the rays, their angle of arrival, and the antenna pattern. The true angle spread is first estimated as described in [10, Annex A], and then the estimation method described in Section 6.3 is applied.

C. LARGE SCALE CORRELATIONS

The correlation coefficient between two variables is defined by the normalized covariance as

$$\rho = \langle a, b \rangle = \frac{\text{E}[ab] - m_a m_b}{\sqrt{(\text{E}[a^2] - m_a^2)(\text{E}[b^2] - m_b^2)}}. \qquad (C.5)$$

At all times when calculating the cross-correlation between LS parameters, even for small subsets of data, like when analyzing the correlation as a function of angular separation between BSs, the mean values are global. Hence the values m_a and m_b are calculated using the full data set of each BSs sector, respectively. If the mean values would be estimated locally, it is equal to assuming that the parameters are locally zero mean, and this is not what we are studying. What we want to investigate is if one parameter is large (or small) given the other.

ACKNOWLEDGMENTS

This work was sponsored partly within the Antenna Center of Excellence (FP6-IST 508009), the WINNER project IST-2003-507581, and wireless@KTH.

REFERENCES

[1] G. Foschini and M. J. Gans, "On limits of wireless communications in a fading environmen when using multiple antennas," *Wireless Personal Communications*, vol. 6, no. 3, pp. 311–335, 1998.

[2] I. E. Telatar, "Capacity of multi-antenna Gaussian channels," *European Transactions on Telecommunications*, vol. 10, no. 6, pp. 585–595, 1999.

[3] D. S. Baum, H. El-Sallabi, T. Jämsä, et al., "IST-WINNER D5.4, final report on link and system level channel models," http://www.ist-winner.org/, October 2005.

[4] D. Chizhik, J. Ling, P. Wolniansky, R. Valenzuela, N. Costa, and K. Huber, "Multiple-input-multiple-output measurements and modeling in manhattan," *IEEE Journal on Selected Areas in Communication*, vol. 21, no. 3, pp. 321–331, 2003.

[5] V. Eiceg, H. Sampath, and S. Catreux-Erceg, "Dual-polarization versus single-polarization MIMO channel measurement results and modeling," *IEEE Transactions on Wireless Communications*, vol. 5, no. 1, pp. 28–33, 2006.

[6] P. Kyritsi, D. C. Cox, R. A. Valenzuela, and P. W. Wolniansky, "Correlation analysis based on MIMO channel measurements in an indoor environment," *IEEE Journal on Selected Areas in Communications*, vol. 21, no. 5, pp. 713–720, 2003.

[7] M. Steinbauer, A. F. Molisch, and E. Bonek, "The double-directional radio channel," *IEEE Antennas and Propagation Magazine*, vol. 43, no. 4, pp. 51–63, 2001.

[8] R. Stridh, K. Yu, B. Ottersten, and P. Karlsson, "MIMO channel capacity and modeling issues on a measured Indoor radio channel at 5.8 GHz," *IEEE Transactions on Wireless Communications*, vol. 4, no. 3, pp. 895–903, 2005.

[9] J. Wallace and M. Jensen, "Time-varying MIMO channels: measurement, analysis, and modeling," *IEEE Transactions on Antennas and Propagation*, vol. 54, no. 11 ,part 1, pp. 3265–3273, 2006.

[10] 3GPP-SCM, "Spatial channel model for multiple input multiple output (MIMO) simulations," *TR.25.966 v.6.10*, http://www.3gpp.org/, September 2003.

[11] M. Gudmundson, "Correlation model for shadow fading in mobile radio systems," *IEEE Electronics Letters*, vol. 27, no. 23, pp. 2145–2146, 1991.

[12] A. Algans, K. I. Pedersen, and E. P. Mogensen, "Experimental analysis of the joint statistical properties of azimuth spread, delay spread, and shadow fading," *IEEE Journal on Selected Areas in Communications*, vol. 20, no. 3, pp. 523–531, 2002.

[13] V. Graziano, "Propagation correlation at 900MHz," *IEEE Transactions on Vehicular Technology*, vol. 27, no. 4, pp. 182–189, 1978.

[14] J. Weitzen and T. J. Lowe, "Measurement of angular and distance correlation properties of log-normal shadowing at 1900 MHz and its application to design of PCS systems," *IEEE Transactions on Vehicular Technology*, vol. 51, no. 2, pp. 265–273, 2002.

[15] A. Mawira, "Models for the spatial correlation functions of the (log)-normal component of the variability of VHF/UHF field strength in urban environment," in *Proceedings of the 3rd IEEE International Symposium on Personal, Indoor and Mobile Radio Communications (PIMRC '92)*, pp. 436–440, Boston, Mass, USA, October 1992.

[16] K. Zayana and B. Guisnet, "Measurements and modelisation of shadowing cross-correlationsbetween two base-stations," in *IEEE International Conference on Universal Personal Communications (ICUPC '98)*, vol. 1, pp. 101–105, Florence, Italy, October 1998.

[17] E. Perahia, D. C. Cox, and S. Ho, "Shadow fading cross correlation between basestations," in *The 53rd IEEE Vehicular Technology Conference (VTC '01)*, vol. 1, pp. 313–317, Rhodes, Greece, May 2001.

[18] H. W. Arnold, D. C. Cox, and R. R. Murray, "Macroscopic diversity performance measured in the 800-MHz portable radio communications environment," *IEEE Transactions on Antennas and Propagation*, vol. 36, no. 2, pp. 277–281, 1988.

[19] T. Klingenbrunn and P. Mogensen, "Modelling cross-correlated shadowing in network simulations," in *The 50th Vehicular Technology Conference (VTC '99)*, vol. 3, pp. 1407–1411, Amsterdam, The Netherlands, September 1999.

[20] N. Jaldén, P. Zetterberg, M. Bengtsson, and B. Ottersten, "Analysis of multi-cell MIMO measurements in an urban macrocell environment," in *General Assembly of International Union of Radio Science (URSI '05)*, New Delhi, India, October 2005.

[21] L. Garcia, N. Jaldén, B. Lindmark, P. Zetterberg, and L. D. Haro, "Measurements of MIMO capacity at 1800MHz with in- and outdoor transmitter locations," in *Proceedings of the European Conference on Antennas and Propagation (EuCAP '06)*, Nice, France, November 2006.

[22] L. Garcia, N. Jaldin, B. Lindmark, P. Zetterberg, and L. D. Haro, "Measurements of MIMO indoor channels at 1800MHz with multiple indoor and outdoor base stations," *EURASIP Journal on Wireless Communication and Networking*, vol. 2007, Article ID 28073, 10 pages, 2007.

[23] P. Zetterberg, "WIreless DEvelopment LABoratory (WIDE-LAB) equipment base," Signal Sensors and Systems (KTH), *iR-SB-IR-0316*, http://www.ee.kth.se/, August 2003.

[24] http://www.hubersuhner.com/.

[25] P. Zetterberg, N. Jaldén, K. Yu, and M. Bengtsson, "Analysis of MIMO multi-cell correlations and other propagation issues based on urban measurements," in *Proceedings of the 14th IST Mobile and Wireless Communications Summit*, Dresden, Germany, June 2005.

[26] T. Rappaport, *Wireless Communications: Principles and Practice*, Prentice-Hall, Upper Saddle River, NJ, USA, 1996.

[27] T. Trump and B. Ottersten, "Estimation of nominal direction of arrival and angular spread using an array of sensors," *Signal Processing*, vol. 50, no. 1-2, pp. 57–69, 1996.

[28] M. Bengtsson and B. Ottersten, "Low-complexity estimators for distributed sources," *IEEE Transactions on Signal Processing*, vol. 48, no. 8, pp. 2185–2194, 2000.

[29] M. Tapio, "Direction and spread estimation of spatially distributed signals via the power azimuth spectrum," in *Proceedings of IEEE International Conference on Acoustics, Speech, and Signal Processing (ICASSP '02)*, vol. 3, pp. 3005–3008, Orlando, Fla, USA, May 2002.

[30] K. I. Pedersen, P. E. Mogensen, and B. H. Fleury, "A stochastic model of the temporal and azimuthal dispersion seen at the base station in outdoor propagation environments," *IEEE Transactions on Vehicular Technology*, vol. 49, no. 2, pp. 437–447, 2000.

[31] N. Jaldén, "Analysis of radio channel measurements using multiple base stations," Licenciate Thesis, Royal Institute of Technology, Stockholm, Sweden, May 2007.

Hindawi Publishing Corporation
EURASIP Journal on Wireless Communications and Networking
Volume 2007, Article ID 51358, 11 pages
doi:10.1155/2007/51358

Research Article

Multiple-Antenna Interference Cancellation for WLAN with MAC Interference Avoidance in Open Access Networks

Alexandr M. Kuzminskiy[1] and Hamid Reza Karimi[1, 2]

[1] Alcatel-Lucent, Bell Laboratories, The Quadrant, Stonehill Green, Westlea, Swindon SN5 7DJ, UK
[2] Ofcom, Riverside House, 2a Southwark Bridge Road, London SE1 9HA, UK

Received 31 October 2006; Accepted 3 September 2007

Recommended by Monica Navarro

The potential of multiantenna interference cancellation receiver algorithms for increasing the uplink throughput in WLAN systems such as 802.11 is investigated. The medium access control (MAC) in such systems is based on carrier sensing multiple-access with collision avoidance (CSMA/CA), which itself is a powerful tool for the mitigation of intrasystem interference. However, due to the spatial dependence of received signal strengths, it is possible for the collision avoidance mechanism to fail, resulting in packet collisions at the receiver and a reduction in system throughput. The CSMA/CA MAC protocol can be complemented in such scenarios by interference cancellation (IC) algorithms at the physical (PHY) layer. The corresponding gains in throughput are a result of the complex interplay between the PHY and MAC layers. It is shown that semiblind interference cancellation techniques are essential for mitigating the impact of interference bursts, in particular since these are typically asynchronous with respect to the desired signal burst. Semiblind IC algorithms based on second- and higher-order statistics are compared to the conventional no-IC and training-based IC techniques in an open access network (OAN) scenario involving home and visiting users. It is found that the semiblind IC algorithms significantly outperform the other techniques due to the bursty and asynchronous nature of the interference caused by the MAC interference avoidance scheme.

1. INTRODUCTION

Interference at the radio receiver is a key source of degradation in quality of service (QoS) as experienced in wireless communication systems. It is for this reason that a great proportion of mobile radio engineering is exclusively concerned with the development of transmitter and receiver technologies, at various levels of the protocol stack, for mitigation of interference.

Multiple-antenna interference cancellation (IC) at the receiver has been the subject of a great deal of research in different application areas including wireless communications [1–3] and others. Despite the considerable interest in this area, IC techniques are typically studied at the physical (PHY) layer and in isolation from the higher layers of the protocol stack, such as the medium access control (MAC). However, it is clear that any gains at the system level are highly dependent on the nature of cross-layer interactions, particularly if multiple layers are designed to contribute to the interference mitigation process. This is indeed the case for the IEEE 802.11 family of wireless local area network (WLAN) systems [4], where the carrier sensing multiple-access with

collision avoidance (CSMA/CA) MAC protocol is itself designed to eliminate the possibility of interference at the receiver from other users of the same system.

Although the MAC layer CSMA/CA protocol may be very effective for avoidance of intrasystem interference in typical conditions, certain applications which experience significant hidden terminal problems and/or interference from coexisting "impolite" systems may also benefit from PHY layer IC. PHY/MAC cross-layer design is clearly required in such situations.

One important example of the above is an open access network (OAN) where visiting users (VUs) are allowed to share the radio resource with home users (HUs) [5]. In many scenarios, VUs typically experience greater distances from an access point (AP) compared to HUs. This means that VUs may interfere with each other with higher probability compared to HUs, leading to throughput reduction for VUs or gaps in coverage. A multiple-antenna AP with IC may be a solution to this problem.

A cross-layer design in such a system is required because the CSMA/CA protocol leads to an asynchronous

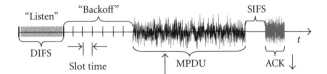

FIGURE 1: Transmission of MPDU and ACK bursts.

interference structure, where interference bursts appear with random delays during the desired signal data burst. One way to account for higher-layer effects is to develop interference models that reflect key features of cross-layer interaction and design PHY-layer algorithms that address these. This is the methodology adopted in [6–11], where semiblind space-time/frequency adaptive second- and higher-order statistic IC algorithms have been developed in conjunction with an asynchronous (intermittent) interference model. The second-order algorithm is based on the conventional least-squares (LS) criterion formulated over the training interval, regularized by means of the covariance matrix estimated over the data interval. This simple analytical solution demonstrates performance that is close to the nonasymptotic maximum likelihood (ML) benchmark [6, 7]. Further analysis is given in [8], which introduces non-stationary interval-based processing and benchmark in the asynchronous interference scenario. The regularized semiblind algorithms can be applied independently or as an initialization for higher-order algorithms that exploit the finite alphabet (FA) or constant modulus (CM) properties of communication signals. The efficiency of these algorithms has been compared to the conventional LS solution [1] by means of PHY simulations. These involve evaluation of metrics such as mean square error (MSE), bit-error rate (BER), or packet-error rate (PER), as a function of signal-to-interference ratio (SIR) for given signal-to-noise ratio (SNR), and a number of independent asynchronous interferers.

Our goal in this paper is to evaluate cross-layer interference avoidance/cancellation effects for different algorithms and estimate the overall system performance in terms of throughput and coverage. The combined performance of different IC algorithms at the PHY layer and the CSMA/CA protocol at the MAC layer is evaluated in the context of an IEEE 802.11a/g-based OAN. This is performed via simulations where the links between all radios are modelled at symbol level based on orthogonal frequency multiplexing (OFDM) as defined in specification [4], subject to path loss, shadowing and multipath fading according to the IEEE 802.11 channel models [12, 13]. Conventional and semiblind multiple-antenna algorithms are assumed at the PHY layer in order to identify possible improvements in system throughput and coverage for different OAN scenarios with VU and HU terminals. Cross-layer effects of continuous and intermittent intersystem interference from a coexisting impolite transmitter are also addressed.

The asynchronous interference model is derived in Section 2 in the context of typical OAN scenarios. The 802.11 CSMA/CA protocol is also briefly reviewed in Section 2. Problem formulation is given in Section 3. This is followed in

Section 4 by a description of the conventional and semiblind IC receiver algorithms, along with a demonstration of their performance at the PHY layer. Section 5 provides a description of the simulation framework and the cross-layer simulation results in typical OAN scenarios with intra- and inter-system interference. Conclusions are presented in Section 6.

2. INTERFERENCE SCENARIOS

The MAC mechanism specified in the IEEE 802.11 family of WLAN standards describes the process by which MAC protocol data units (MPDUs) are transmitted and subsequently acknowledged. Specifically, once a receiver detects and successfully decodes a transmitted MPDU, it responds after a short interframe space (SIFS) period, with the transmission of an acknowledgement (ACK) packet. Should an ACK not be successfully received and decoded after some interval, the transmitter will attempt to retransmit the MPDU.

Each IEEE 802.11 transmitter contends for access to the radio channel based on the CSMA/CA protocol. This is essentially a "listen before talk" mechanism, whereby a radio always listens to the medium before commencing a transmission. If the medium is determined to be already carrying a transmission (i.e., the measured background signal level is above a specified threshold), the radio will not commence transmission. Instead, the radio enters a deferral or back-off mode, where it waits until the medium is determined to be quiet over a certain interval before attempting to transmit. This is illustrated in Figure 1.

A "listen before talk" mechanism may fail in the so-called "hidden" terminal scenario. In this case, a transmitter senses the medium to be idle, despite the fact that a hidden transmitter is causing interference at the receiver, that is, the hidden terminal is beyond the reception range of the transmitter, but within the reception range of the receiver.

A single-cell uplink scenario is illustrated in Figure 2. An AP equipped with N antennas is surrounded by K terminals, uniformly distributed up to a maximum distance D. Terminals located within distance D_v of the AP are referred to as HUs. Terminals located at a distance greater than D_v are referred to as VUs[1] .

One can expect that the extent of possible collisions in this scenario depends on the distance from the AP. HUs located near the AP do not interfere with each other because of the CSMA/CA protocol. Even if signals from certain VUs collide with the signals from the HUs, the VU signal power levels received at the AP are most probably small, and will not result in erroneous decoding of the HUs' data. On the contrary, weaker VU signals are likely to be affected by collisions with stronger "hidden" VU and/or HU signals. This means that without IC, the VU throughput may suffer, leading to reduction or gaps in coverage even if the cell radius is sufficient for reliable reception from individual users.

---

[1] This distinction is made for illustrative purposes only. In practice, location bounds of HU and VU may be more complicated than the concentric rings shown in Figure 2.

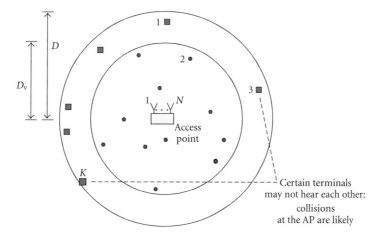

- Home user (HU) terminal
- Visiting user (VU) terminal

FIGURE 2: A single-cell OAN scenario with HUs and VUs.

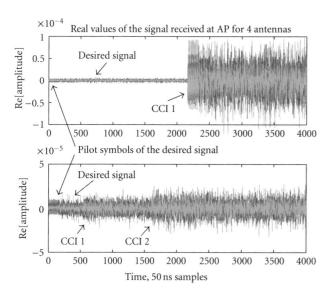

FIGURE 3: Typical collision patterns for $N = 4$.

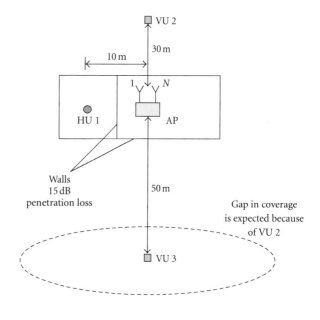

- Home user (HU) terminal
- Visiting user (VU) terminal

FIGURE 4: Residential OAN scenario with home and visiting users.

It is important to emphasize that collisions are typically asynchronous with random overlap between the colliding bursts. Typical collision examples are illustrated in Figure 3, which shows the real values of the received signals for $N = 4$ AP antennas, involving the desired signal and one or two cochannel interference (CCI) components. In both cases, the desired signals correspond to VUs and the interference comes from one HU in the first plot, and from two VUs in the second plot. In both cases, the interference bursts are randomly delayed with respect to the desired signal because of the random back-off intervals of the CSMA/CA protocol. The main consequence of this asynchronous interference structure for IC is that there is no overlap between the pilot symbols of the desired signal (located in the preamble) and the interference bursts.

The single-cell OAN scenario of Figure 2 can be specified for particular home/visitor situations. Figure 4 illustrates a residential scenario with walls that can be taken into account by means of a penetration loss. Home user HU 1 would always get a good connection in this scenario. Visiting users 2 and 3, however, may not hear each other and their transmissions may collide in some propagation conditions. Signals received from VU 2 would typically be much stronger than those from VU 3 due to the shorter distance, resulting in low throughput for VU 3. Another residential scenario with

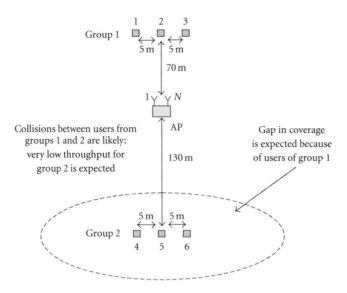

FIGURE 5: Residential scenario with two groups of visiting users.

two groups of three visiting users each is shown in Figure 5. This scenario illustrates the situation, where gaps in coverage can be expected for VUs 4–6 without effective IC at the PHY layer because of another group of strong VUs 1–3. The asynchronous structure of the interference in these scenarios is similar to the one illustrated in Figure 3.

3. PROBLEM FORMULATION

Based on the scenarios in Figures 2, 4, and 5, and other similar OAN scenarios, one may conclude that the MAC layer impact on the interference structure can be taken into account by means of an asynchronous interference model. An example of such model for three interference components is illustrated in Figure 6, where random delays and varying burst durations are assumed. This model can be exploited for developing and comparing different IC algorithms at the PHY layer. After this cross-layer design, the developed PHY IC algorithms can be tested via cross-layer simulations.

The problem formulation, including the main objective, constraints, and system assumptions, as well as the main effects taken into account, is as follows.

Objective

- Increase uplink throughput for VUs in an OAN system based on OFDM WLAN with CSMA/CA.

Constraints and system assumptions

- Single-antenna user terminals.
- Multiple-antenna AP.
- CSMA/CA transmission protocol at the AP and terminals.
- PHY layer interference cancellation at the AP taking into account the asynchronous interference model induced by the MAC layer.

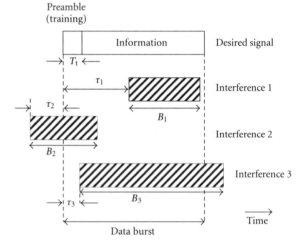

FIGURE 6: Asynchronous interference model.

- OAN scenarios with HUs and VUs as well as external interference from a coexisting system.

Effects taken into account

- MPDU and ACK structures, interleaving, coding, and modulation according to the IEEE 802.11a/g PHY.
- Propagation channels: multipath delay spread; path loss and shadowing; line-of-sight (LoS) and non-LoS (NLoS) conditions; spatial correlation between antenna elements at the AP.

4. INTERFERENCE CANCELLATION

Since training symbols are most reliable for estimation of the desired signal by means of the conventional LS criterion, the main idea here is to apply regularization of the LS criterion by a penalty function associated with the covariance matrix estimated over the data interval. In the narrow-band scenario, that is, for each individual OFDM subcarrier, the modified (regularized) LS criterion can be expressed as follows [6, 7]:

$$\hat{\mathbf{w}} = \arg \min_{\mathbf{w}} \sum_{t \in \tau_t} |s(t) - \mathbf{w}^* \mathbf{x}(t)|^2 + \rho F(\mathbf{w}), \quad (1)$$

where t is the time index, $s(t)$ is the training sequence for the desired signal, $\mathbf{x}(t)$ is the output $N \times 1$ vector from the receiving antenna array, N is the number of antenna elements, $\mathbf{w}$ is the $N \times 1$ weight vector, τ_t is the interval of T_t known training symbols assuming perfect synchronization for the desired signal, $\rho > 0$ is a regularization parameter, $F(\mathbf{w})$ is a regularization function that exploits a priori information for specific problem formulations, and $(\cdot)^*$ is the complex conjugate transpose.

In the considered asynchronous interference scenario, the working interval may be affected by interference components that are not present during the training interval. Thus, selection of the regularization function such that it contains information from the data interval increases the ability to cancel asynchronous interference. For the second-order statistics class of algorithms, this can be achieved by means of the following quadratic function [6, 7]:

$$F(\mathbf{w}) = \mathbf{w}^* \hat{\mathbf{R}}_t \mathbf{w} - \hat{\mathbf{r}}_t^* \mathbf{w} - \mathbf{w}^* \hat{\mathbf{r}}_t, \quad (2)$$

leading to the semiblind (SB) solution

$$\hat{\mathbf{w}}_{\text{SB}} = [(1 - \delta)\hat{\mathbf{R}}_t + \delta \hat{\mathbf{R}}_b]^{-1} \hat{\mathbf{r}}_t, \quad (3)$$

where $\hat{\mathbf{R}}_t = T_t^{-1} \sum_{t \in \tau_t} \mathbf{x}(t)\mathbf{x}^*(t)$ and $\hat{\mathbf{r}}_t = T_t^{-1} \sum_{t \in \tau_t} \mathbf{x}(t)s^*(t)$ are the covariance matrix and cross-correlation vector estimated over the training interval, $\hat{\mathbf{R}}_b = T^{-1} \sum_{t=1}^{T} \mathbf{x}(t)\mathbf{x}^*(t)$ is the covariance matrix estimated over the whole data burst of T symbols, and $0 \leq \delta = \rho/(1 + \rho) \leq 1$ is the regularization coefficient. Selection of the regularization parameter δ has been studied in [6, 11] and will be discussed below.

One can see that the SB estimator (3) contains the conventional LS solution

$$\hat{\mathbf{w}}_{\text{LS}} = \hat{\mathbf{R}}_t^{-1} \hat{\mathbf{r}}_t, \quad (4)$$

as a special case for $\delta = 0$.

An iterative higher-order statistics estimation algorithm with projections onto the FA with SB initialization (SBFA) can be described as follows:

$$\hat{\mathbf{w}}_{\text{SBFA}} = \hat{\mathbf{w}}^{[J]},$$

$$\hat{\mathbf{w}}^{[j]} = (\mathbf{X}\mathbf{X}^*)^{-1}\mathbf{X}\Theta[\mathbf{X}^*\hat{\mathbf{w}}^{[j-1]}], \quad j = 1, \ldots, J, \quad (5)$$

$$\hat{\mathbf{w}}^{[0]} = \hat{\mathbf{w}}_{\text{SB}},$$

$$\hat{\mathbf{w}}^{[J]} = \hat{\mathbf{w}}^{[J-1]}, \quad (6)$$

where $\mathbf{X} = [\mathbf{x}(1), \ldots, \mathbf{x}(T)]$ is the $N \times T$ matrix of input signals, $\hat{\mathbf{w}}^{[j]}$ is the weight vector at the jth iteration, $\Theta[\cdot]$ is the projection onto the FA, and J is the total number of iterations with stopping rule (6).

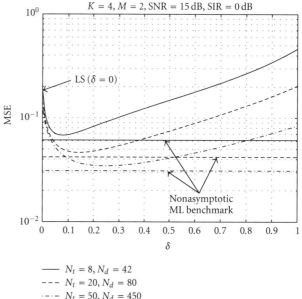

$K = 4, M = 2, \text{SNR} = 15\,\text{dB}, \text{SIR} = 0\,\text{dB}$

— $N_t = 8, N_d = 42$
--- $N_t = 20, N_d = 80$
-·- $N_t = 50, N_d = 450$

FIGURE 7: Typical MSE performance for the SB algorithm for variable regularization parameter.

Efficiency of the SB algorithm (3) is studied in [6, 7] by means of comparison to the especially developed nonasymptotic ML benchmark. Typical estimated MSE performance for different burst structures and variable regularization parameter δ is illustrated in Figure 7 for $N = 4$, $K = 2$, SNR = 15 dB, SIR = 0 dB, QPSK signals, and independent complex Gaussian vectors as propagation channels. The corresponding ML benchmark results from [6] are also shown in Figure 7 for comparison. One can see that the SB performance is very close to the ML benchmark for properly selected regularization parameter. Furthermore, the MSE functions are not very sharp, which means that some fixed parameter δ can be used for a wide range of scenarios. Indeed, the results in Figure 7 suggest that $\delta \approx 0.1$ can be effectively applied for very different slot structures.

The narrowband versions of the LS, SB, and SBFA algorithms can be expanded to the OFDM case. The problem with this expansion is that the available amount of training and data symbols at each subcarrier may not be large enough to achieve desirable performance. Different approaches can be applied to overcome this difficulty, such as grouping (clustering) or other interpolation techniques [14, 15]. According to the grouping technique, subcarriers of an OFDM system are divided into groups, and a single set of parameters is estimated for all subcarriers within a group, using all pilot and information symbols from that group.

Next, we compare the LS, SB, and SBFA algorithms at the PHY layer of an OFDM radio link subject to asynchronous interference. We consider the "D-"channel [13] environment and apply a group-based technique [14] with $Q = 12$ groups of subcarriers. We simulate a single-input multiple-output (SIMO) system ($N = 5$) for IEEE 802.11g time-frequency bursts of 14 OFDM QPSK modulated symbols and 64 subcarriers (only 52 are used for data and pilot transmission).

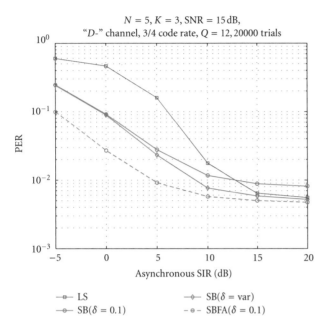

$N = 5, K = 3, \mathrm{SNR} = 15\,\mathrm{dB}$,
"D-" channel, 3/4 code rate, $Q = 12, 20000$ trials

FIGURE 8: Typical PHY-layer OFDM performance for LS, SB, and SBFA.

The transmitted signal is encoded according to the IEEE 802.11g standard with a 3/4 code rate [4]. Each packet contains 54 information bytes. Each time-frequency burst includes two information packets and two preamble blocks of 52 binary pilot symbols. This simulation environment corresponds to an over-the-air data rate of 18 Mbit/s.

Figure 8 presents the packet-error rate (PER) curves for LS, SB, and SBFA with a fixed SNR of 15 dB. The SB algorithm is presented for fixed ($\delta = 0.1$) regularization as well as adaptive ($\delta = var$) regularization parameter selected on a burst-by-burst basis based on the CM criterion:[2]

$$\hat{\delta} = \arg \min_{\delta} \sum_{t=1}^{T} \left[\left| \hat{\mathbf{w}}_{\mathrm{SB}}^{*}(\delta)\mathbf{x}(t) \right|^{2} - 1 \right]^{2}. \tag{7}$$

In Figure 8, the SIR is varied for two asynchronous interference components, and is fixed at 0 dB for a synchronous interference component (note that the latter is still asynchronous on a symbol basis, but always overlaps with the whole data burst of the desired signal including the preamble).

One can see that the regularized SB solution with the fixed regularization parameter significantly outperforms the conventional LS algorithm for low asynchronous SIR. Particularly, it outperforms LS by 4 dB at 3% PER, and by 7 dB at 10% PER. In the high SIR region, the scenario becomes similar to the synchronous case (asynchronous CCI actually disappears), where the LS estimator actually gives the best possible results [16]. Thus, $\delta \to 0$ is required for the best SB performance in this region. Online adaptive selection of the regularization parameter may be adopted in this case, as illustrated in Figure 8. However, one can see in Figure 7 that performance degradation for fixed $\delta = 0.1$ in the synchronous case is small and may well be acceptable. The SBFA algorithm brings additional performance improvement of up to 5 dB for low SIR at the cost of higher complexity.

OFDM versions of the LS, SB, and SBFA algorithms with a fixed regularization parameter, together with the conventional matched filter (no-IC), will be evaluated next via cross-layer simulations.

5. CROSS-LAYER SIMULATION RESULTS

5.1. Simulation assumptions

We simulate the IEEE 802.11g PHY (OFDM) and CSMA/CA subject to the following assumptions:

- 2.4 GHz center frequency,
- 4-QAM, 1/2 rate convolutional coding,
- MPDU burst of 2160 information bits, 50 OFDM symbols, 200 microseconds duration,
- ACK burst of 8 OFDM symbols, 32 microseconds slot duration,
- maximum ratio beamforming at the AP for ACK transmissions,
- trial duration of 10 milliseconds,
- "E"-channel propagation model [13] (100 nanaoseconds delay spread, LOS/NLOS conditions depending on distance),
- 1-wavelength separation between $N = 4$ AP antennas,
- 20 dBm transmit power for the AP and terminals,
- − 92 dBm noise power,
- − 82 dBm clear-channel assessment threshold.

A number of simplifying assumptions are made: ideal channel reciprocity (uplink channel estimates are used for downlink beamforming for ACK transmission); ideal (linear) front-end filters at the AP and terminals; zero frequency offset; perfect receiver synchronization at the AP and terminals; stationary propagation channels during a 10-millisecond trial. The last assumption is applicable in the considered scenario because all channel and weight estimates are derived on a slot-by-slot basis, and channel variations in WLAN environments are normally negligible over these time scales (200 microseconds).

5.2. Single-cell OAN

Typical histograms for collision statistics in the scenario of Figure 2 are shown in Figure 9 for $D = 150$ m. As expected, the average number of colliding MPDUs increases with the total number, K, of users contending for the channel.

The VU throughputs are presented in Figure 10 for variable visitor radius D_v and total number of users. The conventional matched filtering (no-IC), LS, SB, and SBFA ($\delta = 0.1$) algorithms are compared.

---

[2] A simplified switched CM-based selection of the regularization parameter is developed in [11].

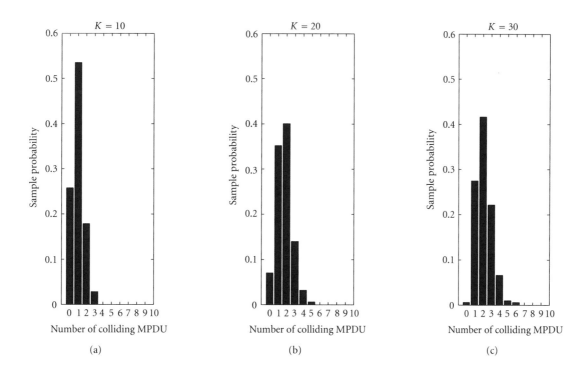

FIGURE 9: Collision statistics for $D = 150$ m with the single-cell scenario of Figure 2.

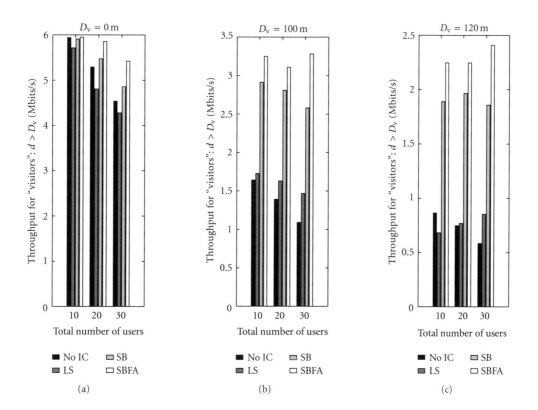

FIGURE 10: Visiting user throughput for $D = 150$ and $N = 4$ for no IC, LS, SB, and SBFA algorithms (left to right for each total number of users) with the single-cell scenario of Figure 2.

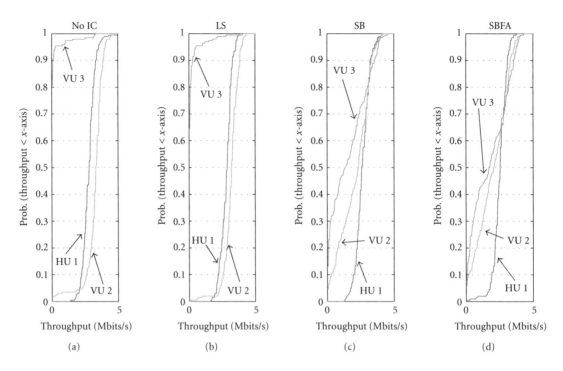

FIGURE 11: Throughput CDF for the residential scenario of Figure 4.

The VU throughput U_v is calculated as follows:

$$U_v = \frac{1}{T_S I} \sum_{i=1}^{I} \sum_{D_v < d_{ik} \leq D} B_{ik}, \qquad (8)$$

where d_{ik} is the distance between the AP and the kth terminal at the ith trial, B_{ik} is the total number of bits from the kth terminal successfully received and acknowledged at the AP at the ith trial, and T_s and I are the duration and number of trials. The throughput results in Figure 10 are averaged over $I = 20$ trials of $T_s = 10$ milliseconds, each with independent user locations and propagation channel realizations[3].

The first plot for $D_v = 0$ actually shows the total cell throughput. One can see that all the algorithms show some performance degradation with growing total number of users in the cell. The SB and SBFA algorithms demonstrate a small improvement over both no-IC and LS for $K = [20, 30]$. The low IC gain is in fact expected in this case since the interference avoidance CSMA/CA protocol dominates for users located close to the AP, making any IC redundant.

The situation is quite different when we consider the throughput of the VUs only. One can see in Figure 10, for

$D_v = [100, 120]$ m, that both semiblind solutions significantly outperform the other two techniques by up to a factor of 4. Furthermore, it appears that the main improvement comes from the second-order SB solution (3). Iterative projections to the FA in SBFA add up to 25% to the SB gain.

5.3. Residential OAN

Cumulative distribution functions (CDFs) of VU throughput over 200 trials are plotted in Figure 11. These corresponds to the residential scenario of Figure 4 with wall penetration loss of 15 dB. As expected, home user (HU) 1 is not affected by VU 2 and 3. On the contrary, visiting user (VU) 3 hardly achieves any throughput unless efficient semiblind IC is utilized. Both semiblind estimators demonstrate significant performance improvement and allow both visiting users (VU) 2 and 3 to share the radio resource almost equally.

The throughput results estimated over 100 trials in the scenario shown in Figure 5 are given in Figure 12. They illustrate the situation, where gaps in coverage because of strong VUs may be significantly reduced by means of the proposed semiblind cancellation at the PHY layer.

5.4. Intersystem interference in residential OAN

As mentioned in Section 2, PHY layer IC may also be effective in scenarios where interference from other systems is not subject to any interference avoidance schemes such as CSMA/CA (i.e., is "impolite"). We illustrate this situation in a residential scenario as presented in Figure 13, which consists of one HU, one VU, and a low-power ($10\,\mu$W) "impolite" interferer located close to the AP. In this scenario,

---

[3] Cross-layer simulations are very computationally demanding. For each trial, we generated $(K+1)K/2$ independent propagation channels for random terminal positions (e.g., for $K = 30$ we generated 465 channels per trial). Typically, during 10 milliseconds we observed around ten collisions between different users (burst duration is 0.2 milliseconds) leading to approximately 200 collisions for 20 trials. This is why we accepted a low number of trials for the single-cell scenario. For particular residential scenarios with low number of terminals, we simulated around 100–200 trials to keep a similar level of averaging over different propagation conditions.

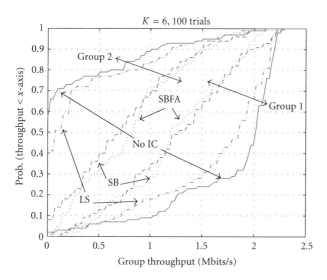

FIGURE 12: Throughput CDF for the residential scenario of Figure 5.

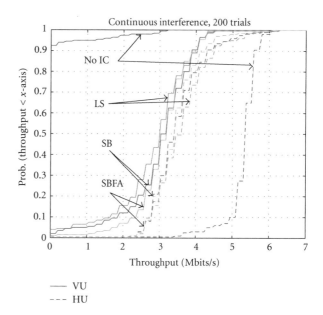

— VU
--- HU

FIGURE 14: Throughput CDF for residential scenario of Figure 12 and a continuous intersystem interferer.

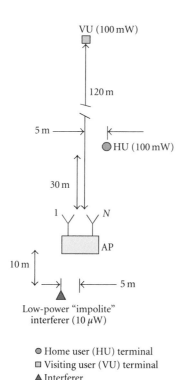

FIGURE 13: Residential scenario with intersystem interference.

CSMA/CA for HU and VU is not affected, because the interference power at the HU and VU locations is normally below the clear-channel assessment threshold. Furthermore, the HU signal received at the AP is much stronger than the interference. So, the HU is also unaffected at the PHY layer. On the contrary, the VU signal received at the AP may be comparable to the interference level, and so, may be significantly affected. Again, IC efficiency depends on the temporal interference structure, as discussed in Section 3.

Figure 14 shows the results for a continuous, white Gaussian, 10-μW "impolite" interferer. One can see that the VU

throughput can be significantly improved by means of all the considered training-based LS and semiblind SB and SBFA IC algorithms. This is because the interference always overlaps with the MPDU pilot symbols, resulting in what we classify in [16] as a synchronous interference.

Again, the situation becomes quite different for intermittent intersystem interference. We simulate this as a stream of 200 microseconds bursts with duty-cycle of 50%. Typical collision patterns between data and interference bursts are plotted in Figure 15. Here, the random MPDU back-offs result in random overlaps between the interference bursts and the training symbols. Figure 16 presents the throughput results for intermittent interference. It is not surprising that both SB and SBFA significantly outperform the conventional no-IC and pilot-based LS IC in this scenario. However, one can see in Figure 16 that LS demonstrates more significant performance improvement over the no-IC solution compared to the intrasystem interference scenario presented above. This is because in the considered intermittent interference scenario, collisions that overlap with the training symbols occur with practically the same probability as those involving no overlap with the training symbols. Both types of collisions are illustrated in Figure 15 in the upper and lower plots, respectively. In the intrasystem interference case, collisions that do not overlap with the training symbols dominate because of the CSMA/CA protocol, as discussed in Section 2, leading to significant semiblind gain over the conventional training-based IC algorithms such as LS.

6. CONCLUSION

The potential gains provided by multiantenna interference cancellation receiver algorithms, in the context of WLAN systems employing CSMA/CA protocols, were evaluated in this paper. Cross-layer interactions were captured via joint

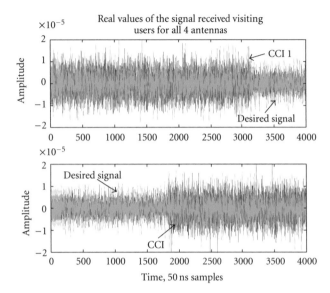

FIGURE 15: Typical received signal patterns in the intersystem intermittent interference scenario.

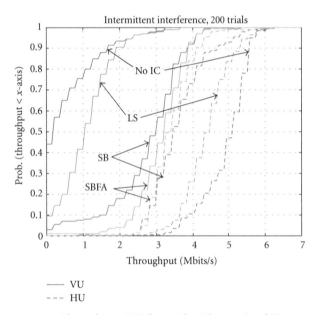

FIGURE 16: Throughput CDF for residential scenario of Figure 12 with an intermittent intersystem interferer (50% duty-cycle).

PHY/MAC simulations involving multiple terminals contending for the opportunity to transmit data to the access point. The impact of impolite cochannel interference from a coexisting system was also accounted for. It was shown that the developed semiblind interference cancellation techniques are essential for addressing the asynchronous interference experienced in WLAN. Significant performance gain has been demonstrated by means of cross-layer simulations in the OAN scenarios. It has been found that the main effect comes from the regularization in the SB algorithm with complexity similar to the conventional LS solution. The more complicated SBFA iterations lead to an additional marginal performance improvement.

ACKNOWLEDGMENTS

The authors would like to thank Professor Y. I. Abramovich for participating in many fruitful discussions on PHY IC in the course of this work. Part of this work has been performed with financial support from the IST FP6 OBAN project and also part of this work has been presented at ICC '07 [17].

REFERENCES

[1] A. J. Paulraj and C. B. Papadias, "Space-time processing for wireless communications," *IEEE Signal Processing Magazine*, vol. 14, no. 6, pp. 49–83, 1997.

[2] J. G. Andrews, "Interference cancellation for cellular systems: a contemporary overview," *IEEE Wireless Communications*, vol. 12, no. 2, pp. 19–29, 2005.

[3] A. M. Kuzminskiy, "Finite amount of data effects in spatio-temporal filtering for equalization and interference rejection in short burst wireless communications," *Signal Processing*, vol. 80, no. 10, pp. 1987–1997, 2000.

[4] IEEE Std 802.11a, "Wireless LAN Medium Access Control (MAC) and Physical Layer (PHY) Specifications," 1999.

[5] Open BroadbAccess Network (OBAN), IST 6FP Contract no. 001889, http://www.ist-oban.org/.

[6] A. M. Kuzminskiy and Y. I. Abramovich, "Second-order asynchronous interference cancellation: regularized semi-blind technique and non-asymptotic maximum likelihood benchmark," *Signal Processing*, vol. 86, no. 12, pp. 3849–3863, 2006.

[7] A. M. Kuzminskiy and Y. I. Abramovich, "Adaptive second-order asynchronous CCI cancellation: maximum likelihood benchmark for regularized semi-blind technique," in *Proceedings of IEEE International Conference on Acoustics, Speech, and Signal Processing (ICASSP '04)*, vol. 4, pp. 453–456, Montreal, Que, Canada, May 2004.

[8] A. M. Kuzminskiy and Y. I. Abramovich, "Interval-based maximum likelihood benchmark for adaptive second-order asynchronous CCI cancellation," in *Proceedings of IEEE International Conference on Acoustics, Speech, and Signal Processing (ICASSP '07)*, vol. 2, pp. 865–868, Honolulu, Hawaii, USA, April 2007.

[9] A. M. Kuzminskiy and C. B. Papadias, "Asynchronous interference cancellation with an antenna array," in *Proceedings of the 13th IEEE International Symposium on Personal, Indoor and Mobile Radio Communications (PIMRC '02)*, vol. 1, pp. 260–264, Lisbon, Portugal, September 2002.

[10] A. M. Kuzminskiy and C. B. Papadias, "Re-configurable semiblind cancellation of asynchronous interference with an antenna array," in *Proceedings of IEEE International Conference on Acoustics, Speech, and Signal Processing (ICASSP '03)*, vol. 4, pp. 696–699, Hong Kong, April 2003.

[11] A. M. Kuzminskiy and Y. I. Abramovich, "Adaptive asynchronous CCI cancellation: selection of the regularization parameter for regularized semi-blind technique," in *Proceedings of the 7th IEEE Workshop on Signal Processing Advances in Wireless Communications (SPAWC '06)*, pp. 1–5, Cannes, France, July 2006.

[12] V. Erceg, L. Schumacher, P. Kyritsi, et al., "TGn Channel Models (IEEE 802.11-03/940r2)," High Throughput Task Group, IEEE P802.11, March 2004.

[13] L. Schumacher, "WLAN MIMO channel MATLAB program," http://www.info.fundp.ac.be/~lsc/Research/IEEE_80211_HTSG_CMSC/distribution_terms.html.

[14] D. Bartolome, X. Mestre, and A. I. Perez-Neira, "Single input multiple output techniques for Hiperlan/2," in *Proceedings of IST Mobile Communications Summit*, Barcelona, Spain, September 2001.

[15] A. M. Kuzminskiy, "Interference cancellation in OFDM with parametric modeling of the antenna array weights," in *Proceedings of the 35th Asilomar Conference on Signals, Systems and Computers*, vol. 2, pp. 1611–1615, Pacific Grove, Calif, USA, November 2001.

[16] Y. I. Abramovich and A. M. Kuzminskiy, "On correspondence between training-based and semiblind second-order adaptive techniques for mitigation of synchronous CCI," *IEEE Transactions on Signal Processing*, vol. 54, no. 6, pp. 2347–2351, 2006.

[17] A. M. Kuzminskiy and H. R. Karimi, "Cross-layer design of uplink multiple-antenna interference cancellation for WLAN with CSMA/CA in open access networks," in *Proceedings of IEEE International Conference on Communications (ICC '07)*, pp. 2568–2573, Glasgow, Scotland, UK, June 2007.

Hindawi Publishing Corporation
EURASIP Journal on Wireless Communications and Networking
Volume 2007, Article ID 60654, 11 pages
doi:10.1155/2007/60654

Research Article

Transmit Diversity at the Cell Border Using Smart Base Stations

Simon Plass, Ronald Raulefs, and Armin Dammann

German Aerospace Center (DLR), Institute of Communications and Navigation, Oberpfaffenhofen, 82234 Wessling, Germany

Received 27 October 2006; Revised 1 June 2007; Accepted 22 October 2007

Recommended by A. Alexiou

We address the problems at the most critical area in a cellular multicarrier code division multiple access (MC-CDMA) network, namely, the cell border. At a mobile terminal the diversity can be increased by using transmit diversity techniques such as cyclic delay diversity (CDD) and space-time coding like Alamouti. We transfer these transmit diversity techniques to a cellular environment. Therefore, the performance is enhanced at the cell border, intercellular interference is avoided, and soft handover procedures are simplified all together. By this, macrodiversity concepts are exchanged by transmit diversity concepts. These concepts also shift parts of the complexity from the mobile terminal to smart base stations.

1. INTRODUCTION

The development of future mobile communications systems follows the strategies to support a single ubiquitous radio access system adaptable to a comprehensive range of mobile communication scenarios. Within the framework of a global research effort on the design of a next generation mobile system, the European IST project WINNER—Wireless World Initiative New Radio—[1] is also focusing on the identification, assessment, and comparison of strategies for reducing and handling intercellular interference at the cell border. For achieving high spectral efficiency the goal of future wireless communications systems is a total frequency reuse in each cell. This leads to a very critical area around the cell borders.

Since the cell border area is influenced by at least two neighboring base stations (BSs), the desired mobile terminal (MT) in this area has to scope with several signals in parallel. On the one hand, the MT can cancel the interfering signals with a high signal processing effort to recover the desired signal [2]. On the other hand, the network can manage the neighboring BSs to avoid or reduce the negative influence of the transmitted signals at the cell border. Due to the restricted power and processing conditions at the MT, a network-based strategy is preferred.

In the region of overlapping cells, handover procedures exist. Soft handover concepts [3] have shown that the usage of two base stations at the same time increases the robustness of the received data and avoids interruption and calling resources for reinitiating a call. With additional information about the rough position of the MT, the network can avoid fast consecutive handovers that consume many resources, for example, the MT moves in a zigzag manner along the cell border.

Already in the recent third generation mobile communications system, for example, UMTS, macrodiversity techniques with two or more base stations are used to provide reliable handover procedures [4]. Future system designs will take into account the advanced transmit diversity techniques that have been developed in the recent years. As the cell sizes decrease further, for example, due to higher carrier frequencies, the cellular context gets more dominant as users switch cells more frequently. The ubiquitous approach of having a reliable link everywhere emphasizes the need for a reliable connection at cell border areas.

A simple transmit diversity technique is to combat flat fading conditions by retransmitting the same signal from spatially separated antennas with a frequency or time offset. The frequency or time offset converts the spatial diversity into frequency or time diversity. The effective increase of the number of multipaths is exploited by the forward error correction (FEC) in a multicarrier system. The elementary method, namely, delay diversity (DD), transmits delayed replicas of a signal from several transmit (TX) antennas [5]. The drawback are increased delays of the impinging signals. By using the DD principle in a cyclic prefix-based system, intersymbol interference (ISI) can occur due to too large delays.

This can be circumvented by using cyclic delays which results in the cyclic delay diversity (CDD) technique [6].

Space-time block codes (STBCs) from orthogonal designs [7] improve the performance in a flat and frequency selective fading channel by coherently adding the signals at the receiver without the need for multiple receive antennas. The number of transmit antennas increases the performance at the expense of a rate loss. The rate loss could be reduced by applying nearly orthogonal STBCs which on the other hand would require a more complex space-time decoder. Generally, STBCs of orthogonal or nearly orthogonal designs need additional channel estimation, which increases the complexity.

The main approach of this paper is the use and investigation of transmit diversity techniques in a cellular environment to achieve macrodiversity in the critical cell border area. Therefore, we introduce cellular CDD (C-CDD) which applies the CDD scheme to neighboring BSs. Also the Alamouti scheme is addressed to two BSs [8] and in the following this scheme is called cellular Alamouti technique (CAT). The obtained macrodiversity can be utilized for handover demands, for example.

Proposals for a next generation mobile communications system design favor a multicarrier transmission, namely, OFDM [9]. It offers simple digital realization due to the fast Fourier transformation (FFT) operation and low complexity receivers. The WINNER project aims at a generalized multicarrier (GMC) [10] concept which is based on a high flexible packet-oriented data transmission. The resource allocation within a frame is given by time-frequency units, so called chunks. The chunks are preassigned to different classes of data flows and transmission schemes. They are then used in a flexible way to optimize the transmission performance [11].

One proposed transmission scheme within GMC is the multicarrier code division multiple access (MC-CDMA). MC-CDMA combines the benefits of multicarrier transmission and spread spectrum and was simultaneously proposed in 1993 by Fazel and Papke [12] and Yee et al. [13]. In addition to OFDM, spread spectrum, namely, code division multiple access (CDMA), gives high flexibility due to simultaneous access of users, robustness, and frequency diversity gains [14].

In this paper, the proposed techniques C-CDD and CAT are applied to a cellular environment based on an MC-CDMA transmission scheme. The structure of the paper is as follows. Section 2 describes the used cellular multicarrier system based on MC-CDMA. Section 3 introduces the cellular transmit diversity technique based on CDD and the application of the Alamouti scheme to a cellular environment. At the end of this section both techniques are compared and the differences are highlighted. A more detailed analytical investigation regarding the influence of the MT position for the C-CDD is given in Section 4. Finally, the proposed schemes are evaluated in Section 5.

2. CELLULAR MULTICARRIER SYSTEM

In this section, we first give an outline of the used MC-CDMA downlink system. We then describe the settings of the cellular environment and the used channel model.

2.1. MC-CDMA system

The block diagram of a transmitter using MC-CDMA is shown in Figure 1. The information bit streams of the N_u active users are convolutionally encoded and interleaved by the outer interleaver Π_{out}. With respect to the modulation alphabet, the bits are mapped to complex-valued data symbols. In the subcarrier allocation block, N_d symbols per user are arranged for each OFDM symbol. The kth data symbol is multiplied by a user-specific orthogonal Walsh-Hadamard spreading code which provides chips. The spreading length L corresponds to the maximum number of active users $L = N_{u,max}$. The ratio of the number of active users to $N_{u,max}$ represents the resource load (RL) of an MC-CDMA system.

An inner random subcarrier interleaver Π_{in} allows a better exploitation of diversity. The input block of the interleaver is denoted as one OFDM symbol and N_s OFDM symbols describe one OFDM frame. By taking into account a whole OFDM frame, a two-dimensional (2D) interleaving in frequency and time direction is possible. Also an interleaving over one dimension (1D), the frequency direction, is practicable by using one by one OFDM symbols. These complex valued symbols are transformed into time domain by the OFDM entity using an inverse fast Fourier transform (IFFT). This results in N_{FFT} time domain OFDM symbols, represented by the samples

$$x_l^{(n)} = \frac{1}{\sqrt{N_{FFT}}} \sum_{i=0}^{N_{FFT}-1} X_i^{(n)} \cdot e^{j(2\pi/N_{FFT})il}, \qquad (1)$$

where l, i denote the discrete time and frequency and n the transmitting BS out of N_{BS} BSs. A cyclic prefix as a guard interval (GI) is inserted in order to combat intersymbol interference (ISI). We assume quasistatic channel fading processes, that is, the fading is constant for the duration of one OFDM symbol. With this quasistatic channel assumption the well-known description of OFDM in the frequency domain is given by the multiplication of the transmitted data symbol $X_{l,i}^{(n)}$ and a complex channel transfer function (CTF) value $H_{l,i}^{(n)}$. Therefore, on the receiver side the lth received MC-CDMA symbol at subcarrier i becomes

$$Y_{l,i} = \sum_{n=0}^{N_{BS}-1} X_{l,i}^{(n)} H_{l,i}^{(n)} + N_{l,i} \qquad (2)$$

with $N_{l,i}$ as an additive white Gaussian noise (AWGN) process with zero mean and variance σ^2, the transmitter signal processing is inverted at the receiver which is illustrated in Figure 2. In MC-CDMA the distortion due to the flat fading on each subchannel is compensated by equalization. The received chips are equalized by using a low complex linear minimum mean square error (MMSE) one-tap equalizer. The resulting MMSE equalizer coefficients are

$$G_{l,i} = \frac{H_{l,i}^{(n)*}}{|H_{l,i}^{(n)}|^2 + (L/N_u)\sigma^2}, \quad i = 1, \dots, N_c. \qquad (3)$$

Furthermore, N_c is the total number of subcarriers. The operator $(\cdot)^*$ denotes the complex conjugate. Further, the symbol demapper calculates the log-likelihood ratio for each bit

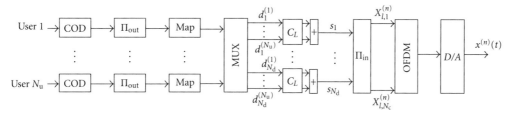

FIGURE 1: MC-CDMA transmitter of the nth base station.

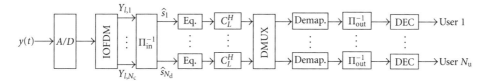

FIGURE 2: MC-CDMA receiver.

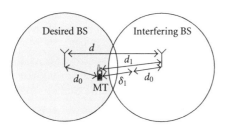

FIGURE 3: Cellular environment.

based on the selected alphabet. The code bits are deinterleaved and finally decoded using soft-decision Viterbi decoding [15].

2.2. Cellular environment

We consider a synchronized cellular system in time and frequency with two cells throughout the paper, see Figure 3. The nth BS has a distance d_n to the desired MT. A propagation loss model is assumed to calculate the received signal energy. The signal energy attenuation due to path loss is generally modeled as the product of the γth power of distance and a log-normal component representing shadowing losses. The propagation loss normalized to the cell radius r is defined by

$$\alpha(d_n) = \left(\frac{d_n}{r}\right)^{-\gamma} \cdot 10^{\eta/10\,\text{dB}}, \tag{4}$$

where the standard deviation of the Gaussian-distributed shadowing factor η is set to 8 dB. The superimposed signal at the MT is given by

$$\begin{aligned} Y_{l,i} &= X_{l,i}^{(0)}\alpha(d_0)H_{l,i}^{(0)} + X_{l,i}^{(1)}\alpha(d_1)H_{l,i}^{(1)} + N_{l,i} \\ &= S_{l,i}^{(0)} + S_{l,i}^{(1)} + N_{l,i}. \end{aligned} \tag{5}$$

Depending on the position of the MT the carrier-to-interference ratio (C/I) varies and is defined by

$$\frac{C}{I} = \frac{E\{|S_{l,i}^{(0)}|^2\}}{E\{|S_{l,i}^{(1)}|^2\}}. \tag{6}$$

3. TRANSMIT DIVERSITY TECHNIQUES FOR CELLULAR ENVIRONMENT

In a cellular network the MT switches the corresponding BS when it is requested by the BS. The switch is defined as the handover procedure from one BS to another. The handover is seamless and soft when the MT is connected to both BSs at the same time. The subcarrier resources in an MC-CDMA system within a spreading block are allocated to different users. Some users might not need a handover as they are (a) in a stable position or (b) away from the cell border. In both cases these users are effected by intercell interference as their resource is also allocated in the neighboring cell. To separate the different demands of the users, users with similar demands are combined within time-frequency units, for example, chunks, in an OFDM frame. The requested parameters of the users combined in these chunks are similar, like a common pilot grid. The spectrum for the users could then be shared between two cells within a chunk by defining a broadcast region. By this the affected users of the two cells would reduce their effective spectrum in half. This would be a price to pay avoiding intercellular interference. Intercellular interference could be tackled by intercellular interference cancellation techniques at complexity costs for all mobile users. Smart BSs could in addition try to balance the needed transmit power by risking an increase of intercellular interference also in neighboring cells. The approach presented in the following avoids intercellular interference by defining the effected area as a broadcast region and applying transmit diversity schemes for a cellular system, like cyclic delay diversity and STBCs. Part of the ineluctable loss of spectrum efficiency are compensated by exploiting additional diversity gains on the physical layer, avoiding the need of high complex intercellular cancellation techniques and decreasing the overall intercellular interference in the cellular network for the common good.

In the following, two transmit diversity techniques are in the focus. The first is based on the cyclic delay diversity (CDD) technique which increases the frequency diversity of the received signal and requires no change at the receiver to

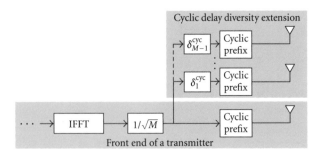

FIGURE 4: Principle of cyclic delay diversity.

exploit the diversity. The other technique applies the Alamouti scheme which flattens the frequency selectivity of the received signal and requires an additional decoding process at the mobile.

3.1. Cellular cyclic delay diversity (C-CDD)

The concept of cyclic delay diversity to a multicarrier-based system, that is, MC-CDMA, is briefly introduced in this section. Later on, the CDD concept will lead to an application to a cellular environment, namely, cellular CDD (C-CDD). A detailed description of CDD can be found in [16]. The idea of CDD is to increase the frequency selectivity, that is, to decrease the coherence bandwidth of the system. The additional diversity is exploited by the FEC and for MC-CDMA also by the spreading code. This will lead to a better error performance in a cyclic prefix-based system. The CDD principle is shown in Figure 4. An OFDM modulated signal is transmitted over M antennas, whereas the particular signals only differ in an antenna specific cyclic shift δ_m^{cyc}. MC-CDMA modulated signals are obtained from a precedent coding, modulation, spreading, and framing part; see also Section 2.1. Before inserting a cyclic prefix as guard interval, the time domain OFDM symbol (cf. (1)) is shifted cyclically, which results in the signal

$$x_{l-\delta_m^{\mathrm{cyc}}\bmod N_{\mathrm{FFT}}} = \frac{1}{\sqrt{N_{\mathrm{FFT}}}} \sum_{i=0}^{N_{\mathrm{FFT}}-1} e^{-j(2\pi/N_{\mathrm{FFT}})i\delta_m^{\mathrm{cyc}}} \cdot X_i \cdot e^{j(2\pi/N_{\mathrm{FFT}})il}.$$

(7)

The antenna specific TX-signal is given by

$$x_l^{(m)} = \frac{1}{\sqrt{M}} \cdot x_{l-\delta_m^{\mathrm{cyc}}\bmod N_{\mathrm{FFT}}},$$

(8)

where the signal is normalized by $1/\sqrt{M}$ to keep the average transmission power independent of the number of transmit antennas. The time domain signal including the guard interval is obtained for $l = -N_{\mathrm{GI}}, \ldots, N_{\mathrm{FFT}} - 1$. To avoid ISI, the guard interval length N_{GI} has to be larger than the maximum channel delay $\tau_{\max}$. Since CDD is done before the guard interval insertion in the OFDM symbol, CDD does not increase the $\tau_{\max}$ in the sense of ISI occurrence. Therefore, the length of the guard interval for CDD does not depend on the cyclic delays δ_m^{cyc}, where δ_m^{cyc} is given in samples.

On the receiver side and represented in the frequency domain (cf. (2)), the cyclic shift can be assigned formally to the channel transfer function, and therefore, the overall CTF

$$H_{l,i} = \frac{1}{\sqrt{M}} \sum_{m=0}^{M-1} e^{-j(2\pi/N_{\mathrm{FFT}})\delta_m^{\mathrm{cyc}} \cdot i} \cdot H_{l,i}^{(m)}$$

(9)

is observed. As long as the effective maximum delay $\tau'_{\max}$ of the resulting channel

$$\tau'_{\max} = \tau_{\max} + \max_m \delta_m^{\mathrm{cyc}}$$

(10)

does not intensively exceed N_{GI}, there is no configuration and additional knowledge at the receiver needed. If $\tau'_{\max} \gg N_{\mathrm{GI}}$, the pilot grid and also the channel estimation process has to be modified [17]. For example, this can be circumvented by using differential modulation [18].

The CDD principle can be applied in a cellular environment by using adjacent BSs. This leads to the cellular cyclic delay diversity (C-CDD) scheme. C-CDD takes advantage of the aforementioned resulting available resources from the neighboring BSs. The main goal is to increase performance by avoiding interference and increasing diversity at the most critical areas.

For C-CDD the interfering BS also transmits a copy of the users' signal as the desired BS to the designated MT located in the broadcast area. Additionally, a cyclic shift δ_n^{cyc} is inserted to this signal, see Figure 5. Therefore, the overall delay in respect to the signal of the desired BS in the cellular system can be expressed by

$$\delta_n = \delta(d_n) + \delta_n^{\mathrm{cyc}},$$

(11)

where $\delta(d_n)$ represents the natural delay of the signal depending on distance d_n. At the MT the received signal can be described by

$$Y_{l,i} = X_{l,i}^{(0)} \left(\alpha(d_0) H_{l,i}^{(0)} e^{-j(2\pi/N_{\mathrm{FFT}})\delta_0 \cdot i} + \alpha(d_1) H_{l,i}^{(1)} e^{-j(2\pi/N_{\mathrm{FFT}})\delta_1 \cdot i} \right).$$

(12)

The transmission from the BSs must ensure that the reception of both signals are within the guard interval. Furthermore, at the MT the superimposed statistical independent Rayleigh distributed channel coefficients from the different BSs sum up again in a Rayleigh distributed channel coefficient. The usage of cyclic shifts prevents the occurrence of additional ISI. For C-CDD no additional configurations at the MT for exploiting the increased transmit diversity are necessary.

Finally, the C-CDD technique inherently provides another transmit diversity technique. If no cyclic shift δ_n^{cyc} is introduced, the signals from the different BSs may arrive at the desired MT with different delays $\delta(d_n)$. These delays can be also seen as delay diversity (DD) [5] for the transmitted MC-CDMA signal or as macrodiversity [19] at the MT. Therefore, an inherent transmit diversity, namely, cellular delay diversity (C-DD), is introduced if the adjacent BSs just transmit the same desired signal at the same time to the designated MT. The C-CDD techniques can be also easily extended to more than 2 BSs.

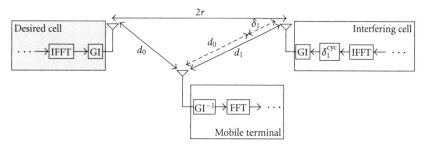

FIGURE 5: Cellular MC-CDMA system with cellular cyclic delay diversity (C-CDD).

3.2. Cellular Alamouti technique (CAT)

In this section, we introduce the concept of transmit diversity by using the space-time block codes (STBCs) from orthogonal designs [7], namely, the Alamouti technique. We apply this scheme to the aforementioned cellular scenario. These STBCs are based on the theory of (generalized) orthogonal designs for both real- and complex-valued signal constellations. The complex-valued STBCs can be described by a matrix

$$\mathbf{B} = \begin{pmatrix} b_{0,0} & \cdots & b_{0,N_{\mathrm{BS}}-1} \\ \vdots & \ddots & \vdots \\ b_{l-1,0} & \cdots & b_{l-1,N_{\mathrm{BS}}-1} \end{pmatrix} \begin{array}{c} \uparrow \\ \text{time} \\ \downarrow \end{array}, \quad (13)$$

where l and N_{BS} are the STBC length and the number of BS (we assume a single TX-antenna for each BS), respectively. The simplest case is the Alamouti code [20],

$$\mathbf{B} = \begin{pmatrix} x_0 & x_1 \\ -x_1^* & x_0^* \end{pmatrix}. \quad (14)$$

The respective assignment for the Alamouti-STBC to the kth block of chips containing data from one or more users is obtained:

$$\vec{y}^{(k)} = \begin{pmatrix} y_0^{(k)} \\ y_1^{(k)*} \end{pmatrix}$$
$$= \begin{pmatrix} h^{(0,k)} & h^{(1,k)} \\ h^{(1,k)*} & -h^{(0,k)*} \end{pmatrix} \cdot \begin{pmatrix} x_0 \\ x_1 \end{pmatrix} + \begin{pmatrix} n_0^{(k)} \\ n_1^{(k)*} \end{pmatrix}. \quad (15)$$

$\vec{y}^{(k)}$ is obtained from the received complex values $y_i^{(k)}$ or their conjugate complex $y_i^{(k)*}$ at the receiver. At the receiver, the vector $\vec{y}^{(k)}$ is multiplied from left by the Hermitian of matrix $\mathbf{H}^{(k)}$. The fading between the different fading coefficients is assumed to be quasistatic. We obtain the (weighted) STBC information symbols

$$\hat{\vec{x}} = \mathbf{H}^{(k)H} \cdot \vec{y}^{(k)} = \mathbf{H}^{(k)H} \cdot \mathbf{H}^{(k)} \vec{x} + \mathbf{H}^{(k)H} \vec{n}^{(k)}$$
$$= \mathbf{H}^{(k)H} \cdot \vec{n}^{(k)} + \vec{x} \cdot \sum_{i=0}^{1} \left| h^{(i,k)} \right|^2, \quad (16)$$

corrupted by noise. For STBCs from orthogonal designs, MIMO channel estimation at the receiver is mandatory, that is, $h^{(n,k)}$, $n = 0, \ldots, N_{\mathrm{BS}} - 1$, $k = 0, \ldots, K - 1$, must be

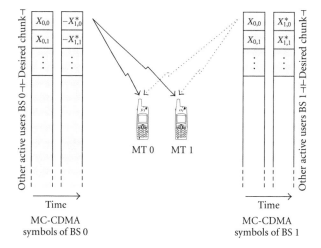

FIGURE 6: MC-CDMA symbol design for CAT for 2 MTs.

estimated. Disjoint pilot symbol sets for the TX-antenna branches can guarantee a separate channel estimation for each BS [8]. Since the correlation of the subcarrier fading coefficients in time direction is decreasing with increasing Doppler spread—that is, the quasistationarity assumption of the fading is incrementally violated—the performance of this STBC class will suffer from higher Doppler frequencies. Later we will see that this is not necessarily true as the stationarity of the fading could also be detrimental in case of burst errors in fading channels.

Figure 6 shows two mobile users sojourning at the cell borders. Both users data is spread within one spreading block and transmitted by the cellular Alamouti technique using two base stations. The base stations exploit information from a feedback link that the two MTs are in a similar location in the cellular network. By this both MTs are served simultaneously avoiding any interference between each other and exploiting the additional diversity gain.

3.3. Résumé for C-CDD and CAT

Radio resource management works perfectly if all information about the mobile users, like the channel state information, is available at the transmitter [21]. This is especially true if the RRM could be intelligently managed by a single genie manager. As this will be very unlikely the described schemes C-CDD and CAT offer an improved performance especially at the critical cell border without the need of any information about the channel state information on the transmitter side. The main goal is to increase performance by avoiding interference and increasing diversity at the most critical environment. In this case, the term C/I is misleading (cf. (6)), as there is no I (interference). On the other hand, it describes the ratio of the power from the desired base station and the other base station. This ratio also indicates where the mobile user is in respect to the base stations. For C/I = 0 dB the MT is directly between the two BSs, for C/I > 0 dB the MT is closer to the desired BS, and for C/I < 0 dB the MT is closer to the adjacent BS. Since the signals of the neighboring BSs for the desired users are not seen as interference, the MMSE equalizer coefficients of (3) need no modification as in the intercellular interfering case [22]. Therefore, the transmit diversity techniques require no knowledge about the intercellular interference at the MT. By using C-CDD or CAT the critical cell border area can be also seen as a broadcast scenario with a multiple access channel.

For the cellular transmit diversity concepts C-CDD and CAT, each involved BS has to transmit additionally the signal of the adjacent cell; and therefore, a higher amount of resources are allocated at each BS. Furthermore, due to the higher RL in each cell the multiple-access interference (MAI) for an MC-CDMA system is increased. There will be always a tradeoff between the increasing MAI and the increasing diversity due to C-CDD or CAT.

Since the desired signal is broadcasted by more than one BS, both schemes can reduce the transmit signal power, and therefore, the overall intercellular interference. Using MC-CDMA for the cellular diversity techniques the same spreading code set has to be applied at the involved BSs for the desired signal which allows simple receivers at the MT without multiuser detection processes/algorithms. Furthermore, a separation between the inner part of the cells and the broadcast area can be achieved by an overlaying scrambling code on the signal which can be also used for synchronization issues as in UMTS [4].

Additionally, if a single MT or more MTs are aware that they are at the cell border, they could already ask for the C-CDD or CAT procedure on the first hand. This would ease the handover procedure and would guarantee a reliable soft handover.

We should point out two main differences between C-CDD and CAT. For C-CDD no changes at the receiver are needed, there exists no rate loss for higher number of transmit antennas, and there are no requirements regarding constant channel properties over several subcarriers or symbols and transmit antenna numbers. This is an advantage over already established diversity techniques [7] and CAT. The Alamouti scheme-based technique CAT should provide a better performance due to the coherent combination of the two transmitted signals [23].

4. RESULTING CHANNEL CHARACTERISTICS FOR C-CDD

The geographical influence of the MT for CAT has a symmetric behavior. In contrast, C-CDD is influenced by the position of the served MT. Due to $\delta_0^{\text{cyc}} \neq \delta_1^{\text{cyc}}$ and the relation in (11), the resulting performance regarding the MT position of C-CDD should have an asymmetric characteristic. Since the influence of C-CDD on the system can be observed at the receiver as a change of the channel conditions, we will investigate in the following this modified channel in terms of its channel transfer functions and fading correlation in time and frequency direction. These correlation characteristics also describe the corresponding single transmit antenna channel seen at the MT for C-CDD.

The frequency domain fading processes for different propagation paths are uncorrelated in the assumed quasistatic channel. Since the number of subcarriers is larger than the number of propagation paths, there exists correlation between the subcarriers in the frequency domain. The received signal at the receiver in C-CDD can be represented by

$$Y_{l,i} = X_{l,i} \cdot \underbrace{\sum_{n=0}^{N_{\text{BS}}-1} e^{-j(2\pi/N_{\text{FFT}})i\delta_n} \alpha(d_n) H_{l,i}^{(n)}}_{H'_{l,i}} + N_{l,i}. \quad (17)$$

Since the interest is based on the fading and signal characteristics observed at the receiver, the AWGN term $N_{l,i}$ is skipped for notational convenience. The expectation

$$\mathbf{R}(l_1, l_2, i_1, i_2) = E\{H'_{l_1,i_1} \cdot H'^*_{l_2,i_2}\} \quad (18)$$

yields the correlation properties of the frequency domain channel fading. Due to the path propagations $\alpha(d_n)$ and the resulting power variations, we have to normalize the channel transfer functions $H_{l,i}^{(n)}$ by the multiplication factor $1/\sqrt{\sum_{n=0}^{N_{\text{BS}}-1} \alpha^2(d_n)}$ which is included for $\mathbf{R}_n(l, i)$.

The fading correlation properties can be divided in three cases. The first represents the power, the second investigates the correlation properties between the OFDM symbols (time direction), and the third examines the correlation properties between the subcarriers (frequency direction).

Case 1. Since we assume uncorrelated subcarriers the autocorrelation of the CTF ($l_1 = l_2 = l$, $i_1 = i_2 = i$) is

$$\mathbf{R}(l, i) = \sum_{n=0}^{N_{\text{BS}}-1} \underbrace{e^{-j(2\pi/N_{\text{FFT}})i\delta_n} \cdot e^{+j(2\pi/N_{\text{FFT}})i\delta_n}}_{=1} \alpha^2(d_n)$$

$$\cdot \underbrace{E\{H_{l,i}^{(n)} \cdot H_{l,i}^{(n)*}\}}_{E\{|H_{l,i}^{(n)}|^2\}=1} = \sum_{n=0}^{N_{\text{BS}}-1} \alpha^2(d_n), \quad (19)$$

and the normalized power is

$$\mathbf{R}_n(l, i) = \sum_{n=0}^{N_{\text{BS}}-1} \alpha^2(d_n) E\left\{\left|\frac{H_{l,i}^{(n)}}{\sqrt{\sum_{n=0}^{N_{\text{BS}}-1} \alpha^2(d_n)}}\right|^2\right\} = 1. \quad (20)$$

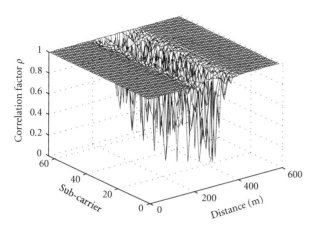

FIGURE 7: Characteristic of correlation factor ρ over the subcarriers depending on the distance d_0.

Case 2. The correlation in time direction is given by $l_1 \neq l_2$, $i_1 = i_2 = i$. Since the channels from the BSs are i.i.d. stochastic processes, $E\{H_{l_1,i}^{(n)} \cdot H_{l_2,i}^{(n)*}\} = E\{H_{l_1,i} \cdot H_{l_2,i}^*\}$ and

$$\mathbf{R}(l_1 \neq l_2, i) = E\{H_{l_1,i} H_{l_2,i}^*\} \sum_{n=0}^{N_{BS}-1} \alpha^2(d_n),$$

$$\mathbf{R}_n(l_1 \neq l_2, i) = E\left\{ \frac{H_{l_1,i} H_{l_2,i}^*}{\sum_{n=0}^{N_{BS}-1} \alpha^2(d_n)} \right\} \sum_{n=0}^{N_{BS}-1} \alpha^2(d_n)$$

$$= E\{H_{l_1,i} H_{l_2,i}^*\}. \tag{21}$$

We see that in time direction, the correlation properties of the resulting channel are independent of the MT position.

Case 3. In frequency direction ($l_1 = l_2 = l, i_1 \neq i_2$) the correlation properties are given by

$$\mathbf{R}(l, i_1 \neq i_2) = E\{H_{l,i_1} H_{l,i_2}^*\} \cdot \underbrace{\sum_{n=0}^{N_{BS}-1} \alpha^2(d_n) e^{-j(2\pi/N_{FFT})(i_1-i_2)\delta_n}}_{\text{C-CDD component}}. \tag{22}$$

For large d_n ($\alpha(d_n)$ gets small) the influence of the C-CDD component vanishes. And there is no beneficial increase of the frequency diversity close to a BS anymore. The normalized correlation properties yield

$$\mathbf{R}_n(l, i_1 \neq i_2) = E\{H_{l,i_1} H_{l,i_2}^*\}$$
$$\cdot \underbrace{\frac{1}{\sum_{n=0}^{N_{BS}-1} \alpha^2(d_n)} \cdot \sum_{n=0}^{N_{BS}-1} \alpha^2(d_n) e^{-j(2\pi/N_{FFT})(i_1-i_2)\delta_n}}_{\text{correlation factor } \rho}. \tag{23}$$

The correlation factor ρ is directly influenced by the C-CDD component and determines the overall channel correlation properties in frequency direction. Figure 7 shows the characteristics of ρ for an exemplary system with $N_{FFT} = 64$, $\gamma = 3.5$, $N_{BS} = 2$, $r = 300\,\mathrm{m}$, $\delta_0^{cyc} = 0$, and $\delta_1^{cyc} = 7$. One

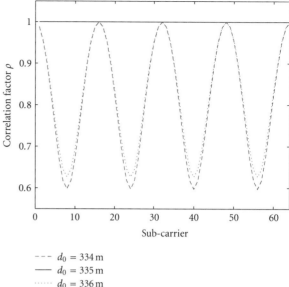

$--- \quad d_0 = 334\,\mathrm{m}$
$\underline{} \quad d_0 = 335\,\mathrm{m}$
$\cdots\cdots \quad d_0 = 336\,\mathrm{m}$

FIGURE 8: Correlation characteristics over the subcarriers for $d_0 = [334\,\mathrm{m}, 335\,\mathrm{m}, 336\,\mathrm{m}]$.

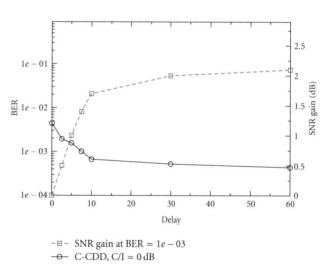

$-\square- \quad$ SNR gain at BER $= 1e - 03$
$-\circ- \quad$ C-CDD, C/I $= 0\,\mathrm{dB}$

FIGURE 9: BER and SNR gains versus the cyclic delay at the cell border (C/I $= 0\,\mathrm{dB}$).

sample of the delay represents 320 microseconds or approximately $10\,\mathrm{m}$, respectively. In the cell border area ($200\,\mathrm{m} < d_0 < 400\,\mathrm{m}$), C-CDD increases the frequency diversity by decorrelating the subcarriers. As mentioned before, there is less decorrelation the closer the MT is to a BS.

A closer look on the area is given in Figure 8 where the inherent delay and the added cyclic delay are compensated, that is, for $d_0 = 335\,\mathrm{m}$ the overall delay is $\delta_1 = \delta(265\,\mathrm{m}) + \delta_1^{cyc} = -70\,\mathrm{m} + 70\,\mathrm{m} = 0$ (cf. (11)). The plot represents exemplarily three positions of the MT ($d_0 = [334\,\mathrm{m}, 335\,\mathrm{m}, 336\,\mathrm{m}]$) and shows explicitly the degradation of the correlation properties over all subcarriers due to the nonexisting delay in the system. These analyses verify the asymmetric and δ^{cyc} dependent characteristics of C-CDD.

TABLE 1: Parameters of the cellular transmission systems.

Bandwidth	B	100.0 MHz
No. of subcarriers	N_c	1664
FFT length	N_{FFT}	2048
Guard interval length	N_{GI}	128
Sample duration	T_{samp}	10.0 ns
Frame length	N_{frame}	16
No. of active users	N_u	$\{1,\ldots,8\}$
Spreading lengh	L	8
Modulation	—	4-QAM, 16-QAM
Interleaving C-CDD	—	2D
Interleaving CAT	—	1D, 2D
Channel coding	—	CC $(561,753)_{oct}$
Channel coding rate	R	1/2
Channel model	—	IEEE 802.11n Model C
Velocity	—	0 mph, 40 mph

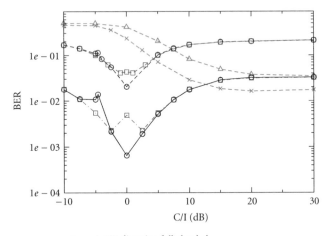

- –△– w/o TX diversity, fully loaded
- –✗– w/o TX diversity, half loaded
- –⊟– C-DD, halved TX power
- –⊖– C-CDD, halved TX power
- –▫– C-DD
- –⊙– C-CDD

FIGURE 10: BER versus C/I for an SNR of 5 dB using no transmit diversity technique, C-DD, and C-CDD for different scenarios.

5. SIMULATION RESULTS

The simulation environment is based on the parameter assumptions of the IST-project WINNER for next generation mobile communications system [24]. The used channel model is the 14 taps IEEE 802.11n channel model C with $\gamma = 3.5$ and $\tau_{max} = 200$ nanoseconds. This model represents a large open space (indoor and outdoor) with non-light-of-sight conditions with a cell radius of $r = 300$ m. The transmission system is based on a carrier frequency of 5 GHz, a bandwidth of 100 MHz, and an FFT length of $N_c = 2048$. One OFDM symbol length (excluding the GI) is 20.48 microseconds and the GI is set to 0.8 microseconds (corresponding to 80 samples). The spreading length L is set to

8. The number of active users can be up to 8 depending on the used RL. 4-QAM is used throughout all simulations and for throughput performances 16-QAM is additionally investigated. For the simulations, the signal-to-noise ratio (SNR) is set to 5 dB and perfect channel knowledge at the receiver is assumed. Furthermore, a $(561,753)_8$ convolutional code with rate $R = 1/2$ was selected as channel code. Each MT moves with an average velocity of 40 mph (only for comparison to see the effect of natural time diversity) or is static. As described in Section 3, users with similar demands at the cell border are combined within time-frequency units. We assume i.i.d. channels with equal stochastic properties from each BS to the MT. If not stated otherwise, a fully loaded system is simulated for the transmit diversity techniques, and therefore, their performances can be seen as upper bounds. All simulation parameters are summarized in Table 1. In the following, we separate the simulation results in three blocks. First, we discuss the performances of CDD; then, the simulation results of CAT are debated; and finally, the influence of the MAI to both systems and the throughput of both systems is investigated.

5.1. C-CDD performance

Figure 9 shows the influence of the cyclic delay δ_1^{cyc} to the bit-error rate (BER) and the SNR gain at the cell border (C/I = 0 dB) for C-CDD. At the cell border there is no influence due to C-DD, that is, ($\delta_1 = 0$). Two characteristics of the performance can be highlighted. First, there is no performance gain for $\delta_1^{cyc} = 0$ due to the missing C-CDD. Secondly, the best performance can be achieved for an existing higher cyclic shift which reflects the results in [25]. The SNR gain performance for a target BER of 10^{-3} depicts also the influence of the increased cyclic delay. For higher delays the performance saturates at a gain of about 2 dB.

The performances of the applied C-DD and C-CDD methods are compared in Figure 10 with the reference system using no transmit (TX) diversity technique. For the reference system both BSs are transmitting independently

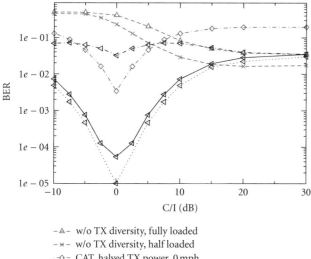

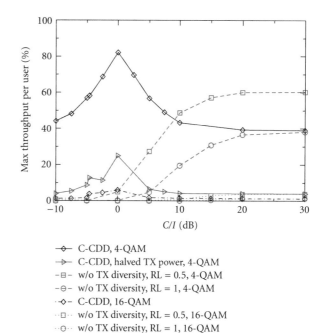

- –△– w/o TX diversity, fully loaded
- –✕– w/o TX diversity, half loaded
- ··◇·· CAT, halved TX power, 0 mph
- ·◁· CAT, 0 mph, 2D interleaving
- –◄– CAT, 0 mph
- ·◁·· CAT, 40 mph

FIGURE 11: BER versus C/I for an SNR of 5 dB using no transmit diversity and CAT for different scenarios.

- –◇– C-CDD, 4-QAM
- –▷– C-CDD, halved TX power, 4-QAM
- –⊟– w/o TX diversity, RL = 0.5, 4-QAM
- –⊖– w/o TX diversity, RL = 1, 4-QAM
- ··◇·· C-CDD, 16-QAM
- ··□·· w/o TX diversity, RL = 0.5, 16-QAM
- ··○·· w/o TX diversity, RL = 1, 16-QAM

FIGURE 13: Throughput per user for 4-QAM versus C/I using no transmit diversity or C-CDD with full and halved transmit power.

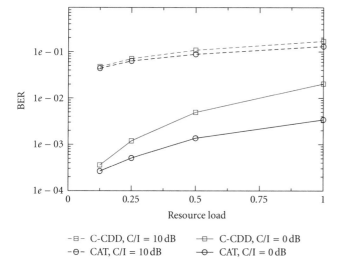

- –⊟– C-CDD, C/I = 10 dB   –□– C-CDD, C/I = 0 dB
- –⊖– CAT, C/I = 10 dB   –○– CAT, C/I = 0 dB

FIGURE 12: Influence of the MAI to the BER performance for varying resource loads at the cell border and the inner part of the cell.

their separate MC-CDMA signal. From Figure 9, we choose $\delta_1^{cyc} = 30$ samples and this cyclic delay is chosen throughout all following simulations. The reference system is half (RL = 0.5) and fully loaded (RL = 1.0). We observe a large performance gain in the close-by area of the cell border (C/I = -10 dB, ..., 10 dB) for the new proposed diversity techniques C-DD and C-CDD. Furthermore, C-CDD enables an additional substantial performances gain at the cell border. The C-DD performance degrades for C/I = 0 dB because $\delta = 0$ and no transmit diversity is available. The same effect can be seen for C-CDD at C/I = -4.6 dB ($\delta_1 = -30$, $\delta_1^{cyc} = 30 \Rightarrow \delta = 0$); see also Section 4. Since both BSs in C-DD and C-CDD transmit the signal with the same power

as the single BS in the reference system, the received signal power at the MT is doubled. Therefore, the BER performance of C-DD and C-CDD at $\delta = 0$ is still better than the reference system performance. For higher C/I ratios, that is, in the inner cell, the C-DD and C-CDD transmit techniques lack the diversity from the other BS and additionally degrade due to the double load in each cell. Thus, the MT has to cope with the double MAI. The loss due to the MAI can be directly seen by comparing the transmit diversity performance with the half-loaded reference system. The fully loaded reference system has the same MAI as the C-CDD system, and therefore, the performances merge for high C/I ratios. To establish a more detailed understanding we analyze the C-CDD with halved transmit power. For this scenario, the total designated received power at the MT is equal to the conventional MC-CDMA system. There is still a performance gain due to the exploited transmit diversity for C/I < 5 dB. The performance characteristics are the same for halved and full transmit power. The benefit of the halved transmit power is a reduction of the intercellular interference for the neighboring cells. In the case of varying channel models in the adjacent cells, the performance characteristics will be the same but not symmetric anymore. This is also valid for the following CAT performances.

5.2. CAT performance

Figure 11 shows the performances of the applied CAT in the cellular system for different scenarios. If not stated otherwise, the systems are using a 1D interleaving. In contrast to the conventional system, the BER can be dramatically improved at the cell border. By using the CAT, the MT exploits the additional transmit diversity where the maximum is given at the

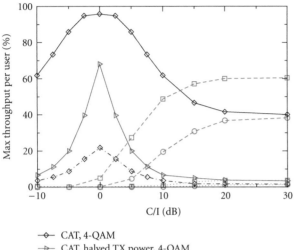

FIGURE 14: Throughput per user for 4-QAM and 16-QAM versus C/I using no transmit diversity or CAT with full and halved transmit power.

cell border. If the MT moves with higher velocity (40 mph), the correlation of the subcarrier fading coefficients in time direction decreases. This incremental violation of the quasistationarity assumption of the fading is profitable compensated by the channel code. The total violation of the aforementioned constraint of CAT (cf. Section 3.2) is achieved by a fully interleaved (2D) MC-CDMA frame. There is a large performance degradation compared to the CAT performance with a noninterleaved frame. Nevertheless, a residual transmit diversity exists, the MT benefits at the cell border, and the performance is improved. The applied CAT is not only robust for varying MT velocities but also for non-quasistatic channel characteristics. Similar to C-CDD, there is still a performance gain due to the exploited transmit diversity for C/I < 5 dB in the case of halved transmit powers at both BSs.

5.3. *MAI and throughput performance of C-CDD and CAT*

The influence of the MAI is shown in Figure 12. The BER performance versus the resource load of the systems is presented. Two different positions of the MT are chosen: directly at the cell border (C/I = 0 dB) and closer to one BS (C/I = 10 dB). Both transmit diversity schemes suffer from the increased MAI for higher resource loads which is in the nature of the used MC-CDMA system. CAT is not influenced by the MAI as much as C-CDD for both scenarios. Both performances merge for C/I = 10 dB because the influence of the transmit diversity techniques is highly reduced in the inner part of the cell.

Since we assume the total number of subcarriers is equally distributed to the maximum number of users per cell,

each user has a maximum throughput of $\eta_{\max}$. The throughput η of the system, by using the probability $P(n)$ of the first correct MC-CDMA frame transmission after $n - 1$ failed retransmissions, is given by

$$\eta = \sum_{n=0}^{\infty} \frac{\eta_{\max}}{n + 1} P(n) \geq \eta_{\max}(1 - \text{FER}). \tag{24}$$

A lower bound of the system is given by the right-hand side of (24) by only considering $n = 0$ and the frame-error rate (FER). Figures 13 and 14 illustrate this lower bound for different modulations in the case of C-CDD and CAT.

C-CDD in Figure 13 outperforms the conventional system at the cell border for all scenarios. Due to the almost vanishing performance for 16-QAM with halved transmit power for an SNR of 5 dB, we do not display this performance curve. For 4-QAM and C-CDD, a reliable throughput along the cell border is achieved. Since C-CDD with halved transmit power still outperforms the conventional system, it is possible to decrease the intercellular interference.

The same performance characteristics as in C-CDD regarding the throughput can be seen in Figure 14 for applying the transmit diversity technique CAT. Due to the combination of two signals in the Alamouti scheme, CAT can provide a higher throughput than C-CDD in the cell border area. The CAT can almost achieve the maximum possible throughput in the cell border area. For both transmit diversity techniques, power and/or modulation adaptation from the BSs opens the possibility for the MT to request a higher throughput in the critical cell border area. All these characteristics can be utilized by soft handover concepts.

6. CONCLUSIONS

This paper handles the application of transmit diversity techniques to a cellular MC-CDMA-based environment. Addressing transmit diversity by using different base stations for the desired signal to a mobile terminal enhances the macrodiversity in a cellular system. Analyses and simulation results show that the introduced cellular cyclic delay diversity (C-CDD) and cellular Alamouti technique (CAT) are capable of improving the performance at the severe cell borders. Furthermore, the techniques reduce the overall intercellular interference. Therefore, it is desirable to use C-CDD and CAT in the outer part of the cells, depending on available resources in adjacent cells. The introduced transmit diversity techniques can be utilized for more reliable soft handover concepts.

ACKNOWLEDGMENTS

This work has been performed in the framework of the IST Project IST-4-027756 WINNER, which is partly funded by the European Union. The authors would like to acknowledge the contributions of their colleagues. The material in this paper was presented in part at the IEEE 64th Vehicular Technology Conference, Montréal, Canada, September 25–28, 2006.

REFERENCES

[1] IST-2003-507581 WINNER Project, https://www.ist-winner .org.

[2] S. Plass, "On intercell interference and its cancellation in cellular multicarrier CDMA systems," *EURASIP Journal on Wireless Communications and Networking*, vol. 2008, Article ID 173645, 11 pages, 2008.

[3] D. Wong and T. J. Lim, "Soft handoffs in CDMA mobile systems," *IEEE Personal Communications*, vol. 4, no. 6, pp. 6–17, 1997.

[4] M. Schinnenburg, I. Forkel, and B. Haverkamp, "Realization and optimization of soft and softer handover in UMTS networks," in *Proceedings of European Personal Mobile Communications Conference (EPMCC '03)*, pp. 603–607, Glasgow, UK, April 2003.

[5] A. Wittneben, "A new bandwidth efficient transmit antenna modulation diversity scheme for linear digital modulation," in *Proceedings of IEEE International Conference on Communications (ICC '93)*, pp. 1630–1634, Geneva, Switzerland, May 1993.

[6] A. Dammann and S. Kaiser, "Performance of low complex antenna diversity techniques for mobile OFDM systems," in *Proceedings of International Workshop on Multi-Carrier Spread Spectrum (MC-SS '01)*, pp. 53–64, Oberpfaffenhofen, Germany, September 2001.

[7] V. Tarokh, H. Jafarkhani, and A. R. Calderbank, "Space-time block codes from orthogonal designs," *IEEE Transactions on Information Theory*, vol. 45, no. 5, pp. 1456–1467, 1999.

[8] M. Inoue, T. Fujii, and M. Nakagawa, "Space time transmit site diversity for OFDM multi base station system," in *Proceedings of the 4th EEE International Workshop on Mobile and Wireless Communication Networks (MWCN '02)*, pp. 30–34, Stockholm, Sweden, September 2002.

[9] S. B. Weinstein and P. M. Ebert, "Data transmission by frequency-division multiplexing using the discrete Fourier transform," *IEEE Transactions on Communications*, vol. 19, no. 5, pp. 628–634, 1971.

[10] Z. Wang and G. B. Giannakis, "Wireless multicarrier communications: where Fourier meets Shannon," *IEEE Signal Processing Magazine*, vol. 17, no. 3, pp. 29–48, 2000.

[11] M. Sternad, T. Svensson, and G. Klang, "The WINNER B3G system MAC concept," in *Proceedings of IEEE Vehicular Technology Conference (VTC '06)*, pp. 3037–3041, Montreal, Canada, September 2006.

[12] K. Fazel and L. Papke, "On the performance of concolutionally-coded CDMA/OFDM for mobile communications systems," in *Proceedings of IEEE International Symposium on Personal, Indoor and Mobile Radio Communications (PIMRC '93)*, pp. 468–472, Yokohama, Japan, September 1993.

[13] N. Yee, J.-P. Linnartz, and G. Fettweis, "Multi-carrier CDMA for indoor wireless radio networks," in *Proceedings of IEEE International Symposium on Personal, Indoor and Mobile Radio Communications (PIMRC '93)*, pp. 109–113, Yokohama, Japan, September 1993.

[14] K. Fazel and S. Kaiser, *Multi-Carrier and Spread Spectrum Systems*, John Wiley & Sons, San Francisco, Calif, USA, 2003.

[15] A. Viterbi, "Error bounds for convolutional codes and an asymptotically optimum decoding algorithm," *IEEE Transactions on Information Theory*, vol. 13, no. 2, pp. 260–269, 1967.

[16] A. Dammann and S. Kaiser, "Transmit/receive-antenna diversity techniques for OFDM systems," *European Transactions on Telecommunications*, vol. 13, no. 5, pp. 531–538, 2002.

[17] G. Auer, "Channel estimation for OFDM with cyclic delay diversity," in *Proceedings of IEEE International Symposium on Personal, Indoor and Mobile Radio Communications (PIMRC '04)*, vol. 3, pp. 1792–1796, Barcelona, Spain, September 2004.

[18] G. Bauch, "Differential modulation and cyclic delay diversity in orthogonal frequency-division multiplex," *IEEE Transactions on Communications*, vol. 54, no. 5, pp. 798–801, 2006.

[19] G. L. Stüber, *Principles of Mobile Communication*, Kluwer Academic Publishers, Norwell, Mass, USA, 2001.

[20] S. M. Alamouti, "A simple transmit diversity technique for wireless communications," *IEEE Journal on Selected Areas in Communications*, vol. 16, no. 8, pp. 1451–1458, 1998.

[21] D. Tse and P. Viswanath, *Fundamentals of Wireless Communication*, Cambridge University Press, New York, NY, USA, 2005.

[22] S. Plass, X. G. Doukopoulos, and R. Legouable, "On MC-CDMA link-level inter-cell interference," in *Proceedings of the 65th IEEE Vehicular Technology Conference (VTC '07)*, pp. 2656–2660, Dublin, Ireland, April 2007.

[23] H. Schulze, "A comparison between Alamouti transmit diversity and (cyclic) delay diversity for a DRM+ system," in *Proceedings of International OFDM Workshop*, Hamburg, Germany, August 2006.

[24] IST-2003-507581 WINNER, "D2.10: final report on identified RI key technologies, system concept, and their assessment," December 2005.

[25] G. Bauch and J. S. Malik, "Cyclic delay diversity with bit-interleaved coded modulation in orthogonal frequency division multiple access," *IEEE Transactions on Wireless Communications*, vol. 5, no. 8, pp. 2092–2100, 2006.

Hindawi Publishing Corporation
EURASIP Journal on Wireless Communications and Networking
Volume 2007, Article ID 98186, 15 pages
doi:10.1155/2007/98186

Research Article

SmartMIMO: An Energy-Aware Adaptive MIMO-OFDM Radio Link Control for Next-Generation Wireless Local Area Networks

Bruno Bougard,[1, 2] Gregory Lenoir,[1] Antoine Dejonghe,[1] Liesbet Van der Perre,[1] Francky Catthoor,[1, 2] and Wim Dehaene[2]

[1] IMEC, Department of Nomadic Embedded Systems, Kapeldreef 75, 3001 Leuven, Belgium
[2] K. U. Leuven, Department of Electrical Engineering, Katholieke Universiteit Leuven, ESAT, 3000 Leuven, Belgium

Received 15 November 2006; Revised 12 June 2007; Accepted 8 October 2007

Recommended by Monica Navarro

Multiantenna systems and more particularly those operating on multiple input and multiple output (MIMO) channels are currently a must to improve wireless links spectrum efficiency and/or robustness. There exists a fundamental tradeoff between potential spectrum efficiency and robustness increase. However, multiantenna techniques also come with an overhead in silicon implementation area and power consumption due, at least, to the duplication of part of the transmitter and receiver radio frontends. Although the area overhead may be acceptable in view of the performance improvement, low power consumption must be preserved for integration in nomadic devices. In this case, it is the tradeoff between performance (e.g., the net throughput on top of the medium access control layer) and average power consumption that really matters. It has been shown that adaptive schemes were mandatory to avoid that multiantenna techniques hamper this system tradeoff. In this paper, we derive *smartMIMO*: an adaptive multiantenna approach which, next to simply adapting the modulation and code rate as traditionally considered, decides packet-per-packet, depending on the MIMO channel state, to use either space-division multiplexing (increasing spectrum efficiency), space-time coding (increasing robustness), or to stick to single-antenna transmission. Contrarily to many of such adaptive schemes, the focus is set on using multiantenna transmission to improve the link energy efficiency in real operation conditions. Based on a model calibrated on an existing reconfigurable multiantenna transceiver setup, the link energy efficiency with the proposed scheme is shown to be improved by up to 30% when compared to nonadaptive schemes. The average throughput is, on the other hand, improved by up to 50% when compared to single-antenna transmission.

1. INTRODUCTION

The performance of wireless communication systems can drastically be improved when using multiantenna transmission techniques. Specifically, multiantenna techniques can be used to increase antenna gain and directionality (beamforming, [1]), to improve link robustness (space-time coding [2, 3]), or to improve spectrum efficiency (space division multiplexing [4]). Techniques where multiple antennas are considered both at transmit and receive sides can combine those assets and are referred to as multiple-input multiple-output (MIMO). On the other hand, because of its robustness in harsh frequency selective channel combined with a low implementation cost, orthogonal frequency division multiplexing (OFDM) is now pervasive in broadband wireless communication. Therefore, MIMO-OFDM schemes turn out to be excellent candidates for next generation broadband wireless standards.

Traditionally, the benefit of MIMO schemes is characterized in terms of *multiplexing gain* (i.e., the increase in spectrum efficiency) and *diversity gain* (namely, the increase in immunity to the channel variation, quantified as the order of the decay of the bit-error rate as a function of the signal-to-noise ratio). In [5], it is shown that, given a multiple-input multiple-output (MIMO) channel and assuming a high signal-to-noise ratio, there exists a fundamental tradeoff between how much of these gains a given coding scheme can extract. Since then, the merit of a new multiantenna scheme is mostly evaluated with regard to that tradeoff. However, from the system perspective, one has also to consider the impact on implementation cost such as silicon area and energy efficiency. When multiantenna techniques are integrated in battery-powered nomadic devices, as it is mostly the case for wireless systems, it is the tradeoff between the effective link performance (namely, the net data rate on top of the medium access control layer) and the link energy efficiency

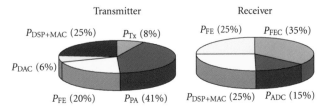

FIGURE 1: Power consumption breakdown of typical single-antenna OFDM transceivers [6, 7]. At the transmit part, the power amplifier contribution $(P_{Tx} + P_{PA})$, which can scale with the transmit power and linearity if specific architectures are considered [8], accounts for 49%. At the receiver, the digital baseband processing (P_{DSP}), forward error correction (P_{FEC}), and medium access control (P_{MAC}) units are dominant and do not scale with the transmit power. The power of the analog/digital and digital/analog converters $(P_{ADC} + P_{DAC})$ and the fixed front-end power (P_{FE}) are not considered because they are constant.

(total energy spent in the transmission and the reception per bit of data) that really matter. Characterizing how a diversity gain, a multiplexing gain, and/or a coding gain influence that system-level tradeoff remains a research issue.

The transceiver power consumption is generally made of two terms. The first corresponds to the power amplifier(s) consumption and is a function of the transmit power, inferred from the link budget. The second corresponds to the other electronics consumption and is independent of the link budget. We refer, respectively, to dynamic and static power consumption. The relative contribution of those terms is illustrated in Figure 1 where the typical power consumption breakdown of single-antenna OFDM transceivers is depicted.

The impact on power consumption of multiantenna transmission (MIMO), when compared with traditional single-antenna transmission (SISO), is twofold. On the one hand, the general benefit in spectral efficiency versus signal-to-noise ratio can be exploited either to reduce the required transmit power, with impact on the dynamic power consumption, or to reduce the transceiver duty cycle with impact on both dynamic and static power contributions. On the other hand, the presence of multiple antennas requires duplicating part of the transceiver circuitry, which increases both the static and dynamic terms.

The question whether multiantenna transmission techniques increase or decrease the transceiver energy efficiency has only recently been addressed in the literature [9–11]. Interestingly, it has been shown that for narrow-band single-carrier transmission, multiantenna techniques basically decrease the energy efficiency if they are not combined with adaptive modulation [9]. It has also been shown, in the same context, that energy efficiency improvement is achievable by adapting the type of multiantenna encoding to the transmission condition [10, 11].

The purpose of this paper is first to extend previously mentioned system-level energy efficiency studies to the case of broadband links based on MIMO-OFDM. Therefore, we investigate the performance versus energy efficiency tradeoff of two typical multiantenna techniques—space-time block code (STBC) [3] and space-division multiplexing (SDM)

[4]—and compare it to the single-antenna case. Both are implemented on top of a legacy OFDM transmission chain as used in IEEE 802.11a/g/n and proceed to spatial processing at the receiver only. The IEEE 802.11 MAC has been adapted to accommodate those transmission modes. For the sake of clarity, without hampering the generality of the proposed approach, we limit the study to 2 × 2 antennas systems.

Second, we propose *smartMIMO*, a coarsely adaptive MIMO-OFDM scheme that, on a packet-per-packet basis, switches between STBC, SDM, and SISO depending on the channel conditions to simultaneously secure the throughput and/or robustness improvement provided by the multiantenna transmission and guarantees an energy-efficiency improved compared with the current standards. Contrarily in other adaptive scheme [12–16], using SISO still reveal effective in many channel condition because of the saving in static power consumption.

The remainder of the paper is structured as follows. In Section 2, we present some related work. The MIMO-OFDM physical (PHY) and medium access control (MAC) layers are described in Section 3. The unified performance and energy models used to investigate the average throughput versus energy-efficiency tradeoff are presented in Section 4. The impact of SDM and STBC on the net throughput versus energy-efficiency tradeoff is discussed in Section 5. Finally, in Section 6, we present the *smartMIMO* scheme and evaluate its benefit on the aforementioned tradeoff.

2. RELATED WORK

The question whether multiantenna techniques increase or decrease the energy efficiency has only very recently been addressed. Based on comprehensive first order energy and performance models targeted to narrow-band single carrier transceivers (as usually considered in wireless microsensor), Shuguang et al. have evaluated, taking both static and dynamic power consumption into account, the impact on energy efficiency of single-carrier space-time block coding (STBC) versus traditional single antenna (SISO) transmission [9]. Interestingly, it is shown that in short-/middle-range applications such as sensor networks—and by extension, wireless local area networks (WLANs)—nonadaptive STBC actually degrades the system energy efficiency at same data rate. However, when combined with adaptive modulation in so-called adaptive multiantenna transmission, energy-efficiency can be improved. Liu and Li have extended those results by showing that energy-efficiency can further be improved by adaptively combining multiplexing and diversity techniques [10, 11]. Adaptive schemes are hence mandatory to achieve both high-throughput and energy-efficient transmissions.

In the context of broadband wireless communication, many adaptive multiantenna schemes have also been proposed and are often combined with orthogonal frequency division multiplexing (OFDM). Adaptation is most often carried out to minimize the bit-error (BER) probability or maximize the throughput. In [12], for instance, a scheme is proposed to switch between diversity and multiplexing codes based on limited channel state information (CSI) feedback.

In [13], a pragmatic *coarse grain* adaptation scheme is evaluated. Modulation, forward error correction (FEC) coding rate and MIMO encoding are adapted according to CSI estimator—specifically, the average signal-to-noise ratio (SNR) and packet error rate (PER)—to maximize the effective throughput. More recently, *fine grain* adaptive schemes have been proposed [14, 15]. The modulation and multiantenna encoding are here adapted on a carrier-per-carrier basis. The main challenge with such schemes is however to provide the required CSI to the transmitter with minimal overhead. This aspect is tackled, for instance in [16].

The approaches mentioned above have been proven to be effective to improve net throughput and/or bit-error rate (BER). Some are good candidate to be implemented in commercial chipset. However, in none of those contributions, the electronics power consumption is considered in the optimization. Moreover, most adaptive policies are designed to maximize gross data-rate and/or minimize (uncoded) bit-error rate without taking into account the coupling between physical layer data rate and bit-error rate incurred in medium access control (MAC) layer [17].

In this paper, adaptive MIMO-OFDM schemes are looked at with as objective to jointly optimize the average link throughput (on top of the medium access control layer) and the average transceiver energy efficiency. The total transceiver power consumption is considered, including the terms that vary with the transmit power and the fixed term due the radio electronics.

To enable low-complexity policy-based adaptation and to limit the required CSI feedback, coarsely adaptive schemes, as defined in [13], are considered. For this, pragmatic empirical performance, energy, and channel state information models are developed based on observation and measurements collected on the reconfigurable MIMO-OFDM setup previously described in [18].

3. MIMO PHYSICAL AND MAC LAYERS

3.1. MIMO-OFDM physical layer

The multiantenna schemes considered here are orthogonal space-time block coding (STBC) [2, 3] and space-division multiplexing with linear spatial processing at the receiver (SDM-RX) [4]. Both are combined with OFDM so that multiantenna encoding and/or receive processing is performed on a per-carrier basis. The N OFDM carriers are QAM-modulated with a constellation size set by the link adaptation policy presented in [17]. The same constellation is considered for the different carriers of a given symbol, therefore we refer to "coarse grain" adaptation in opposition to fine grain adaptation where the subcarriers can receive different constellation.

Figure 2 illustrates the general setup for MIMO-OFDM on which either SDM or STBC can be implemented.

For SDM processing, a configuration with U transmit antenna and A receive antenna is considered. The multiantenna preprocessing reduces to demultiplexing the input stream in sub-streams that are transmitted in parallel. Vertical encoding is considered: the original bit stream is FEC encoded,

interleaved, and demultiplexed between the OFDM modulators. The MIMO processors at the receiver side take care of the spatial interference mitigation on a per-subcarrier basis. We consider a minimum mean-square–error-(MMSE-) based detection algorithm. Although the MMSE algorithm is outperformed by nonlinear receiving algorithm such as successive interference cancellation [4], its implementation ease keeps it attractive in low-cost, high-throughput solution such as wireless local area networks.

In STBC mode, space-time block codes from orthogonal designs [2, 3] are considered. Such scheme reduces to an equivalent diagonal system that can be interpreted as a SISO model where the channel is the quadratic average of the MIMO sub-channels [3].

Channel encoding and OFDM modulation are done according to the IEEE 802.11a standard specifications, transmission occurs in the 5 GHz ISM band [19].

As mentioned previously, from the transceiver energy-efficiency perspective, it may still be interesting to operate transmission in single-antenna mode. In that case, a single finger of the transmitter and receiver are activated and the MIMO encoder and receive processor are bypassed.

3.2. Medium access control layer

The multiantenna medium access control (MAC) protocol we consider is a direct extension of the IEEE 802.11 distributed coordination function (DCF) standard [19]. A carrier sense multiple access/collision avoidance (CSMA/CA) medium access procedure performs automatic medium access sharing. Collision avoidance is implemented by mean of the exchange of request-to-send (RTS) and clear-to-send (CTS) frames. The data frames are acknowledged (ACK). The IEEE 802.11 MAC can easily be tuned for adaptive multiantenna systems. We assume that the basic behavior of each terminal is single-antenna transmission. Consequently, single-antenna exchange establishes the multiantenna features prior to the MIMO exchange. This is made possible via the RTS/CTS mechanism and the data header. A signaling relative to the multiple-antenna mode is added to the physical layer convergence protocol (PLCP) header.

Further, the transactions required for channel estimation need to be adapted. In the considered 2×2 configuration, not only one but four-channel path must be identified. Therefore, the preamble structure is adapted as sketched in Figure 3. The transmitter sends preambles consecutively on antennas 1 and 2. The receiver can then easily identify all channel paths. The complete protocol transactions to transmit a packet of data in the SDM and STBC modes are detailed in Figures 4 and 5, respectively. For SISO transmission, one relies on the standard 802.11 CSMA/CA transaction and on the standard preambles.

4. UNIFIED PERFORMANCE AND ENERGY MODEL

The physical (PHY) and medium access control (MAC) layers being known, one wants to compute the net throughput (on top of the MAC) and the energy per bit as functions of the transmission parameters, including the type of

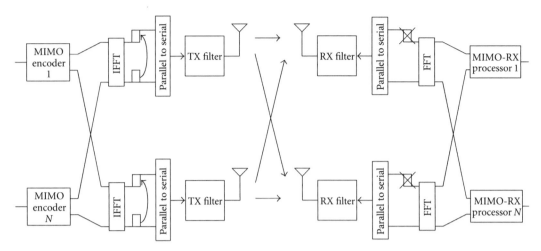

FIGURE 2: Reconfigurable multiantenna transceiver setup supporting SDM, STBC, and SISO transmissions.

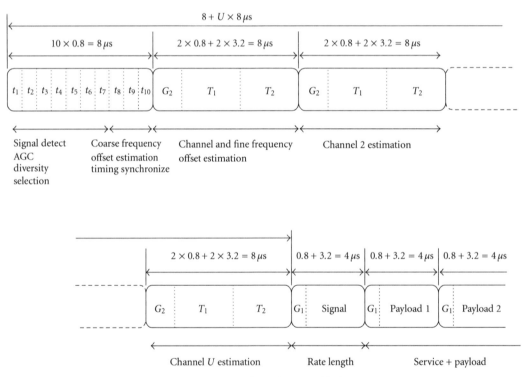

FIGURE 3: Channel estimations from the preamble for a 2×2 system.

multiantenna processing, and the channel state. To enable simple policy-based adaptation scheme and limit the required CSI feedback, a coarse channel state model is needed to capture the CSI in a synthetic way. The proposed model should cover the system performance and energy consumption of the different multiantenna techniques under consideration.

An important aspect is to identify tractable channel state parameters that dominate the instantaneous packet error probability. The average packet errors rate (PER) as tradi-

tionally evaluated misses that instantaneous dimension. In narrow-band links affected by Rayleigh fading, the signal-to-noise ratio (SNR) suffices to track the channel state. In MIMO-OFDM, however, the impact of the channel is more complex. With spatial multiplexing, for instance, the error probability for a given modulation and SNR still depends on the rank of the channel. Moreover, not all the subcarriers experience the same MIMO channel. Finally, a given channel instance can be good for a specific MIMO mode while being bad for another one.

(1) TX sends RTS in SISO. the PLCP header contains:

    (i) 2 bits specifying the MIMO exchange type (here channel state extraction at RX)

    (ii) 2 bits for the number of TX antennas that will be activated for the next MIMO exchange (nTX) and that have to be considered in the channel extraction

    (iii) 2 bits for the minimum numbers of RX antennas to activate for next reception.

(2) RX sends CTS.

(3) TX send DATA preamples in time division: 10 short training sequences for coarse synchronization ($t_1 \cdots t_{10} - 8\,\mu s$) and $2 \times nTX$ long training sequences ($G_2 T_1 T_2 - 8\,\mu s$, called C-C sequence). Each TX antenna transmits after each other its C-C sequence. RX activates its antennas and extracts information about each channel that each antenna sees. PLCP header is sent from the last TX antenna and conveys 2 bits for the mode (here SDM-RX) and 2 bits for the number of streams. finally, TX sends the DATA in MIMO.

(4) RX sends ACK.

FIGURE 4: Considered SDM protocol extension.

(1) TX sends RTS in SISO. the PLCP header contains:

    (i) 2 bits specifying the MIMO exchange type (here channel state extraction at RX)

    (ii) 2 bits for the number of TX antennas that will be activated for the next MIMO exchange (nTX) and that have to be considered in the channel extraction

    (iii) 2 bits for the minimum numbers of RX antennas to activate for next reception.

(2) RX sends CTS.

(3) TX send DATA preamples in TDMA: 10 short training sequences for coarse synchronization ($t_1 \cdots t_{10} - 8\,\mu s$) and $2 \times nTX$ long training sequences ($G_2 T_1 T_2 - 8\,\mu s$, called C-C sequence). Each TX antenna transmits after each other its C-C sequence. RX activates its antennas and extracts information about each channel that each antenna sees. PLCP header is sent from the last TX antenna and conveys 2 bits for the mode (here STBC) and 4 bits for the code used. Finally, TX sends the DATA in MI-SO/MO depending on the number of recieved antenna (nRX).

(4) RX sends ACK.

FIGURE 5: Considered STBC protocol extension.

Possible coarse channel state information (CSI) indicators for MIMO-OFDM are discussed in [13]. An empirical approach based on multiple statistics of the postprocessing SNR (the SNR after MIMO processing) and running-average PER monitoring is proposed. Yet, it is difficult to define such SNR-based indicators consistently across different MIMO schemes. Moreover, relying on PER information

results in a tradeoff between accuracy and feedback latency, both with potential impact on stability.

As already proposed in [20], based on the key observation that energy efficiency and net throughput are actually weak functions of the packet error probability [21], we prefer to use the outage probability—that is, the probability that the channel instantaneous capacity is lower that the link spectrum efficiency—as indicator of the packet error probability. The instantaneous capacity depends on the average signal-to-noise ratio (SNR), the normalized instantaneous channel response $\mathbf{H}$, and the multiantenna encoding. The instantaneous capacity can be easily derived for the different multiantenna encoding. Practically, it is convenient to derive the capacity-over-bandwidth ratio that can be compared to the transmission spectrum efficiency η instead of absolute rate.

In the remainder of this session, we first derive the instantaneous capacity expressions for the different transmission mode considered (Section 4.1). Then, we derive the condition for quasi-error-free packet transmission (Section 4.2). Based on that, we compute the expressions of the net throughput and the energy per bit (Sections 4.3 and 4.4). Finally, we discuss the derivation of the coarse channel model required to develop policy-based radio link control strategies (Section 4.5).

4.1. Instantaneous capacity

Let $\mathbf{H} = (h_{ua}^n)$ be a normalized MIMO-OFDM channel realization. The coefficient h_{11}^n corresponds to the (flat) channel response between the single active transmit antenna and the single active receive antenna for the subcarrier n (including transmit and receive filters). The *instantaneous capacity* of the single-antenna channel is given by (1), where W is the signal bandwidth and N is the number of subcarriers. SNR is the average link signal-to-noise ratio. If SNR is high compared to 1, the capacity relative to the bandwidth can be decomposed in a term proportional to SNR and independent of $\mathbf{H}$ and a second term function of $\mathbf{H}$ only:

$$C = W \cdot \frac{1}{N} \sum_{n=1}^{N} \log_2 (1 + h_{11}^{n2} \mathrm{SNR}), \tag{1}$$

$$\frac{C}{W} \cong \frac{\mathrm{SNR}|_{\mathrm{dB}}}{10 \log_{10} 2} + \frac{1}{N} \sum_{n-1}^{N} \log_2 (h_{11}^{n2}). \tag{2}$$

In the STBC case, as mentioned in Section 3, the MIMO channel can be reduced to equivalent SISO channel corresponding to the quadratic average of the subchannels between each pairs of transmit and receive antennas [3]. The *instantaneous capacity* can then be computed just as for SISO:

$$\frac{C}{W} \cong \frac{\mathrm{SNR}|_{\mathrm{dB}}}{10 \log 2} + \frac{1}{N} \sum_{n-1}^{N} \log_2 \left(\frac{1}{UA} \sum_{u=1}^{U} \sum_{a=1}^{A} h_{ua}^{n2} \right). \tag{3}$$

In the SDM case finally, the compound channel results from the concatenation of the transmission channel with the interference cancellation filter. The instantaneous capacity can be computed based on the *postprocessing SNR's (γ)*— that is, for each stream, the signal-to-noise-and-interference ratio at the output of the interference cancellation filter. Let $\mathbf{H}^n$ and $\mathbf{F}^n$, respectively, denote the MIMO channel realization for the subcarrier n, and the corresponding MMSE filter (4):

$$\mathbf{F}^n = \mathbf{H}^{nH} \cdot \left(\mathbf{H}^n \mathbf{H}^{nH} + \sigma^2 \mathbf{I}_{AxA}\right)^{-1}. \tag{4}$$

In the considered 2×2 case, let us assume an equal transmit power at both transmit antennas $p_1 = p_2 = p/2$ and let us denote with $\mathbf{f}_1^n$, $\mathbf{f}_2^n$, $\mathbf{h}_1^n$, $\mathbf{h}_2^n$, respectively, the first row, second row, first column, second column of the matrices $\mathbf{H}^n$ and $\mathbf{F}^n$. The substream postprocessing SNRs γ_1 and γ_2 can then be computed as

$$\gamma_1^n = \frac{\left|\mathbf{f}_1^n \mathbf{h}_1^n\right|^2 \times P_1}{\left|\mathbf{f}_1^n \mathbf{h}_2^n\right|^2 \times P_2 + \left|\mathbf{f}_1^n\right|^2 \times \sigma^2} = \frac{\left|\mathbf{f}_1^n \mathbf{h}_1^n\right|^2}{\left|\mathbf{f}_1^n \mathbf{h}_2^n\right|^2 + \left|\mathbf{f}_1^n\right|^2 \times 2/\mathrm{SNR}},$$

$$\gamma_2^n = \frac{\left|\mathbf{f}_2^n \mathbf{h}_2^n\right|^2 \times P_2}{\left|\mathbf{f}_2^n \mathbf{h}_1^n\right|^2 \times P_1 + \left|\mathbf{f}_2^n\right|^2 \times \sigma^2} = \frac{\left|\mathbf{f}_2^n \mathbf{h}_2^n\right|^2}{\left|\mathbf{f}_2^n \mathbf{h}_1^n\right|^2 + \left|\mathbf{f}_2^n\right|^2 \times 2/\mathrm{SNR}}. \tag{5}$$

SNR is again the average link signal-to-noise ratio. The instantaneous *capacity* can then be computed in analogy with (2) and (3) as follows:

$$\frac{C}{W} = \frac{1}{N} \sum_{n=1}^{N} \left[\log_2\left(1 + \gamma_1^n\right) + \log_2\left(1 + \gamma_2^n\right)\right]. \tag{6}$$

This development can easily be extended to more than 2×2 antenna setups.

4.2. Condition for quasi-error-free packet transmission on a given channel

Because the link throughput and energy efficiency (our objective functions) are weak functions of the packet error probability [21], one does not need to estimate the latter accurately in order to define adaptation policies that optimize the formers. It is sufficient to derive a condition under which the packet error rate is sufficiently low in order not to significantly affect the aforementioned objective functions. It is easy to verify that the accuracy obtained on the throughput on top of the MAC and on the energy efficiency is of the same order of magnitude that the packet error rate.

Based on Monte Carlo simulations, we have verified that, for the purpose of computing the average throughput on top of the MAC and the transceiver energy consumption per bit, the packet error event probability can be approximated by the outage without significant prejudice to the accuracy (Figure 6). We hence assume that, the channel being known, P_e, equals 1 if the spectrum efficiency η exceeds C/W. To account for the nonoptimality of the coding chain, we apply an empirical margin $\delta = 0.5$ bit/s/Hz, calibrated from simulation:

$$P_e = \begin{cases} 1 & \text{if } \dfrac{C}{W} < \eta + \delta, \\ \cong 0 & \text{otherwise.} \end{cases} \tag{7}$$

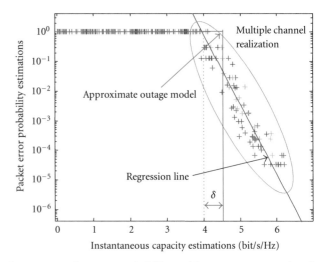

FIGURE 6: Packet error probability and instantaneous capacity observation on a link with spectrum efficiency 4 for a large set of channel realizations. One can observe the strong correlation and the steep descend of the regression line beyond the point where the instantaneous capacity breaks the spectrum efficiency line. The net throughput and energy per bit are weak function of the packet error probability; these observations motivate considering an outage model to derive policy-based adaptive schemes.

4.3. Net throughput

Assuming that the channel capacity criteria is met and, hence, the PER is close to zero, knowing the physical layer throughput (R_{phy}) and the details of the protocols, the *net throughput* (R_{net}) can be computed as follows:

$$R_{\mathrm{net}}$$
$$\cong \frac{L_d}{T_{\mathrm{DIFS}} + 3 \cdot T_{\mathrm{SIFS}} + T_{CW} + \left((L_d + L_h)/R_{\mathrm{phy}}^d\right) + 4 \cdot T_{\mathrm{plcp}} + L_{\mathrm{ctrl}}/R_{\mathrm{phy}}^b}. \tag{8}$$

To better understand that expression, refer to Figure 7 and notice that the denominator corresponds to the total time required for the transmission of one packet of data size L_d with a L_h-bit header according to the 802.11 DCF protocol [19]. T_{SIFS} is the so-called short interframe time. R_{phy}^d is the physical layer data rate. L_{ctrl} corresponds to the aggregate length of all control frames (RTS, CTS, and ACK) transmitted at the basic rate (R_{phy}^b) and T_{plcp} is the transmission time of the PLCP header. T_{DIFS} is the minimum carrier sense duration and T_{CW} holds for the average contention time due to the CSMA procedure. The physical layer data rate R_{phy}^d can be expressed as a function of modulation order (N_{mod}) and the code rate (R_c), considering the number of data carrier per OFDM symbol (N) and the symbol rate (R_s), in addition to the number of streams U (9). In our study, one has $U = 1$ for SISO and $U = 2$ for both SDM and STBC. In the MIMO case ($U > 1$), one of the four T_{PLCP}'s in (8) must be replaced by $T_{\mathrm{PLCP\_MIMO}}$ given in (10) with $T_{\mathrm{CC\_Seq}}$ being equal 8 μs:

$$R_{\mathrm{phy}}^d = U \cdot N \cdot N_{\mathrm{mod}} \cdot R_c \cdot R_s, \tag{9}$$

$$T_{\mathrm{PLCP\_MIMO}} = T_{\mathrm{PLCP\_SISO}} + (U - 1) \times T_{\mathrm{CC\_Seq}}. \tag{10}$$

4.4. Energy per bit

To compute the energy efficiency, the system power consumption needed to sustain the required average SNR must be assessed. The latter consists of a fixed term due to the electronics, and a variable term, function of the power consumption

$$P_{\text{system}} = P_{\text{elec}} + \frac{P_{\text{Tx}}}{\varepsilon}, \tag{11}$$

where ε denotes the efficiency of the transmitter power amplifier (PA), that is, the ratio of the output power (P_{Tx}) by the power consumption (P_{PA}). In practical OFDM transmitters, class A amplifiers are typically used. The power consumption of the latter component only depends on its maximum output power (P_{max}) (12). Next, the transmitter signal-to-distortion ratio (S/D_{Tx}) can be derived as a function of the sole backoff (OBO) of the actual PA output power (P_{Tx}) to P_{max} (13)-(14). The latter relation is design dependent and usually not analytical. In this study, we consider an empirical curve-fitted model calibrated on the energy-scalable transmission chain design presented in [8], which has as key feature to enable both variable output power (P_{Tx}) and variable linearity (S/D_{Tx}) with a monotonic impact on the power consumption;

$$P_{\text{PA}} = \frac{P_{\text{max}}}{2}, \tag{12}$$

$$\text{OBO} = \frac{P_{\text{max}}}{P_{\text{Tx}}}, \tag{13}$$

$$(S/D)_{\text{Tx}} = f(\text{OBO}). \tag{14}$$

The path-loss being known, SNR can then be computed as a function of OBO and P_{Tx} (15). P_N is the thermal noise level depending of the temperature (T), the receiver bandwidth (W), and noise factor (N_f), k is the Boltzmann constant:

$$\frac{1}{\text{SNR}} = \frac{1}{(S/D)_{\text{Tx}}} + \frac{P_N \times P_L}{P_{\text{Tx}}}, \tag{15}$$

$$P_N = k \cdot T \cdot W \cdot N_f. \tag{16}$$

The PA power can be expressed as a function of those two parameters (17):

$$P_{\text{PA}} = 2 \times (P_{\text{TX}} \times \text{OBO}). \tag{17}$$

The achievable P_{PA} versus SNR tradeoff obtained with the design as presented in [8] is illustrated for different average link path-loss values in Figure 8. Notice that in case the output power has to stay constant, the proposed reconfigurable architecture still has the possibility to adapt to the linearity requirements. This has less but still significant impact on the power consumption.

4.5. Coarse channel model

At this point in the development, we have relations to compute the link throughput and the transceiver energy consumption per bit for given multiantenna encoding, modulation and code rate, provided that the link SNR is sufficient to satisfy the quasi error free transmission condition.

The latter condition depends also on the actual channel response **H** (equations (2), (3), (5), (6), and (7)). Extensive work has already been done to model broadband channel at that level of abstraction (physical level). In the case of MIMO-OFDM WLAN as considered here, a reference channel model is standardized by the IEEE [22]. However, to be able to derive simple policy-based adaptation schemes that take that channel state information into account, one has to derive a model that captures this information in a more compact way. A valid approach is to operate an empirical classification of the channel merit. This can easily be done based on the instantaneous capacity indicators.

As an example, let us consider the second term of the instantaneous SISO capacity. According to [22], the values of the carrier fading h_{ua}^n are Rayleigh distributed. However, due to the averaging across the carriers, which are only weakly correlated, the distribution of the second term of the instantaneous capacity is almost normal distributed (see Figure 9). Since the first term of the capacity indicator is independent of **H** and therefore not stochastic, the capacity indicator can then also be approximated as normal-distributed. One can verify that the same observation holds also for the STBC and SDM instantaneous capacity indicators (see Figure 9).

Let us, respectively, denote the average and variance of the instantaneous capacity for a given mode as μ_{mode} and σ^2_{mode}. These quantities depend only on SNR. Their evolution in a function of SNR is plotted in Figure 10 for the different multiantenna encoding. It can be observed that for a given mode and a sufficiently large SNR, μ_{mode} grows linearly with SNR in dB while σ^2_{mode} stays sensibly constant. Therefore, a linear regression can be operated. The parameters of the extracted linear model are summarized in Table 1.

Based on the normal distribution of the instantaneous capacity and the linear models for the parameters of that distribution, a channel merit scale can be defined. A given channel instance receives a merit index for a given multi-antenna mode and a given SNR depending on how its actual instantaneous capacity compared to the capacity distribution for that mode and that SNR. The empiric scale we consider goes from 1 (worst) to 5 (best) with the class boundaries as defined in Table 2 (3 first columns).

For each class index (channel merit), a worst-case error-free transmission condition is defined, comparing the signaling spectrum efficiency (η) to the upper bound of the instantaneous capacity class for this channel merit (Table 2, fourth column).

4.6. Usage of the model

One can now compute, for a given channel merit as defined above, what will be the link throughput and the transceiver energy consumption per bit for a given multiantenna mode, a given modulation and a given code rate. The computation occurs as follows.

Step 1. Knowing the modulation and code rate, hence the signaling spectrum efficiency, the minimum link SNR to

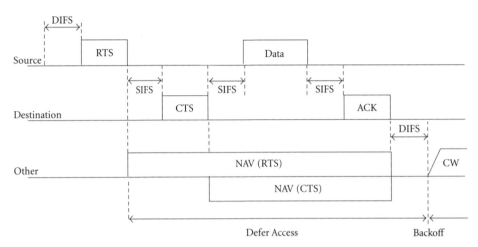

FIGURE 7: Packet transmission transaction according to the IEEE 802.11 protocol modified to support multiantenna operation.

TABLE 1: Instantaneous capacity indicator average and standard deviation as a function of the average SNR.

	$\mu = A \times SNR + B$		σ
	A	B	
SISO	0.33	-0.84	1.41
SDM	0.6	-2.54	2.41
STBC	0.33	-0.24	0.73

satisfy the worst-case quasi-error-free transmission condition (for the given multi-antenna mode and channel merit) is computed using the appropriate inequality from Table 2, fourth column, and the linear model exposed in Table 1.

Step 2. Knowing the multi-antenna mode, the modulation and the code rate, assuming quasi-packet-error-free transmission, the link throughput is computed according to (8). The condition is calibrated for a packet error rate of <1%, yielding an accuracy of 1% of the estimated throughput and energy efficiency.

Step 3. Based on the power model exposed in Section 4.4, assuming a given average path loss, the transmitter parameters (output power, backoff) to achieve the link SNR computed in Step 1, and subsequently the transmitter power consumption, are computed.

Step 4. From the transmitter power and the net throughput, the energy per bit can be computed.

5. IMPACT OF MIMO ON THE AVERAGE RATE VERSUS AVERAGE POWER TRADEOFF

The proposed performance and energy models enable computing the net throughputand energy efficiency as functions of the system-level parameters (mode, modulation, code rate, transmit power, and power amplifier backoff). The

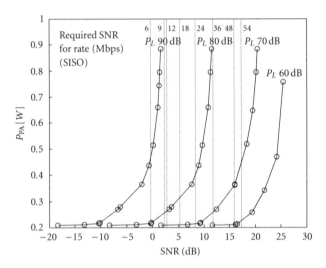

FIGURE 8: Power consumption versus link SNR tradeoff achieved with the energy scalable transmitter for average path-loss, PL = 60 dB, 70 dB, 80 dB, 90 dB. The tradeoff curves are compared to the SNR level required for 20 MHz SISO-OFDM transmission with PER < 10% at various rates.

considered settings for those parameters are summarized in Table 3. Capitalizing on those models and using the techniques already proposed in [17], one can derive a set of close-to-optimal transmission adaptation policies that optimize the average energy efficiency for a range of average throughput targets and, then, analyze the resulting tradeoff.

In this section, we derive these tradeoffs separately for the STBC and SDM modes and compare with SISO. This is done in two steps.

Step 1. For each channel merit, the optimum tradeoff between net throughput and energy per bit (Section 5.1) is derived. This results in a Pareto optimal [23] set of working points (settings of the system-level parameters) for each possible channel merit.

TABLE 2: Definition of the channel merit.

| Channel merit | Channel instance instantaneous capacity indicator | | Maximum spectrum efficiency for quasi error free transmission |
	Min	Max	
1	$-\infty$	$\mu_{\mathrm{mode}} - 2 \times \sigma_{\mathrm{mode}}$	—
2	$\mu_{\mathrm{mode}} - 2 \times \sigma_{\mathrm{mode}}$	$\mu_{\mathrm{mode}} - \sigma_{\mathrm{mode}}$	$\eta < \mu_{\mathrm{mode}}(\mathrm{SNR}) - 2 \times \sigma_{\mathrm{mode}}(\mathrm{SNR})$
3	$\mu_{\mathrm{mode}} - \sigma_{\mathrm{mode}}$	$\mu_{\mathrm{mode}} + \sigma_{\mathrm{mode}}$	$\eta < \mu_{\mathrm{mode}}(\mathrm{SNR}) - \sigma_{\mathrm{mode}}(\mathrm{SNR})$
4	$\mu_{\mathrm{mode}} + \sigma_{\mathrm{mode}}$	$\mu_{\mathrm{mode}} + 2 \times \sigma_{\mathrm{mode}}$	$\eta < \mu_{\mathrm{mode}}(\mathrm{SNR}) + \sigma_{\mathrm{mode}}(\mathrm{SNR})$
5	$\mu_{\mathrm{mode}} + 2 \times \sigma_{\mathrm{mode}}$	$+\infty$	$\eta < \mu_{\mathrm{mode}}(\mathrm{SNR}) + 2 \times \sigma_{\mathrm{mode}}(\mathrm{SNR})$

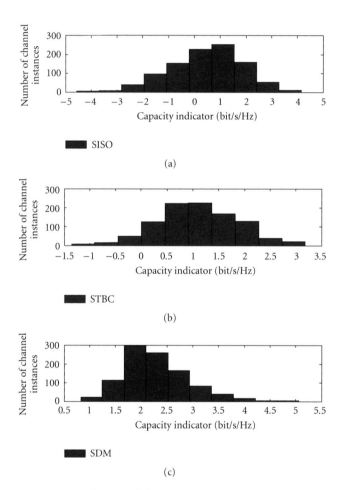

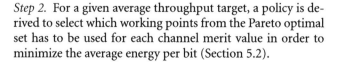

FIGURE 9: Distribution of the instantaneous capacity observation for the different modes (SISO, STBC, SDM) over a large set of channel instances generated with the physical channel model.

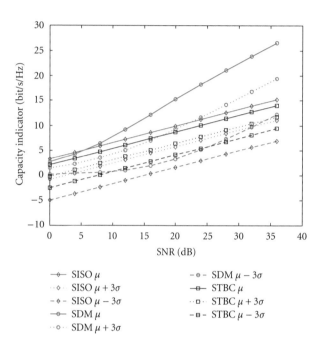

FIGURE 10: Capacity indicator mean and standard deviation as a function of SNR for the various modes.

TABLE 3: System-level parameters considered.

MIMO mode	SISO, SDM2 × 2, STBC2 × 2
N_{mod}	BPSK, QPSK, 16QAM, 64QAM
R_c	1/2, 2/3, 3/4
P_{Tx} [dBm]	0, 5, 10, 15, 20, 23
OBO [dB]	6, 8, 10, 12, 14

Step 2. For a given average throughput target, a policy is derived to select which working points from the Pareto optimal set has to be used for each channel merit value in order to minimize the average energy per bit (Section 5.2).

The resulting average throughput versus energy-per-bit tradeoffs is finally analyzed in Section 5.3. It should be noticed that this approach assume that the channel merit is known at the transmitter (limited CSI at transmit). That information can be acquired during the reception of the CTS frame or piggy-backed in the CTS, assuming that the channel is stable during RTS/CTS/packet transaction. The assumption is valid in nomadic scenario as considered in case of WLAN (typical coherence time of 300 milliseconds).

5.1. Net throughput versus energy-per-bit

To derive the optimal net throughput versus energy-per-bit tradeoff for a given mode in a given channel merit, a multiobjective optimization problem has to be solved: from

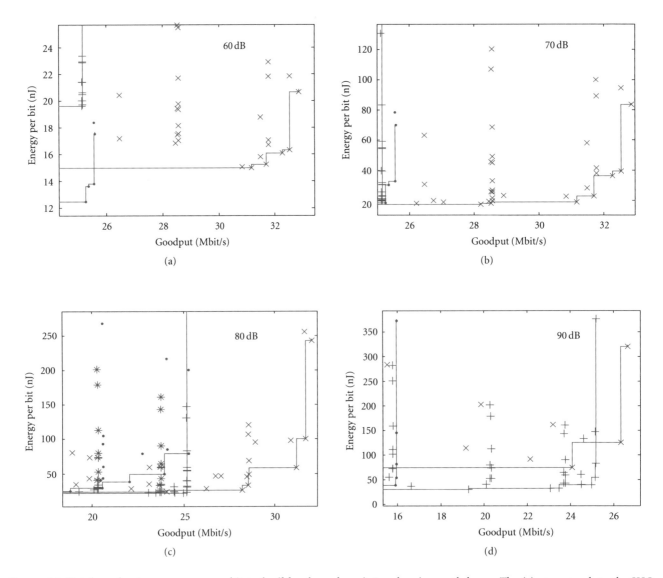

FIGURE 11: Net throughput versus energy-per-bit tradeoff for channel merit 3 and various path-losses. The ($\cdot$) corresponds to the SISO working points, the (+) corresponds to the STBC, and the ($\times$) corresponds to the SDMs. In each case, the Pareto optimal set is interpolated with a step curve.

all system-level parameter combinations, the ones bounding the tradeoff have to be derived. The limited range of the functional parameters still allows us to proceed efficiently to this search with simple heuristics [17]. This optimization can be proceeded to at design time, which limits to a great extend the complexity of the adaptation scheme.

The resulting tradeoff points are plotted in Figure 11 for different path losses and an average channel merit (which is 3). For each mode, we only keep the nondominated tradeoff points, leading to Pareto optimal sets, which are interpolated by step curves. We generally observe that SDM enables reaching higher throughput but that SISO stays more energy efficient for lower rates. STBC becomes attractive in case of large path losses. Similar tradeoff shapes can be observed for the other channel merit values.

5.2. Derivation of the control policies

From the knowledge of the Pareto optimal net throughput versus energy-per-bit tradeoff and the channel merit probabilities, which can be obtained from the Monte Carlo analysis of the physical-level channel model, given an average throughput constraint, we applied the technique presented in [17] to derive the adaptation policy that minimizes the energy per transmitted bit. Such a policy, valid for a given multiantenna mode, a given average path loss, and a given average throughput constraints, maps the possible channel merit to the appropriate setting of the transmission parameters.

Let (r_{ij}, e_{ij}) denote the coordinates of the ith Pareto point in the set corresponding to channel merit j. The average

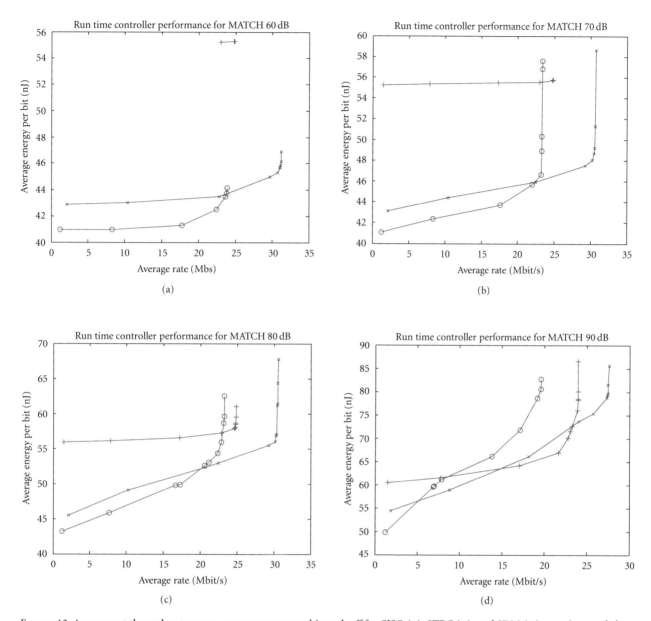

FIGURE 12: Average net throughput versus average energy-per-bit tradeoff for SISO (o), STBC (+), and SDM (×) at various path-loss.

power $\overline{P}$ and rate $\overline{R}$ corresponding to a given control policy—that is, the selection of one point on each throughput energy efficiency tradeoff—can then be expressed by (18). In these equations, x_{ij} is l if the corresponding point is selected, 0 otherwise, and ψ_j is the probability of the channel merit j. The energy per bit can be computed as $\overline{P} / \overline{R}$:

$$\overline{P} = \sum_j \psi_j \sum_i x_{ij} e_{ij} r_{ij} = \sum_i \sum_j x_{ij} \psi_j e_{ij} r_{ij} \widehat{=} \sum_i \sum_j x_{ij} p'_{ij},$$

$$\overline{R} = \sum_j \psi_j \sum_i x_{ij} r_{ij} = \sum_i \sum_j x_{ij} \psi_j r_{ij} \widehat{=} \sum_i \sum_j x_{ij} r'_{ij}.$$

$$(18)$$

We introduce the notation p'_{ij} and r'_{ij} corresponding, respectively, to the power and rate when the channel merit is j and the ith point is selected on the corresponding curve, both

weighted by the probability to be in that channel state. Only one tradeoff point can be selected for a given channel merit, resulting in the following constraints:

$$\sum_i x_{ij} = 1 \quad \forall j, \quad x_{ij} \in \{0, 1\}. \tag{19}$$

For a given average rate constraint R, the optimal control policy is the solution of the following problem:

$$\min \sum_i \sum_j x_{ij} p'_{ij} \quad \text{subject to} \sum_i \sum_j x_{ij} r'_{ij} > R. \tag{20}$$

This is the classical multiple choice knapsack problem. We are interested in the family of control policies corresponding to R ranging from 0 to $R_{\max}$, $R_{\max}$ being the maximum average rate achievable on the link. We call this family the *control*

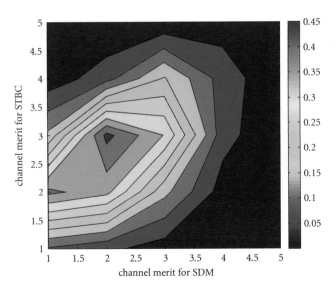

FIGURE 13: Histogram of the channel merit for STBC and SDM.

strategy. Let us denote as k_j the index of the point selected on the jth Pareto curve. Formally, $k_j = i \Leftrightarrow x_{ij} = 1$. A control policy can be represented by the vector $\mathbf{k} = \{k_j\}$. The *control strategy*, denoted $\{\mathbf{k}^{(n)}\}$ corresponds to the set of points $\{(\overline{R}^{(n)}, \overline{P}^{(n)})\}$ in the average throughput versus average power plane. A good approximation of the optimal control strategy (i.e., that bounds the tradeoff between $\overline{R}$ and $\overline{P}$) can be derived iteratively with the greedy heuristic explained in [17].

5.3. Average throughput versus average energy-per-bit

From the knowledge of the Pareto optimal net throughput versus energy-per-bit tradeoff for each channel merit, next to the channel merit probabilities one can now derive, given an average rate constraint $\overline{R}$, the control policy that minimizes the energy per bit. By having the constraint $\overline{R}$ ranging from 0 to its maximum achievable value $R_{\max}$, the average throughput versus average energy-per-bit tradeoff, when applying the proposed policy-based adaptive transmission, can be studied. The tradeoff (for each mode separately) is depicted in Figure 12 for path losses 60, 70, 80, and 90 dB.

The results for SDM and STBC are compared with the tradeoff achieved with a SISO system. One can observe that for low path loss (60–70 dB), SISO reveals, on the average, to be the most energy efficient in almost the whole range it spans. SDM enables, however, a significant increase of the maximum average rate. STBC is irrelevant in this situation. At average path loss (80 dB), a *breakpoint* rate (around 20 Mbps) exists above which both SDM and STBC are more energy efficient than SISO, although SDM is still better than STBC. At high path loss (90 dB), STBC is the most efficient between 20 and 25 Mbps. It is though still beaten by SDM for data rate beyond 25 Mbps and by SISO for smaller data rate.

6. SMARTMIMO

In the previous section, we have observed that STBC or SDM enable a significant average rate and/or range extension but hardly improve the energy efficiency. This is especially true when the average data rate is lower than 50% of the ergodic capacity of the MIMO channel.

Based on that observation, in this section, we propose to extend the policy-based adaptive scheme not only to adapt the transmission parameters with a fixed multiantenna encoding, but also to vary the latter encoding on a packet-per-packet basis. Beside, since it has been observed that SISO transmission is still most energy-efficient in certain condition, it is also considered as a possible transmission mode in the adaptive scheme.

Observing the histogram of the channel merits for the reference 802.11n channel model (Figure 13), one can notice that the merit indexes of a given channel for STBC or SDM are weakly correlated. Since the energy efficiency of a given mode is obviously better on a channel with a high merit, an average energy-efficiency improvement can be expected by letting the adaptation policy select one or the other transmission mode depending on the channel state.

6.1. Extended adaptation policy

The approach followed in Section 5 can be generalized to handle multiantenna mode adaptation, besides the other transmission parameters. For a given average path-loss and a given average rate target, the adaptation policy will now map a compound channel merit (namely, the triplets of channel merit values for the three possible multiantenna mode) to the system-level parameter settings, extended with the decision on which multiantenna mode to use.

As previous, the adaptation policies are derived in two steps.

Step 1. For each possible compound channel merit combination (with our scale, $5 \times 5 \times 5 = 125$ combinations), the Pareto optimal tradeoff between throughput and energy per bit is derived. This tradeoff can be derived by combining the single-mode Pareto tradeoffs for the corresponding single-mode channel merits. This combined Pareto set corresponds basically to the subset of nondominated points in the union of the Pareto sets to be combined.

Step 2. Based on the throughput versus energy-per-bit tradeoff for each compound channel merits and the knowledge of the compound channel merit probabilities, obtained again by Monte Carlo analysis of the physical-level channel model, one can derive the adaptation policy that minimizes the average energy per bit for a given average throughput target. This derivation is identical as in Section 5.

In the remainder, we analyze the average throughput versus energy-per-bit tradeoff achieved by an extended adaptive transmission scheme and compare to the results from Section 5.

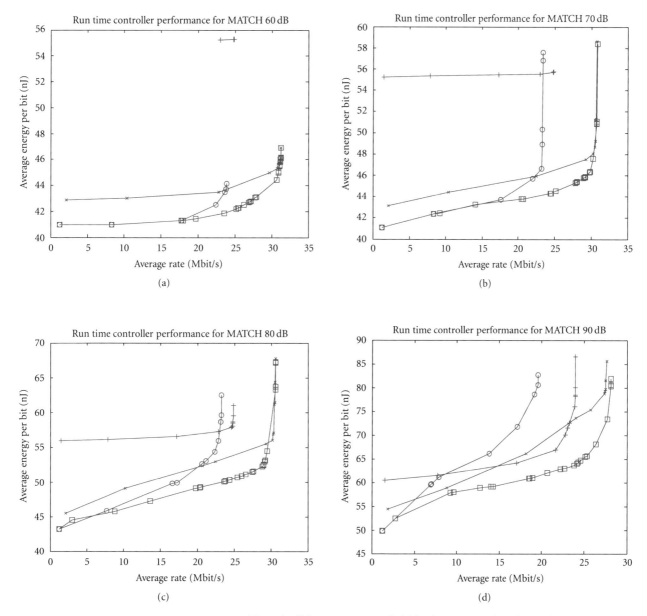

FIGURE 14: Average rate versus average energy-per-bit tradeoff for *smartMIMO* (bold line), superposed to the single-mode results. The energy efficiency is improved by up to 30% when compared to the single-mode results.

6.2. *Average rate versus average energy-per-bit*

By varying the average throughput constraint from 0 to the maximum achievable value ($R_{\max}$), one can derive the set of extended control policies that lead to the Pareto optimal tradeoff between average throughput and average energy per bit. This tradeoff is depicted in Figure 14 for different path losses. The results of the multiantenna mode specific tradeoff curves are superposed for the sake of comparison.

Globally, it can be observed that the tradeoff achieved with the extended control policies always dominate the tradeoff achieved with the multiantenna mode-specific policies. An average power reduction up to 30% can be observed. The resulting throughput-energy tradeoff even dominates SISO in the whole range, meaning that *smartMIMO* always brings a better energy per bit than any single mode. Moreover, this improvement does not affect the maximum throughput and range extension provided, respectively, by SDM and STBC. The energy benefit comes from a better adaptation to the channel conditions.

7. CONCLUSIONS

Multiantenna transmission techniques (MIMO) are being adopted in most broadband wireless standard to improve wireless links spectrum efficiency and/or robustness. There exists a well-documented tradeoff between potential spectrum efficiency and robustness increase. However, at architecture level, multiantenna techniques also come with an overhead in power consumption due, at least, to the duplication of part of the transmitter and receiver radio front ends. Therefore, from a system perspective, it is the trade-off between performance (e.g., the net throughput on top of the medium access control layer) and the average power consumption that really matters. It has been shown, in related works, that, in the case of narrow band single-carrier transceivers, adaptive schemes were mandatory to avoid that multiantenna techniques hamper this system-level tradeoff. In the broadband case, orthogonal frequency division multiplexing (OFDM) is usually associated with multiantenna processing. Adaptive schemes proposed so far for MIMO-OFDM optimize either the baseline physical layer throughput or the robustness in terms of bit-error rate. Energy efficiency is generally disregarded as well as the effects introduced by the medium access control (MAC) layer.

In this paper, the impact of adaptive SDM-OFDM and STBC-OFDM on the net data rate (on top of the MAC layer) versus energy-per-bit tradeoff has been analyzed and compared to adaptive SISO-OFDM. It has been shown that depending on the channel conditions, the one or the other scheme can lead to the best tradeoff. Up to a path loss of 80 dB, SISO always leads to the best energy efficiency up to a breakpoint rate (depending on the path loss) from where SDM is the most energy efficient. STBC improves the energy efficiency in a significant range of data rates only in case of large path loss (>90 dB).

Next, we derived and discussed *SmartMIMO*, an adaptive multiantenna scheme that controls, packet-per-packet, the basic OFDM links parameters (carrier modulation, forward error correction coding rate) as well as the type of multiple-antenna encoding (SISO, SDM, or STBC) in order to optimize the link net data rate (on top of the MAC) versus energy efficiency tradeoff. Based on a model calibrated on an existing multiantenna transceiver setup, the link energy efficiency with the proposed scheme is shown to be improved by up to 30% when compared to nonadaptive schemes. The average rate is, on the other hand, improved by up to 50% when compared to single-antenna transmission.

ACKNOWLEDGMENTS

This work has been partially published in the Proceeding of the 20th IEEE Workshop on Signal Processing Systems (SiPS06) in October 2006. The project has been partially supported by Sony Corporation, the Samsung Advanced Institute of Technology (SAIT), and the Flemish Institute for BroadBand Telecommunication (IBBT, Ghent, Belgium). Bruno Bougard was Research Assistant delegated by the Belgian Foundation for Scientific Research (FWO) until October 2006.

REFERENCES

[1] A. B. Gershman, "Robust adaptive beamforming: an overview of recent trends and advances in the field," in *Proceedings of the 4th International Conference on Antenna Theory and Techniques (ICATT '03)*, vol. 1, pp. 30–35, Sevastopol, Ukraine, September 2003.

[2] S. M. Alamouti, "A simple transmit diversity technique for wireless communications," *IEEE Journal on Selected Areas in Communications*, vol. 16, no. 8, pp. 1451–1458, 1998.

[3] V. Tarokh, H. Jafarkhani, and A. R. Calderbank, "Space-time block codes from orthogonal designs," *IEEE Transactions on Information Theory*, vol. 45, no. 5, pp. 1456–1467, 1999.

[4] G. J. Foschini, G. D. Golden, R. A. Valenzuela, and P. W. Wolniansky, "Simplified processing for high spectral efficiency wireless communication employing multi-element arrays," *IEEE Journal on Selected Areas in Communications*, vol. 17, no. 11, pp. 1841–1852, 1999.

[5] L. Zheng and D. N. C. Tse, "Diversity and multiplexing: a fundamental tradeoff in multiple-antenna channels," *IEEE Transactions on Information Theory*, vol. 49, no. 5, pp. 1073–1096, 2003.

[6] M. Zargari, D. K. Su, C. P. Yue, et al., "A 5-GHz CMOS transceiver for IEEE 802.11a wireless LAN systems," *IEEE Journal of Solid-State Circuits*, vol. 37, no. 12, pp. 1688–1694, 2002.

[7] A. Behzad, L. Lin, Z. M. Shi, et al., "Direct-conversion CMOS transceiver with automative frequency control for 802.11a wireless LANs," in *Proceedings of IEEE International Solid-State Circuits Conference (ISSCC '03)*, vol. 1, pp. 356–357, San Francisco, Calif, USA, February 2003.

[8] B. Debaillie, B. Bougard, G. Lenoir, G. Vandersteen, and F. Catthoor, "Energy-scalable OFDM transmitter design and control," in *Proceedings of the IEEE 43rd Design Automation Conference (DAC '06)*, pp. 536–541, San Francisco, Calif, USA, July 2006.

[9] C. Shuguang , A. J. Goldsmith, and A. Bahai, "Energy-constrained modulation optimization for coded systems," in *Proceedings of IEEE Global Telecommunications Conference (Globecom '03)*, vol. 1, pp. 372–376, San Francisco, Calif, USA, December 2003.

[10] W. Liu, X. Li, and M. Chen, "Energy efficiency of MIMO transmissions in wireless sensor networks with diversity and multiplexing gains," in *Proceedings of IEEE International Conference on Acoustics, Speech and Signal Processing (ICASSP '05)*, vol. 4, pp. 897–900, Philadelphia, Pa, USA, March 2005.

[11] X. Li, M. Chen, and W. Liu, "Application of STBC-encoded cooperative transmissions in wireless sensor networks," *IEEE Signal Processing Letters*, vol. 12, no. 2, pp. 134–137, 2005.

[12] R. W. Heath Jr. and A. J. Paulraj, "Switching between diversity and multiplexing in MIMO systems," *IEEE Transactions on Communications*, vol. 53, no. 6, pp. 962–972, 2005.

[13] S. Catreux, V. Erceg, D. Gesbert, and R. W. Heath Jr., "Adaptive modulation and MIMO coding for broadband wireless data networks," *IEEE Communications Magazine*, vol. 40, no. 6, pp. 108–115, 2002.

[14] M. H. Halmi and D. C. H. Tze, "Adaptive MIMO-OFDM combining space-time block codes and spatial multiplexing," in *Proceedings of the 8th IEEE International Symposium on Spread Spectrum Techniques and Applications (ISSSTA '04)*, pp. 444–448, Sydney, Australia, August 2004.

[15] M. Codreanu, D. Tujkovic, and M. Latva-aho, "Adaptive MIMO-OFDM systems with estimated channel state information at TX side," in *Proceedings of IEEE International Conference on Communications (ICC '05)*, vol. 4, pp. 2645–2649, Seoul, Korea, May 2005.

[16] P. Xia, S. Zhou, and G. B. Giannakis, "Adaptive MIMO-OFDM based on partial channel state information," *IEEE Transactions on Signal Processing*, vol. 52, no. 1, pp. 202–213, 2004.

[17] B. Bougard, S. Pollin, A. Dejonghe, F. Catthoor, and W. Dehaene, "Cross-layer power management in wireless networks

and consequences on system-level architecture," *Signal Processing*, vol. 86, no. 8, pp. 1792–1803, 2006.

[18] M. Wouters, T. Huybrechts, R. Huys, S. De Rore, S. Sanders, and E. Umans, "PICARD: platform concepts for prototyping and demonstration of high speed communication systems," in *Proceedings of the 13th IEEE International Workshop on Rapid System Prototyping (IWRSP '02)*, pp. 166–170, Darmstadt, Germany, July 2002.

[19] IEEE Std 802.11a, "Part 11: wireless LAN medium access control (MAC) and physical layer (PHY) specifications," IEEE, 1999.

[20] S. Valle, A. Poloni, and G. Villa, "802.11 TGn proposal for PHY abstraction in MAC simulators," IEEE 802.11, 04/0184, February, 2004, ftp://ieee:wireless@ftp.802wirelessworld.com/11/04/11-04-0184-00-000n-proposal-phy-abstraction-in-mac-simulators.doc.

[21] C. Schurgers, V. Raghunathan, and M. B. Srivastava, "Power management for energy-aware communication systems," *ACM Transactions on Embedded Computing Systems*, vol. 2, no. 3, pp. 431–447, 2003.

[22] V. Erceg, L. Schumacher, P. Kyritsi, et al., "TGn channel models," Tech. Rep. IEEE 802.11-03/940r4, Wireless LANs, 2004.

[23] K. M. Miettinen, *Non-Linear Multi-Objective Optimization*, Kluwer Academic, Boston, Mass, USA, 1999.

Hindawi Publishing Corporation
EURASIP Journal on Wireless Communications and Networking
Volume 2007, Article ID 14562, 15 pages
doi:10.1155/2007/14562

Research Article

Cross-Layer Admission Control Policy for CDMA Beamforming Systems

Wei Sheng and Steven D. Blostein

Department of Electrical and Computer Engineering, Queen's University, Walter Light Hall (19 Union Street), Kingston, Ontario, Canada K7L 3N6

Received 31 October 2006; Revised 24 June 2007; Accepted 1 August 2007

Recommended by Robert W. Heath Jr.

A novel admission control (AC) policy is proposed for the uplink of a cellular CDMA beamforming system. An approximated power control feasibility condition (PCFC), required by a cross-layer AC policy, is derived. This approximation, however, increases outage probability in the physical layer. A truncated automatic retransmission request (ARQ) scheme is then employed to mitigate the outage problem. In this paper, we investigate the joint design of an AC policy and an ARQ-based outage mitigation algorithm in a cross-layer context. This paper provides a framework for joint AC design among physical, data-link, and network layers. This enables multiple quality-of-service (QoS) requirements to be more flexibly used to optimize system performance. Numerical examples show that by appropriately choosing ARQ parameters, the proposed AC policy can achieve a significant performance gain in terms of reduced outage probability and increased system throughput, while simultaneously guaranteeing all the QoS requirements.

1. INTRODUCTION

In a code division multiple access (CDMA) system, quality-of-service (QoS) requirements rely on interference mitigation schemes and resource management, such as power control, multiuser detection, and admission control (AC) [1–3]. Recently, the problem of ensuring QoS by integrating the design in the physical layer and the admission control (AC) in the network layer is receiving much attention. In [4, 5], an optimal semi-Markov decision process (SMDP)-based AC policy is presented based on a linear-minimum-mean-square-error (LMMSE) multiuser receiver for constant bit rate traffic and circuit-switched networks. In [6], optimal admission control schemes are proposed in CDMA networks with variable bit rate packet multimedia traffic.

The above algorithms [4–6] integrate the optimal AC policy with a multiuser receiver, and as a result, are able to optimize the power control and the AC across the physical and network layers. However, [4–6] only consider single antenna systems, which lack the tremendous performance benefits provided by multiple antenna systems [7–17]. Furthermore, [4–6] rely on an asymptotic signal-to-interference ratio (SIR) expression proposed in [18] which requires a large number of users and a large processing gain. This specific

signal model limits the application of the proposed AC policies. Motivated by these facts, in this paper, we investigate cross-layer AC design for an arbitrary-size CDMA system with multiple antennas at the base station (BS).

To derive an optimal AC policy, a feasible state space and exact power controllability are required but are hard to evaluate for the case of multiple antenna systems. This motivates an approximated power control feasibility condition (PCFC) proposed for admission control of a multiple antenna system. This approximation, however, introduces outage in the physical layer, for example, a nonzero probability that a target signal-to-interference ratio (SIR) cannot be satisfied. To reduce the outage probability in the physical layer, a truncated ARQ-based reduced-outage-probability (ROP) algorithm can be employed. Truncated ARQ is an error-control protocol which retransmits an error packet until correctly received or a maximum number of retransmissions is reached. It is well known that retransmissions can significantly improve transmission reliability, and as a result, can reduce the outage probability. Although retransmissions increase the transmission duration of a packet and thus degrade the network layer performance, this degradation can be controlled to an arbitrarily small level by appropriately choosing the parameters of a truncated ARQ scheme, such as the maximum number of allowed retransmissions and target packet-error rate (PER).

To date, there is no research on cross-layer AC design which considers both link-layer error control schemes and multiple antennas. We remark that this paper differs from prior investigations, for example, [4–6], in the following aspects: (a) here multiple antenna systems are investigated which provide a large capacity gain, while in [4–6], only single antenna systems are discussed; (b) in this paper, a cross-layer AC policy is designed by including error-control schemes, while in [4–6], no such error control schemes are exploited; (c) prior investigations in [4–6] rely on a large system analysis which requires an infinite number of users and infinite length spreading sequences, while here, no such requirements are imposed. In summary, this paper provides a framework for joint optimization across physical, data-link, and network layers, and as a result, is capable of providing a flexible way to handle QoS requirements.

We remark that in the current third generation (3G) system, the application of more efficient methods for packet data transmission such as high-speed uplink packet access (HSUPA) has become more important [19]. In HSUPA, a threshold-based call admission control (CAC) policy is employed, which admits a user request if the load reported is below the CAC threshold. Although the CAC decision can be improved upon by taking advantage of resource allocation information [19], and it is simple to implement, it is well known that the threshold-based CAC policy cannot satisfy QoS requirements in the network layer [5]. Our proposed AC policy provides a solution to guarantee the QoS requirements in both physical and network layers.

The proposed AC policy can be derived offline and then stored in a lookup table. Whenever an arrival or departure occurs, an optimal action can be obtained by table lookup, resulting in low enough complexity for admission control at the packet level. Similar to call/connection level admission control, in a packet-switched system, a packet admission control policy decides if an incoming packet can be accepted or blocked in order to meet quality-of-service (QoS) requirements. In a packet-switched network, blocking a packet instead of blocking the whole user connection can be more spectrally efficient. In this paper, we consider the packet level AC problem.

The rest of this paper is organized as follows. In Section 2, we present the signal model. In Section 3, an approximated PCFC and ARQ-based ROP algorithm are discussed. The formulation and solution of Markov-decision-process (MDP)-based AC policies are proposed in Section 4. Section 5 summarizes the cross-layer design of ARQ parameters. Simulation results are then presented in Section 6.

We will use the following notation: $\ln x$ is the natural logarithm of x, and $*$ denotes convolution. The superscripts $(\cdot)^H$ and $(\cdot)^t$ denote hermitian and transpose, respectively; $\mathrm{diag}(a_1, \ldots, a_n)$ denotes a diagonal matrix with elements $a_1, \ldots, a_n$, and $\mathbf{I}$ denotes an identity matrix. For a random variable X, $E[X]$ is its expectation. The notation and definitions used in this paper are summarized in Table 1.

TABLE 1: Notation and definitions.

Notation	Definition
M	Number of antennas at the BS
K	Number of users
J	Number of classes
R_i	Data rate for packet i
p_i	Transmitted power for packet i
B	Bandwidth
G_i	Link gain for packet i
$\mathbf{a}_i$	Array response vector for packet i
λ_j	Arrival rate for class j
μ_j	Departure rate for class j
Ψ_j	Blocking probability constraint for class j
D_j	Connection delay constraint for class j
L_j	Maximum number of retransmissions for class j
ρ_j	Target PER for class j
$\mathrm{PER}_{\mathrm{overall}}^j$	Achieved overall PER for class j
$\mathrm{PER}_{\mathrm{in}}^j$	Achieved instantaneous PER for class j
γ_j	Target SIR for class j
B_j	Buffer size for class j
$\mathbf{w}_i$	Beamformer weight for packet i
η_0	One-sided power spectral density of additive white Gaussian noise (AWGN)

2. SIGNAL MODEL AND PROBLEM FORMULATION

2.1. Signal model at the physical layer

We consider an uplink CDMA beamforming system, in which M antennas are employed at the BS and a single antenna is employed for each packet. There are K accepted packets in the system, and a channel with slow fading is assumed.

To highlight the design across physical and upper layers considered in this paper, the effects due to multipath are neglected. However, the proposed schemes in this paper can be extended straightforwardly to the case where multipath exists, provided multipath delay profile information is available.

The received vector at the BS antenna array can be written as

$$\mathbf{x}(t) = \sum_{i=1}^{K} \sqrt{P_i} G_i \mathbf{a}_i s_i(t - \tau_i) + \mathbf{n}(t), \qquad (1)$$

where P_i and G_i denote the transmitted power and link gain for packet i, respectively; $\mathbf{a}_i$ is defined as the array response vector for packet i, which contains the relative phases of the received signals at each array element, and depends on the array geometry as well as the angle of arrival (AoA); $s_i(t)$ is the transmitted signal, given by $s_i(t) = \sum_n b_i(n) c_i(t - nT)$, where $b_i(n)$ is the information bit stream, and $c_i(t)$ is the spreading sequence; τ_i is the corresponding time delay and $\mathbf{n}(t)$ is the thermal noise vector at the input of antenna array.

It has been shown that the output of a matched filter sampled at the symbol interval is a sufficient statistic for the estimation of the transmitted signal [14]. The matched filter for a desired packet k is given by $c_k^H(-t)$. The output of the matched filter is sampled at $t = nT$, where T denotes symbol interval. Hence, the received signal at the output of the matched filter is given by [14]

$$
\begin{aligned}
\mathbf{x}_k(n) &= \mathbf{x}(t) * c_k^H(-t)|_{t=nT} \\
&= \sum_{i=1}^{K} \sqrt{P_i} G_i \mathbf{a}_i \int_{(n-1)T+\tau_k}^{nT+\tau_k} \sum_m b_i(m) c_i(t - mT - \tau_i) \\
&\qquad \times c_k(t - nT - \tau_k) dt + \mathbf{n}_k(n),
\end{aligned}
\tag{2}
$$

where $\mathbf{n}_k(n) = \mathbf{n}(t) * c_k^H(-t)|_{t=nT}$.

In order to reduce the interference, we employ a beamforming weighting vector $\mathbf{w}_k$ for a desired packet k. We can write the output of the beamformer as

$$
\begin{aligned}
y_k(n) &= \mathbf{w}_k^H \mathbf{x}(n) \\
&= \sum_{i=1}^{K} \sqrt{P_i} G_i \mathbf{w}_k^H \mathbf{a}_i \int_{(n-1)T+\tau_k}^{nT+\tau_k} \sum_m b_i(m) c_i(t - mT - \tau_i) \\
&\qquad \times c_k(t - nT - \tau_k) dt + \mathbf{w}_k^H \mathbf{n}_k(n).
\end{aligned}
\tag{3}
$$

We assume the signature sequences of the interfering users appear as mutually uncorrelated noise. As shown in [14], the received signal-to-interference ratio (SIR) for a desired packet k can be written as

$$
\mathrm{SIR}_k = \frac{B}{R_i} \frac{p_k \phi_{kk}^2}{\sum_{l \neq i} p_l \phi_{il}^2 + \eta_0 B},
\tag{4}
$$

where B and R_i denote the bandwidth and data rate for packet i, respectively, and the ratio B/R_i represents the processing gain; $p_i = P_i G_i^2$ denotes the received power for packet i, and η_0 denotes the one-sided power spectral density of background additive white Gaussian noise (AWGN); the parameters ϕ_{ii}^2 and ϕ_{ik}^2 are defined as

$$
\phi_{ik}^2 = |\mathbf{w}_k^H \mathbf{a}_i|^2
\tag{5}
$$

which capture the effects of beamforming. In the following, we consider a spatially matched filter receiver, for example, $\mathbf{w}_k = \mathbf{a}_k$.

QoS requirements in the physical layer

In a wireless communication network, we must allow for outage, defined as the probability that a target SIR, or equivalently, a target packet-error rate (PER), cannot be satisfied. The QoS requirement in the physical layer can be represented by a target outage probability.

In this paper, we rely on a relationship between a target SIR and a target PER. Although an exact relationship may not be available, we can obtain the target SIR according to

an approximate expression of PER. As discussed in [20], in a system with packet length N_p (bits), the target SIR for a desired packet i, denoted by γ_i, can be approximated by

$$
\gamma_i = \frac{1}{g} \big[\ln a - \ln \rho_i \big]
\tag{6}
$$

for $\gamma_i \geq \gamma_0$ dB, where ρ_i denotes the overall target PER; a, g, and γ_0 are constants depending on the chosen modulation and coding scheme. In the above expression, the interference is assumed to be additive white Gaussian noise, which is reasonable in a system with enough interferers.

2.2. Signal model in data-link and network layers

We consider a single-cell CDMA system which supports J classes of packets, characterized by different target PERs ρ_j, different blocking probability requirements Ψ_j, and different connection delay requirements D_j, where $j = 1, \ldots, J$. Requests for packet connections of class j are assumed to be Poisson distributed, with arrival rates λ_j, $j = 1, \ldots, J$.

The admission control (AC) is performed at the BS. An AC policy is derived offline, and stored in a lookup table. When a packet is generated at the mobile station (MS), the MS sends an access request to the BS. In this request, the class of this packet is indicated. After receiving the request, the BS makes a decision, which is then sent back to the MS, on whether the incoming packet should be either accepted, queued in the buffer, or blocked. Similarly, whenever a packet departs, the BS decides whether the packet in the queue can be served (transmitted).

Once a packet is accepted, its first transmission round will be performed, and then the receiver will send back an acknowledgement (ACK) signal to the transmitter. A positive ACK indicates that the packet is correctly received while a negative ACK indicates an incorrect transmission.

If a positive ACK is received or the maximum number of retransmissions, denoted by L, is reached, the packet releases the server and departs. Otherwise, the packet will be retransmitted. Therefore, the service time of a packet can comprise at most $L + 1$ transmission rounds. Each transmission round includes the actual transmission time of the packet and the waiting time of an ACK signal (positive or negative). The duration of a transmission round for a packet in class j is assumed to have an exponential distribution with mean duration $1/\mu_j$, $j = 1, \ldots, J$. However, in this paper, a sub-optimal solution is also provided for a generally distributed duration.

If the packet is not accepted by the AC policy, it will be stored in a queue buffer provided that the queue buffer is not full. Otherwise, the packet will be blocked. Each class of packets shares a common queue buffer, and B_j denotes the queue buffer size of class j.

The QoS requirements in the network layer can be represented by the target blocking probability and connection delay, denoted by Ψ_j and D_j for class j, respectively. For each class j, where $j = 1, \ldots, J$, there are K_j packets physically present in the system, which have the same target packet-error-PER, blocking probability, and connection delay constraints.

We note that there are two types of buffers in the system: queue buffers and server buffers. The queue buffer accommodates queued incoming packets, while the server buffer accommodates transmitted packets in the server in case any packet in the server requires retransmission. For simplicity, we assume that the size of the server buffer is large enough such that all the packets in the server can be stored. In the following, the generic term "buffer" refers to the queue buffer.

2.3. Problem formulation

The AC policy considered in this paper is for the uplink only. However, with an appropriate physical layer model for power allocation, the methodology can be extended straightforwardly to the downlink AC problem. The uplink AC is performed at the BS, and the following information is necessary to derive an admission control policy: traffic model in the system, such as arrival and departure rate, and QoS requirements in both physical and network layers.

The overall system throughput is defined as the number of correctly received packets per second, given by

$$\text{Throughput} = \sum_{j=1}^{J} \left(1 - P_b^j\right)\left(1 - \rho_j\right)\left(1 - P_{\text{out}}^j\right)\lambda_j, \quad (7)$$

where P_b^j, ρ_j and P_{out}^j denote the blocking probability, target PER, and outage probability for class j packets, respectively.

In this paper, we aim to derive an optimal AC policy which incorporates the benefits provided by multiple antennas and ARQ schemes. The objective is to maximize the overall system throughput given in (7), while simultaneously guaranteeing QoS requirements in terms of outage probability, blocking probability, and connection delay.

The above optimization problem can be formulated as a Markov decision process (MDP). With a required power control feasibility condition (PCFC), combined with an ARQ-based reduced-outage-probability (ROP) algorithm, a target outage probability constraint can be satisfied. Blocking probability and connection delay requirements can be guaranteed by the constraints of this MDP.

In the following, we first derive an approximate PCFC combined with an ARQ-based reduced-outage-probability (ROP) algorithm that can guarantee the outage probability constraint. Based on these results, we then formulate the AC problem as a Markov decision process. Afterward, we discuss how to design ARQ parameters optimally in order to achieve a maximum system throughput.

3. PHYSICAL LAYER INVESTIGATION: PCFC DERIVATION AND OUTAGE REDUCTION

To investigate the physical layer performance, we must derive an approximate PCFC, which ensures a positive power solution to achieve target SIRs. Due to the approximation of the derived PCFC, we then propose an ARQ-based ROP algorithm to reduce the resulting outage probability.

3.1. PCFC

In the physical layer, the SIR requirements of packet i can be written as

$$\text{SIR}_i \geq \gamma_i \quad (8)$$

for $i = 1, \ldots, K$, where SIR_i is given in (4).

Inserting the SIR expression in (4) into (8), and letting SIR_i achieve its target value, γ_i, we have the matrix form [15]

$$[\mathbf{I} - \mathbf{QF}]\mathbf{p} = \mathbf{Qu}, \quad (9)$$

where $\mathbf{I}$ is the identity matrix, $\mathbf{p} = [p_1, \ldots, p_K]^t$, $\mathbf{u} = \eta_0 B[1, \ldots, 1]^t$,

$$\mathbf{Q} = \text{diag}\left\{\frac{\gamma_1 R_1/B}{1 + \gamma_1 R_1/B}, \ldots, \frac{\gamma_K R_K/B}{1 + \gamma_K R_K/B}\right\},$$

$$\mathbf{F} = \begin{bmatrix} F_{1,1} & F_{1,2} & \cdots & F_{1,K} \\ F_{2,1} & F_{2,2} & \cdots & F_{2,K} \\ \cdots & \cdots & \cdots & \cdots \\ F_{K,1} & F_{K,2} & \cdots & F_{K,K} \end{bmatrix} \quad (10)$$

in which $F_{ij} = \phi_{ij}^2/\phi_{ii}^2$.

To ensure a positive solution for power vector $\mathbf{p}$, we require the following power control feasibility condition [15],

$$\rho(\mathbf{QF}) < 1, \quad (11)$$

where $\rho(\cdot)$ denotes the maximum eigenvalue.

The outage probability can be obtained as the probability that the above condition is violated. Although the state space, required by an optimal AC policy, can be formulated by evaluating the above outage probability, this evaluation relies on the number of packets as well as the distribution of AoAs for all the packets in the system, and thus results in a very high computation complexity. An approach to evaluate the above outage probability with reasonably low complexity is currently under investigation.

In this paper, we propose an alternative solution, which employs an approximated PCFC, and as a result can dramatically simplify the formulation of the state space.

Without loss of generality, we consider an arbitrary packet i in class 1, where $i = 1, \ldots, K_1$. By considering specific traffic classes and letting SIR achieve its target value, the expression in (4) can be written as

$$\gamma_i = \frac{p_i \phi_{ii}^2 (B/R_1)}{\sum_{l=1, l \neq i}^{K_1} p_l \phi_{il}^2 + \sum_{l=1}^{K_2} p_l \phi_{il}^2 + \cdots \sum_{l=1}^{K_J} p_l \phi_{il}^2 + \sigma^2}, \quad (12)$$

where $\sigma^2 \triangleq \eta_0 B$ denotes noise variance, and p_i represents received power for packet i.

It is not difficult to show that packets in the same class have the same received power. By denoting the received power in class j as p_j, where $j = 1, \ldots, J$, the above expression can be written as

$$\gamma_i = \frac{p_1 \phi_{ii}^2 (B/R_1)}{\sum_{l=1, l \neq i}^{K_1} p_1 \phi_{il}^2 + \cdots + \sum_{l=1}^{K_J} p_J \phi_{il}^2 + \sigma^2}$$

$$= \frac{p_1 \phi_{ii}^2 (B/R_1)}{p_1 (K_1 - 1)\beta_1 + \sum_{j=2}^{J} p_j K_j \beta_j + \sigma^2}, \quad (13)$$

where $\beta_1 = (1/(K_1 - 1)) \sum_{l=1, l \neq i}^{K_1} \phi_{il}^2$ and $\beta_j = (1/K_j) \sum_{l=1}^{K_j} \phi_{il}^2$, in which $j = 2, \ldots, J$.

By exchanging the numerator and denominator, (13) is equivalent to

$$\frac{p_1 (K_1 - 1)\beta_1 + \sum_{j=2}^{J} p_j K_j \beta_j + \sigma^2}{p_1 (B/\gamma_1 R_1)} = \phi_{ii}^2, \quad (14)$$

where $i = 1, \ldots, K_1$.

Summing the above K_1 equations, and calculating the sample average, we obtain

$$\frac{p_1 (K_1 - 1)\alpha_1 + \sum_{j=2}^{J} K_j p_j \alpha_j + \sigma^2}{p_1 (B/\gamma_1 R_1)} = \frac{1}{K_1} \sum_{i=1}^{K_1} \phi_{ii}^2, \quad (15)$$

where $\alpha_1 = (1/K_1) \sum_{i=1}^{K_1} \beta_1$ and $\alpha_j = (1/K_1) \sum_{i=1}^{K_1} \beta_j$.

When the number of packets is large enough, by the weak law of large numbers, the above $\alpha_1, \ldots, \alpha_J$ can be approximated by their mean values, and (15) can be further simplified as

$$\frac{p_1 (K_1 - 1) E_{11}[\phi_{\text{int}}] + \sum_{j=2}^{J} K_j p_j E_{1j}[\phi_{\text{int}}] + \sigma^2}{p_1 (B/\gamma_1 R_1)} = E_1[\phi_{\text{des}}] \quad (16)$$

in which $E_{mn}[\phi_{\text{int}}]$ is the expected fraction of an interferer packet in class n passed by a beamforming weight vector for a desired packet in class m, where $m, n = 1, \ldots, J$, while $E_j[\phi_{\text{des}}]$ is the expected fraction of a desired packet in class j passed by its beamforming weight vector, where $j = 1, \ldots, J$.

The AoAs of active packets in the system are assumed to be independent and identically distributed, that are independent of a packet's specific class. Therefore, it is reasonable to assume that $E_{mn}[\phi_{\text{int}}]$ is also independent of specific classes m and n, which can be denoted by $E[\phi_{\text{int}}]$. Similarly, $E_j[\phi_{\text{des}}]$ is independent of class j, and can be denoted by $E[\phi_{\text{des}}]$. $E[\phi_{\text{des}}]$ and $E[\phi_{\text{int}}]$ represent the expected fractions of the desired packet's power and interference, respectively.

From the above discussion, (16) can be written as

$$\frac{p_1 (K_1 - 1) E[\phi_{\text{int}}] + \sum_{j=2}^{J} K_j p_j E[\phi_{\text{int}}] + \sigma^2}{p_1 (B/\gamma_1 R_1)} = E[\phi_{\text{des}}]. \quad (17)$$

By exchanging the numerator and denominator of the above equation, we have

$$p_1 \frac{B}{\gamma_1 R_1} \bigg/ \bigg(p_1 (K_1 - 1) \frac{E[\phi_{\text{int}}]}{E[\phi_{\text{des}}]} + \sum_{j=2}^{J} K_j p_j \frac{E[\phi_{\text{int}}]}{E[\phi_{\text{des}}]} + \frac{\sigma^2}{E[\phi_{\text{des}}]} \bigg) = 1. \quad (18)$$

The QoS requirement for class 1 in (18) can be extended to any class j,

$$p_j \frac{B}{\gamma_j R_j} \bigg/ \bigg(p_j (K_j - 1) \frac{E[\phi_{\text{int}}]}{E[\phi_{\text{des}}]} + \sum_{m=1, m \neq j}^{J} K_m p_m \frac{E[\phi_{\text{int}}]}{E[\phi_{\text{des}}]} + \frac{\sigma^2}{E[\phi_{\text{des}}]} \bigg) = 1, \quad (19)$$

where $j = 1, \ldots, J$.

The power allocation solution can be obtained by solving the above J equations [21]

$$p_j = \frac{\sigma^2}{E[\phi_{\text{int}}]} \bigg/ \bigg(\bigg(1 + \frac{B}{\gamma_j R_j (E[\phi_{\text{int}}]/E[\phi_{\text{des}}])} \bigg) \times \bigg[1 - \sum_{j=1}^{J} \frac{K_j}{1 + (B/\gamma_j R_j (E[\phi_{\text{int}}]/E[\phi_{\text{des}}]))} \bigg] \bigg), \quad (20)$$

where $j = 1, \ldots, J$.

Positivity of the power solution implies the following power control feasibility condition:

$$\sum_{j=1}^{J} \frac{K_j}{1 + (B/\gamma_j R_j (E[\phi_{\text{int}}]/E[\phi_{\text{des}}]))} < 1. \quad (21)$$

As shown in [22], $E[\phi_{\text{int}}]$ and $E[\phi_{\text{des}}]$ can be determined numerically from (5) for a beamforming system.

We note that the above approximated power control feasibility condition is independent of the angle of arrivals, and thus can provide a less-complicated offline AC policy, which does not require estimation of the current AoA realizations of each packet. However, due to the randomness of the actual SIR, this deterministic power control feasibility condition introduces outage. In the next section, we discuss how to mitigate the outage.

3.2. ARQ-based ROP

We first define two types of PERs. The overall achieved PER, denoted by $\text{PER}_{\text{overall}}^j$, is defined as the probability that a class j packet is incorrectly received after its maximum number of ARQ retransmissions is reached, for example, an error occurs in each of the $L_j + 1$ transmission rounds, where L_j denotes the maximum number of retransmissions. The achieved instantaneous PER, denoted as $\text{PER}_{\text{in}}^j(l)$, is defined as the probability that an error occurs in a single transmission round l for a class j packet.

Under the assumption that each retransmission round is independent from the others, by using an ARQ scheme with a maximum of L_j retransmissions for class j, the achieved overall PER is constrained by [20]

$$\text{PER}_{\text{overall}}^j = \prod_{l=1}^{L_j+1} \text{PER}_{\text{in}}^j(l), \quad (22)$$

$$\leq \rho_j,$$

where ρ_j denotes the target overall PER for class j.

The achieved outage probability for class j, denoted by P_{out}^j, can be written as

$$
\begin{aligned}
P_{\text{out}}^j &= \text{Prob}\left\{ \text{PER}_j^{\text{overall}} > \rho_j \right\} \\
&= \text{Prob}\left\{ \prod_{l=1}^{L_j+1} \text{PER}_{\text{in}}^j(l) > \rho_j \right\},
\end{aligned}
\tag{23}
$$

where $\text{Prob}\{A\}$ denotes the probability of event A. By maintaining PCFC, $\text{PER}_j^{\text{in}}(l)$ remains unchanged. Therefore, by increasing L_j, the outage probability in the above equation can be reduced.

4. AC PROBLEM FORMULATION BY INCLUDING ARQ

In the previous section, we have derived an approximated PCFC combined with an ARQ-based ROP algorithm in the physical layer. In the following, we discuss how to derive an AC policy in the network layer.

An optimal semi-Markov decision process (SMDP)-based AC policy as well as a low-complexity generalized-Markov decision process (GMDP)-based AC policy is discussed.

4.1. SMDP-based AC policy

Traditionally, the decision epoches are chosen as the time instances that a packet arrives or departs. In the system under consideration, the duration of each packet may include several transmission rounds due to ARQ retransmissions, and as a result, the time duration until next system state may not be exponentially distributed. Therefore, the SMDP formulation approach discussed in [4–6], which assumes an exponentially distributed duration, cannot be applied here.

In the following, we propose a novel formulation in which the decision epoch is chosen as the arrival and departure of each transmission round. Based on these decision epoches, the time duration until the next state remains exponentially distributed. The components of a Markov decision process, such as state space, action space, and dynamic statistics, are modified accordingly to represent the characteristics of different transmission rounds. The formulation of this SMDP as well as its LP solution are now described.

State space and action space

Class j packets are divided into L_j+1 subclasses, in which the state of the ith subclass can be represented by the number of packets which are under the ith round transmission, that is, the $(i-1)$th retransmission, where $i = 1, \ldots, L_j+1$.

In admission problems, the discrete-value (finite) state at time t, $s(t)$, can be written as

$$
\begin{aligned}
s(t) = \Big[&\underbrace{n_q^1(t), k^{1,1}(t), \ldots, k^{1,L_1+1}(t)}, \ldots, \\
&\underbrace{n_q^J(t), k^{J,1}(t), \ldots, k^{J,L_J+1}(t)} \Big]^T,
\end{aligned}
\tag{24}
$$

where $k^{j,i}(t)$ represents the number of active packets in class j and subclass i served in the system, and $n_q^j(t)$ denotes the number of packets in the queue buffer of class j. Since the arrival and departure of packets are random, $\{s(t), t > 0\}$ represents a finite state stochastic process [4]. From here on, we will drop the time index.

The state space S is comprised of any state vector s, in which SIR requirements can be satisfied or, equivalently, the power control feasibility condition (PCFC) holds,

$$
\begin{aligned}
S = \Big\{ s : n_q^j &\leq B_j, j = 1, \ldots, J; \\
&\sum_{j=1}^J \frac{\left(\sum_{l=1}^{L_j+1} k^{j,l} \right)}{1 + \left(B/\gamma_j R_j (E[\phi_{\text{int}}]/E[\phi_{\text{des}}]) \right)} < 1 \Big\},
\end{aligned}
\tag{25}
$$

where B_j denotes the buffer size of class j. We have mentioned that the PCFC for the case of no ARQ is used in our AC problem, no matter how many retransmissions are allowed.

At each state s, an action is chosen that determines how the admission control will perform at the next decision moment [4]. In general, an action, denoted as a, can be defined as a vector of dimension $\sum_{j=1}^J L_j + 2J$,

$$
a = \Big[\underbrace{a_1, d_1^1, \ldots, d_1^{L_1+1}} \cdots \underbrace{a_J, d_J^1, \ldots, d_J^{L_J+1}} \Big]^T,
\tag{26}
$$

where a_j denotes the action for class j if an arrival occurs, $j = 1, \ldots, J$. If $a_j = 0$, the new arrival is placed in the buffer provided that the buffer is not full or is blocked if the buffer is full; if $a_j = 1$, the arrival is admitted as an active packet, and the number of servers of class j is incremented by one.

The quantity d_j^i, where $1 \leq i \leq L_j$, denotes the action for class j packet if the ith transmission round is finished, and is received correctly. If $d_j^i = 0$, where $1 \leq i \leq L_j$, $k^{j,i}$ is decremented by one, and no packets that are queued in the buffer are made active; if $d_j^i = 1$, the number of servers is maintained by admitting a packet at the buffer as an active packet.

The quantity $d_j^{L_j+1}$ denotes the action for class j packet if a connection has finished its (L_j+1)th transmission round. If $d_j^{L_j+1} = 0$, no packets that are queued in the buffer are made active, and k^{j,L_j+1} is decremented by one; if $d_j^{L_j+1} = 1$, the number of servers is maintained by admitting a packet at the buffer as an active packet.

The admissible action space for state s, denoted by A_s, can be defined as the set of all feasible actions. A feasible action ensures that after taking this action, the next transition state is still in space S [4].

State dynamics $p_{sy}(a)$ and $\tau_s(a)$

The state dynamics of an SMDP are completely specified by stating the transition probabilities of the embedded chain $p_{sy}(a)$ and the expected holding time $\tau_s(a)$: $p_{sy}(a)$ is defined as the probability that the state at the next decision epoch is

Table 2: Expression of transition probability p_{sy}.

y	$p_{sy}(a)$
$y = s + q^j$	$\lambda_j a_j \tau_s(a)$
$y = s + b^j$	$\lambda_j (1 - a_j) \delta(B_j - n_q^j) \tau_s(a)$
$y = s + c_i^j$	$(1 - \rho_j)[\mu_j k^{j,i}(1 - d_j^i) \tau_s(a)]$ $+ (1 - \rho_j)[\mu_j k^{j,i} d_j^i (1 - \delta(n_q^j)) \tau_s(a)]$
$y = s + r^j$	$\sum_{i=1}^{L_j+1} (1 - \rho_j) \mu_j k^{j,i} d_j^i \tau_s(a) \delta(n_q^j)$
$y = s + e_i^j$	$\rho_j \mu_j k^{j,i} \tau_s(a)$
$y = s + f^j$	$\mu_j k^{j,L_j+1} d_j^{L_j+1} \delta(n_q^j) \tau_s(a)$
$y = s + g^j$	$\mu_j k^{j,L_j+1}(1 - d_j^{L_j+1}) \tau_s(a)$ $+ \mu_j k^{j,L_j+1} d_j^{L_j+1}(1 - \delta(n_q^j)) \tau_s(a)$
Otherwise	0

Table 3: Definition of vectors in Table 2: each vector defined in this table has a dimension of $\sum_{j=1}^{J} L_j + 2J$, which contains only zeros except for the specified positions.

Vector	Nonzero positions
q^j	Position $2(j - 1) + \sum_{t=1}^{j-1} L_t + 2$ contains a 1
b^j	Position $2(j - 1) + \sum_{t=1}^{j-1} L_t + 1$ contains a 1
c_i^j	Position $2(j - 1) + \sum_{t=1}^{j-1} L_t + i + 1$ contains a -1
r^j	Position $2(j - 1) + \sum_{t=1}^{j-1} L_t + 1$ contains a -1
e_i^j	Position $2(j - 1) + \sum_{t=1}^{j-1} L_t + i + 1$ contains a -1 and position $2(j - 1) + \sum_{t=1}^{j-1} L_t + i + 2$ contains a 1
f^j	Position $2(j - 1) + \sum_{t=1}^{j-1} L_t + 1$ contains a -1
g^j	Position $2(j - 1) + \sum_{t=1}^{j-1} L_t + L_j + 2$ contains a -1

Table 4: Representation of vectors in Table 2: each defined vector represents a possible state transition from current state s.

Notation	State transition
$s + q^j$	An increase in subclass 1 of class j by 1
$s + b^j$	An increase in queue j by 1
$s + c_i^j$	A decrease in subclass i of class j by 1
$s + r^j$	A decrease in queue j by 1
$s + e_i^j$	An increase of subclass $i + 1$ by 1, and a decrease in subclass i of class j by 1
$s + f^j$	A decrease in queue j by 1
$s + g^j$	A decrease in subclass $L_j + 1$ of class j by 1

y if action a is selected at the current state s, while $\tau_s(a)$ is the expected time until the next decision epoch after action a is chosen in the present state s [4].

Derivations of $\tau_s(a)$ and $p_{sy}(a)$ rely on the statistical properties of arrival and departure processes [4]. Since the arrival and departure processes are both Poisson distributed and mutually independent, it follows that the cumulative process is also Poisson, and the cumulative event rate is the sum of the rates for all constituent processes [4]. Therefore, the expected sojourn time, $\tau_s(a)$, can be obtained as the inverse of the event rate,

$$
\begin{aligned}
\tau_s(a)^{-1} = &\ \lambda_1 a_1 + \lambda_1 (1 - a_1) \delta(B_1 - n_q^1) \\
&+ \sum_{i=1}^{L_1+1} \mu_1(k^{1,i}) + \cdots + \lambda_J a_J + \lambda_J (1 - a_J) \delta(B_J - n_q^J) \\
&+ \sum_{i=1}^{L_J+1} \mu_J(k^{J,i}),
\end{aligned}
\tag{27}
$$

where

$$
\delta(z) = \begin{cases} 1 & \text{if } z > 0, \\ 0 & \text{if } z = 0. \end{cases}
\tag{28}
$$

To derive the transition probabilities, we employ the decomposition property of a Poisson process, which states that an event of a certain type occurs with a probability equal to the ratio between the rate of that particular type of event and the total cumulative event rate $1/(\tau_s(a))$ [4]. Transition probability $p_{sy}(a)$ is shown in Table 2, where ρ_j denotes the target packet-error rate for class j packets. The set of vectors $\{q_j, b^j, c_i^j, r^j, e_i^j, f^j, g_j\}$ represents the possible state transitions from current state s. Each vector in this set has a dimension of $\sum_{j=1}^{J} L_j + 2J$, and contains only zeros except for one or two positions. The nonzero positions of this set of vectors, as well as the possible state transitions represented by these vectors, are specified in Tables 3 and 4, respectively.

Policy and cost criterion

For any given state $s \in S$, an action a, which decides if the new packet at the next decision epoch will be blocked or accepted, is selected according to a specified policy R. A stationary policy R is a function that maps the state space into the admissible action space.

We consider average cost criterion [4]. The cost criterion for a given policy R and initial state s_0, which includes blocking probability as a special case, is given as follows:

$$
J_R(s_0) = \lim_{t \to \infty} \frac{1}{T} E \left\{ \int_0^T c(s(t), a(t)) dt \right\},
\tag{29}
$$

where $c(s(t), a(t))$ can be interpreted as the expected cost until the next decision epoch and is selected to meet the network layer performance criteria [4].

In the system under investigation, we are interested in blocking probability and connection delay constraints. If the cost criterion $J_R(s_0)$ represents blocking probability, we have $c(s, a) = (1 - a_j)(1 - \delta(B_j - n_q^j))$, and if the cost criterion $J_R(s_0)$ represents connection delay, we have $c(s, a) = n_q^j$.

An optimal policy R^* that minimizes an average cost criterion $J_R(s_0)$ for any initial state s_0 exists,

$$
J_{R^*}(s_0) = \min_{R \in \mathbf{R}} J_R(s_0), \quad \forall s_0 \in S
\tag{30}
$$

under the weak unichain assumption [23], where $\mathbf{R}$ is the class of admissible AC policies.

Solving the AC policy by linear programming (LP)

The optimal AC policy, which can minimize the blocking probability, can be obtained by using the decision variables $z_{sa}, s \in S, a \in A_s$.

The optimal AC policy R^* in (30) can be obtained by solving the following linear programming (LP):

$$\min_{z_{sa} \geq 0, s, a} \sum_{s \in S} \sum_{a \in A_s} \sum_{j=1}^{J} \eta_j (1 - a_j)(1 - \delta(B_j - n_q^j)) \tau_s(a) z_{sa} \tag{31}$$

subject to

$$\begin{aligned}
&\sum_{a \in A_y} z_{ya} - \sum_{s \in S} \sum_{a \in A_s} p_{sy}(a) z_{sa} = 0, \quad y \in S, \\
&\sum_{s \in S} \sum_{a \in A_s} \tau_s(a) z_{sa} = 1, \\
&\sum_{s \in S} \sum_{a \in A_s} (1 - a_j)(1 - \delta(B_j - n_q^j)) \tau_s(a) z_{sa} \leq \Psi_j, \\
&\sum_{s \in S} \sum_{a \in A_s} n_q^j \tau_s(a) z_{sa} \leq D_j,
\end{aligned} \tag{32}$$

where D_j and Ψ_j denote the connection delay and blocking probability constraints, respectively, and η_j is the coefficient representing the weighting of the cost function for a particular class j, where $j = 1, \ldots, J$.

The optimal policy will be a randomized policy: the optimal action $a^* \in A_s$ for state s, where A_s is the admissible action space, is chosen probabilistically according to the probabilities $z_{sa}/\sum_{a \in A_s} z_{sa}$.

We remark that the above randomized AC policy can optimize the long-run performance. The decision variables, z_{sa}, where $s \in S$ and $a \in A_x$, act as the long-run fraction of decision epochs at which the system is in state s and action a. At each state s, there exists a set of feasible actions, and each action induces a different cost $c(s, a)$. The long-run performance can be optimized by appropriately allocating these time fractions, and the allocation leads to a randomized AC policy. When a deterministic policy is desired, a constraint regarding the decision variables z_{sa} should be imposed into the above optimization problem, in order to ensure that at each state s, there is one and only one nonzero decision variable. It is obvious that the more constraints we impose, the worse the achieved performance becomes. We choose a randomized AC policy in order to achieve long-run optimal performance.

4.2. GMDP-based AC policy

In the above, we provide an optimal SMDP formulation. The state space has dimension of $2J + \sum_{j=1}^{J} L_j$ for J classes of traffic. For large J and retransmission number, this leads to a computation problem of excessive size.

In order to reduce complexity, we consider the decision epoch as the time instances that a packet arrives or departs. As we discussed in the previous section, based on these decision epochs, the time interval until the next state is not exponentially distributed. Therefore, we have a generalized Markov decision process (GMDP). While an optimal solution for this GMDP problem is hard to obtain, a linear programming approach provides a suboptimal solution [5].

We remark that the formulation of a GMDP is very similar to the AC problem formulation employed in [4–6], except that the state space has been modified to include beamforming and the mean duration of a packet is modified to consider the impact of ARQ schemes.

In the formulated GMDP, decision epochs are chosen as the time instances that a packet arrives or departs. The arrival process for class j is assumed to have a Poisson distribution with arrival rate λ_j. The duration of the class j packets may have a general distribution, with mean $(1/\mu_j)(1 + \rho_j + \cdots + \rho_j^{L_j})$, where μ_j denotes the departure rate for each transmission round for the class j packets.

The state space S is comprised of any state vector s, which satisfies SIR requirements,

$$S = \left\{ s = [n_q^1, k^1, \ldots, n_q^J, k^J]^T : n_q^j \leq B_j, \right.$$
$$\left. j = 1, \ldots, J; \sum_{j=1}^{J} \frac{k^j}{1 + (B/\gamma_j R_j (E[\phi_{int}]/E[\phi_{des}]))} < 1 \right\}, \tag{33}$$

where k^j denotes the number of active packets for class j.

At each decision epoch, an action is chosen as $a = [a_1, d_1 \ldots, a_J, d_J]^T$, where a_j denotes the action for class j if an arrival occurs, $j = 1, \ldots, J$ and d_j denotes the action for class j packet if a packet in this class departs. The admissible action space for state s, denoted by A_s, can be defined as the set of all feasible actions.

The state dynamics of a SMDP are completely specified by stating the transition probabilities of the embedded chain $p_{sy}(a)$ and the expected holding time $\tau_s(a)$, which are given in [4, 5].

After formulating the AC problem as a GMDP, the AC policy, which minimizes the blocking probability, can be obtained by using the decision variables $z_{sa}, s \in S, a \in A_s$ from linear programming which is presented in (31).

In a low instantaneous PER region, the suboptimal solution proposed in the above is very close to the SMDP-based AC policy. Intuitively, when the PER is very low, retransmission occurs only occasionally, and the duration of a packet would be very close to an exponential distribution. In this case, the LP approach would provide an optimal solution to the above GMDP.

We remark that unlike the SMDP-based AC policy in which the transmission round is assumed to have an exponential distribution, the GMDP-based AC policy discussed in the subsection can be applied to a system with a generally distributed transmission round.

4.3. Complexity

SMDP or GMDP-based AC policies are always calculated offline and stored in a lookup table. Whenever an arrival or departure occurs, an optimal action can be obtained by table lookup using the current system state. This facilitates the implementation of packet-level admission control.

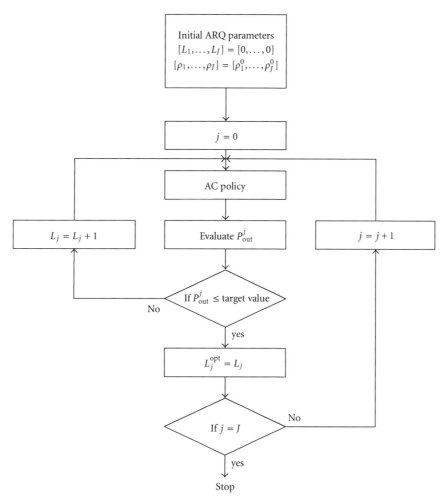

FIGURE 1: Search procedure of the optimal number of retransmissions.

Once system parameters change, an updated policy is required. However, in the system we investigate, the policy only depends on buffer sizes, long-term traffic model, and QoS requirements. These parameters are generally constant for the provision of a given profile of offered services. Therefore, an SMDP or GMDP-based policy has a very reasonable computation complexity.

5. CROSS-LAYER DESIGN OF ARQ PARAMETERS

In the previous sections, we discuss how to derive the PCFC in the physical layer and how to derive admission control in the network layer. These derivations assume that ARQ parameters such as L_j and ρ_j, where $j = 1, \ldots, J$, are already known. In this section, we discuss how to choose these parameters in order to guarantee outage probability constraints and optimize overall system throughput.

The search procedures for optimal ARQ parameters, denoted as vectors $L^{\text{opt}} = [L_1^{\text{opt}}, \ldots, L_J^{\text{opt}}]$ and $\rho^{\text{opt}} = [\rho_1^{\text{opt}}, \ldots, \rho_J^{\text{opt}}]$, are demonstrated in Figures 1 and 2, respectively. The initial parameters are set to $[L_1, \ldots, L_J] = [0, \ldots, 0]$ and $[\rho_1, \ldots, \rho_J] = [\rho_1^0, \ldots, \rho_J^0]$, where ρ_j^0 represents the upper bound target PER for class j, which can be specified for the system. In Figure 2, Δ_j represents the adjustment step size.

From the search procedures presented in Figures 1 and 2, it is observed that the number of allowed retransmissions L_j^{opt}, which can achieve a target outage probability, is minimized; and as a result, the network layer performance degradation can be minimized. Thus, network layer QoS requirements in terms of blocking probability and connection delay can be guaranteed by formulating the AC problem as an SMDP or GMDP.

Summing above, by choosing ARQ parameters in a cross-layer context, QoS requirements in the physical and network layers can be guaranteed, and the overall system throughput can be maximized.

6. SIMULATION RESULTS

We consider a 3-element circular antenna array, for example, $M = 3$, with a uniformly distributed angle of arrival (AoA) over $[0, 2\pi)$ [22]. Numerical values of parameters $E[\phi_{\text{des}}]$ and $E[\phi_{\text{int}}]$ in (21), derived in [22], are shown in Table 5. We remark that the proposed AC policies can be applied to any other array geometry and AoA distribution. Without loss of

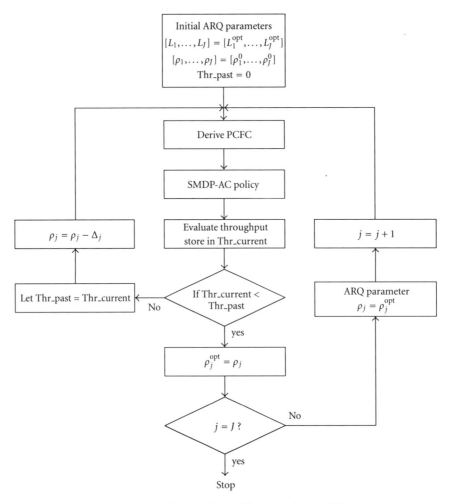

FIGURE 2: Search procedure of the optimal target PER.

TABLE 5: Numerical values of $E[\phi_{\text{des}}]$ and $E[\phi_{\text{int}}]$ in (20) and (21).

M	1	2	3	4	5	6
$E[\phi_{\text{des}}]$	1.0	1.0	1.0	1.0	1.0	1.0
$E[\phi_{\text{int}}]$	1.0	0.5463	0.3950	0.3241	0.2460	0.2058

TABLE 6: Simulation parameters.

B	3.84 MHz	a	90.2514
g	3.4998	γ_0	1.0942 dB
R_1	144 kbps	R_2	384 kbps
λ_1	1	λ_2	0.5
μ_1	0.25	μ_2	0.1375
Ψ_1	0.1	Ψ_2	0.2
D_1	2.25	D_2	0.5360
M	3	η_0	10^{-6}
η_1	0.5	η_2	0.5

generality, we consider a single-path channel and a two-class system with a QPSK and convolutionally coded modulation scheme with rate 1/2 and a packet length $N_p = 1080$. Under this scheme, the parameters of a, g, and γ_0 in (6) can be obtained from [20]. For simplicity, no buffer is employed in the simulation. Simulation parameters are presented in Table 6.

6.1. Performance of SMDP-based AC policies

Here, we investigate how the ARQ scheme can reduce outage probability while only slightly degrading the network layer performance.

We examine the case in which only the class 2 packets can be retransmitted once, for example, $L_1 = 0$ and $L_2 = 1$, and an optimal SMDP-based AC policy is employed. The target PER for the class 1 packets is set to 10^{-4}, while different target

PERs for class 2 are evaluated. We focus on the performance for the class 2 packets since only these packets are allowed retransmission. Figure 3 presents the analytical and simulated blocking probabilities as a function of ρ_2. It is observed that the simulation results are very close to the analytical results. Figure 4 presents the outage probability and throughput for the class 2 packets. It is observed that at a reasonably low PER, the outage probability can be reduced dramatically, and overall system throughput can be significantly improved by allowing only one retransmission. Figure 5, which presents

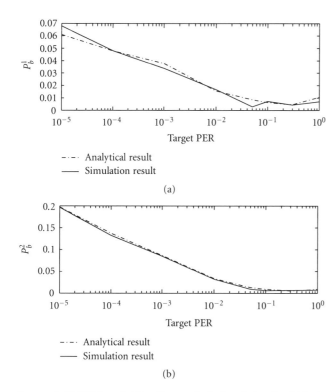

(a)

(b)

Figure 3: SMDP-based AC policy: analytical and simulated blocking probabilities as a function of ρ_2 in which $L_1 = 0$, $L_2 = 1$, and $\rho_1 = 10^{-4}$.

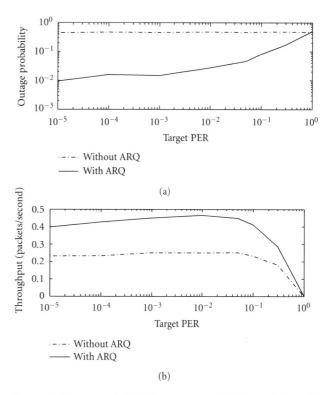

(a)

(b)

Figure 4: SMDP-based AC policy: outage probability and throughput for class 2 packets as a function of ρ_2 in which $L_1 = 0$, $L_2 = 1$, and $\rho_1 = 10^{-4}$.

the network layer performance degradation by employing ARQ, shows that the degradation can be ignored in a low PER region.

6.2. Performance of GMDP-based AC policies

In the above, we discussed the performance of SMDP-based AC policies, which require high computation. To reduce complexity, a GMDP-based AC policy can be employed. The target PER for class 1 is set to 10^{-4}, while different target PER requirements for class 2 are considered.

Figure 6 shows the analytical and simulated blocking probabilities as a function of target PER for the class 2 packets. The gap between the simulated and analytical results is due to the non-exponential distribution of the packet duration.

Figure 7 demonstrates that for a small number of retransmissions, SMDP and GMDP-based AC policies have similar performance. Although performance comparison for large L_j is not presented here since an SMDP-based AC policy would involve excessive computation, it is expected that for low PER, these two AC policies would still have similar performance. For a high PER, however, the packet duration is far from exponentially distributed, and thus linear programming cannot provide an optimal solution to a GMDP and its performance would be inferior to that of SMDP. In summary, GMDP-based AC policy provides a simplified approach which is capable of achieving a near-optimal system

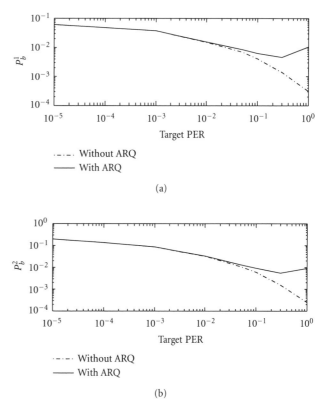

(a)

(b)

Figure 5: SMDP-based AC policy: blocking probability degradation as a function of ρ_2 in which $L_1 = 0$, $L_2 = 1$, and $\rho_1 = 10^{-4}$.

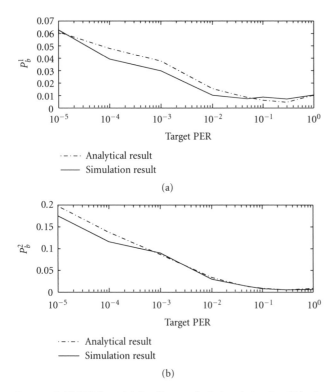

FIGURE 6: GMDP-based AC policy: analytical and simulated blocking probabilities as a function of ρ_2 in which $L_1 = 0$, $L_2 = 1$, and $\rho_1 = 10^{-4}$.

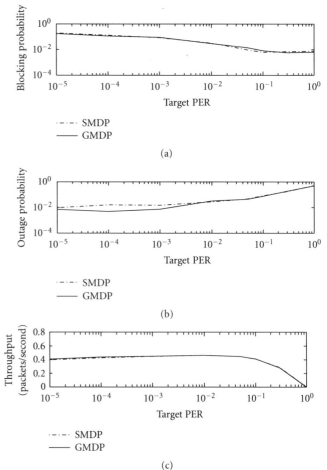

FIGURE 7: Performance comparison between SMDP and GMDP-based AC policies as a function of ρ_2 in which $L_1 = 0$, $L_2 = 1$, and $\rho_1 = 10^{-4}$.

performance for a system with low PER or a small number of retransmissions.

Figures 8–10 compare the performance among different numbers of retransmissions in which $\rho_1 = \rho_2$, and $L_1 = L_2$. From here on, L_j is denoted by L in the figures. We investigate the performance for $L_j = 0, 1$, and 2, respectively. The results for large L_j can be extended straightforwardly. It is observed that in a low PER region, for example, $\rho_j \leq 0.01$, with an increased L_j, outage can dramatically be reduced, while the blocking probability is only slightly degraded. With only one retransmission allowed, the throughput can be improved by 100%. However, when L_j is increased beyond a certain level, for example, $L_j = 2$ in the system under consideration, the outage reduction and throughput improvement are not significant. Beyond this threshold, further increasing L_j may even lead to a performance degradation due to a degraded network layer performance. From Figures 8–10, we also conclude that at high PER, the proposed ARQ-based ROP algorithm is not as efficient as in low PER.

6.3. Performance of a complete-sharing-based admission control policy

For a complete-sharing (CS)-based policy, whenever a packet arrives, the power control feasibility condition in (21) is evaluated by incorporating information of this newly arrived packet. If this condition is satisfied, the incoming packet can be accepted, otherwise, the packet is stored in a buffer or blocked if the buffer is full. CS-based AC policy provides a simple admission control algorithm but ignores the QoS requirements in the network layer.

We now provide a simple example for complete-sharing (CS)-based AC policy. For comparison purposes, the simulation results for a GMDP-based AC policy is also presented. In this example, both classes of packets are allowed to retransmit twice, for example, $L_1 = L_2 = 2$.

We note that in a system with relaxed blocking probability constraints, even a CS-based AC policy can satisfy all the QoS requirements. To illustrate the shortcoming of a CS-based AC policy, we now restrict the blocking probability constraint for class 2 to 0.05 without loss of generality, and all the other parameters in Table 6 remain unchanged.

The results for a GMDP-based AC policy and a CS-based AC policy are shown in Table 7, in which P_b^j denotes the blocking probability for class j packets, where $j = 1, 2$ and P_b denotes the overall blocking probability. It is observed that for a CS-based AC policy, the blocking probability constraint cannot be guaranteed. For example, when the buffer size is $[0, 3]$, the blocking probability for class 1 packets is 0.1185, which exceeds its constraint 0.1. When the buffer size

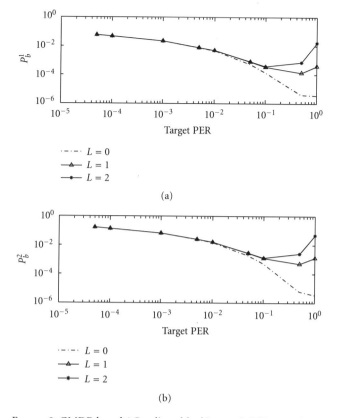

(a)

(b)

FIGURE 8: GMDP-based AC polices: blocking probability as a function of target PER in which $\rho_1 = \rho_2$ and $L_1 = L_2$.

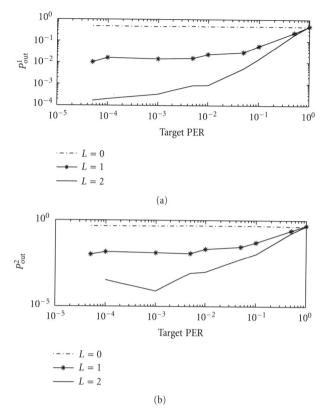

(a)

(b)

FIGURE 9: GMDP-based AC polices: outage probability as a function of target PER in which $\rho_1 = \rho_2$ and $L_1 = L_2$.

TABLE 7: Comparison between CS-based and GMDP-based AC policies, in which $L_1 = L_2 = 2$, $\rho_1 = \rho_2 = 10^{-4}$. The blocking probability constraint is set to $[0.1, 0.05]$, and the connection delay constraint is $[2.25, 0.5360]$.

$[B_1, B_2]$	GMDP: P_b^1	GMDP: P_b^2	GMDP: P_b	CS: P_b^1	CS: P_b^2	CS: P_b
$[0, 2]$	0.0714	0.0359	0.0537	0.0978	0.0390	0.0684
$[0, 3]$	0.0764	0.0280	0.0522	0.1185	0.0171	0.0678
$[1, 2]$	0.0434	0.0412	0.0423	0.0505	0.040	0.0452
$[3, 2]$	0.0179	0.0379	0.0279	0.0210	0.0569	0.0389

is $[3, 2]$, the blocking probability for class 2 packets is 0.0569, which exceeds its blocking probability constraint 0.05. However, for the same buffer sizes, GMDP-based AC policy can always guarantee blocking probability constraints for both classes.

6.4. Choosing ARQ parameters

As discussed in Section 5, ARQ parameters, such as L_j and ρ_j, should be chosen appropriately in order to achieve maximum throughput while simultaneously satisfying the QoS requirements in the physical and network layers.

We now provide a simple example to illustrate how to obtain optimal ARQ parameters by using the algorithm proposed in Section 5. The initial target PERs $\rho_j^0 = 0.05$, where

$j = 1, 2$, are given by the system which represents the upper bound of the target PER.

Using the algorithm presented in Section 5, the optimal ARQ parameters are derived as $L_1^{\text{opt}} = 1$, $L_2^{\text{opt}} = 1$, $\rho_1^{\text{opt}} = 0.005$, and $\rho_2^{\text{opt}} = 0.005$, respectively, for outage probability constraint $[0.01, 0.01]$ and blocking probability constraints $[0.1, 0.2]$. If the blocking probability constraint remains unchanged, and the outage probability constraint is reduced to $[10^{-3}, 10^{-3}]$, the optimal ARQ parameters can be derived as $L_1^{\text{opt}} = 2$, $L_2^{\text{opt}} = 2$, $\rho_1^{\text{opt}} = 0.01$, and $\rho_2^{\text{opt}} = 0.01$, respectively.

6.5. Sensitivity of the proposed algorithm to traffic load

In this subsection, we study the sensitivity of the proposed AC policy to different traffic loads. Traffic load can be represented by the packet occupancy ratio, defined as $[\lambda_1/\mu_1, \lambda_2/\mu_2]$. The following traffic loads are investigated: $[(1, 1/2); (2, 1(1/2)); (3, 2(1/2)); (4, 3(1/2)); (5, 4(1/2))]$.

Let λ and μ denote the overall arrival rate and the average departure rate, respectively, which can be expressed as $\lambda = \lambda_1 + \lambda_2$ and $\mu = (\lambda_1/(\lambda_1 + \lambda_2))\mu_1 + (\lambda_2/(\lambda_1 + \lambda_2))\mu_2$. The overall traffic load is represented by λ/μ. In the following examples, the target PER is assumed to be 10^{-3} for both classes, and a GMDP-based AC policy is employed, which would achieve a very similar performance to an optimal SMDP-based AC policy due to the low target PER under investigation.

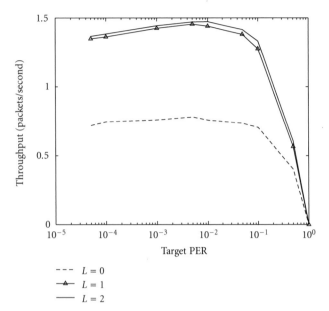

FIGURE 10: GMDP-based AC polices: throughput as a function of target PER in which $\rho_1 = \rho_2$ and $L_1 = L_2$.

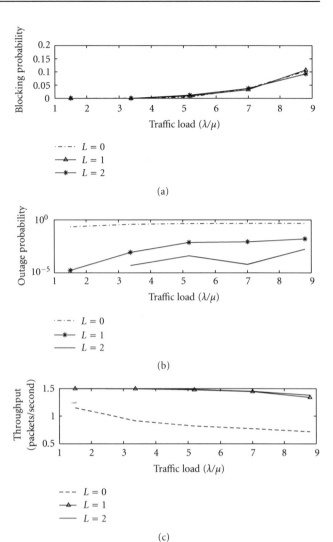

FIGURE 11: Blocking probability, outage probability, and throughput as a function of overall traffic load in which $\rho_1 = \rho_2 = 10^{-3}$.

Figure 11 presents the average blocking probability, outage probability and throughput as a function of overall traffic load. With an increased traffic load, there will be an increased interfering power and thus the performance is degraded. We remark that for all the traffic loads investigated, the proposed ARQ-based ROP algorithm is able to reduce the outage probability significantly at the cost of a slightly degraded network layer performance. Therefore, the proposed ARQ-based ROP algorithm can be applied to a wide variety of traffic conditions.

7. CONCLUSIONS

This paper provides a novel framework which exchanges information among physical, data-link, and network layers, and as a result provides a flexible way to handle the QoS requirements as well as the overall system throughput. In this paper, we propose a cross-layer AC policy combined with an ARQ-based ROP algorithm for a CDMA beamforming system. Both optimal and suboptimal admission control policies are investigated. We conclude that in a low PER region, for example, less than 10^{-2}, the proposed AC policies are capable of achieving significant performance gain while simultaneously satisfying all QoS requirements. Numerical examples show that the throughput can be improved by 100% by employing only one retransmission. Although ARQ schemes may degrade network layer performance, this degradation can be adequately controlled by appropriately choosing ARQ parameters. Furthermore, the proposed AC policy and ARQ-based ROP algorithm can be applied to any traffic load.

ACKNOWLEDGMENT

The support of the Natural Sciences and Engineering Research Council of Canada, discovery Grant 41731, is gratefully acknowledged.

REFERENCES

[1] M. Andersin, Z. Rosberg, and J. Zander, "Soft and safe admission control in power-controlled mobile systems," *IEEE/ACM Transactions on Networking*, vol. 5, no. 2, pp. 255–265, 1997.

[2] Y. Bao and A. S. Sethi, "Performance-driven adaptive admission control for multimedia applications," in *Proceedings of IEEE International Conference on Communications (ICC '99)*, vol. 1, pp. 199–203, Vancouver, BC, Canada, June 1999.

[3] T.-K. Liu and J. Silvester, "Joint admission/congestion control for wireless CDMA systems supporting integrated services," *IEEE Journal on Selected Areas in Communications*, vol. 16, no. 6, pp. 845–857, 1998.

[4] C. Comaniciu and H. V. Poor, "Jointly optimal power and admission control for delay sensitive traffic in CDMA networks

with LMMSE receivers," *IEEE Transactions on Signal Processing*, vol. 51, no. 8, pp. 2031–2042, 2003.

[5] S. Singh, V. Krishnamurthy, and H. V. Poor, "Integrated voice/data call admission control for wireless DS-CDMA systems," *IEEE Transactions on Signal Processing*, vol. 50, no. 6, pp. 1483–1495, 2002.

[6] F. Yu, V. Krishnamurthy, and V. C. M. Leung, "Cross-layer optimal connection admission control for variable bit rate multimedia traffic in packet wireless CDMA networks," *IEEE Transactions on Signal Processing*, vol. 54, no. 2, pp. 542–555, 2006.

[7] I. E. Telatar, "Capacity of multi-antenna Gaussian channels," *European Transactions on Telecommunications*, vol. 10, no. 6, pp. 585–595, 1999.

[8] A. Yener, R. D. Yates, and S. Ulukus, "Combined multiuser detection and beamforming for CDMA systems: filter structures," *IEEE Transactions on Vehicular Technology*, vol. 51, no. 5, pp. 1087–1095, 2002.

[9] S. M. Alamouti, "A simple transmit diversity technique for wireless communications," *IEEE Journal on Selected Areas in Communications*, vol. 16, no. 8, pp. 1451–1458, 1998.

[10] B. Hassibi and B. M. Hochwald, "High-rate codes that are linear in space and time," *IEEE Transactions on Information Theory*, vol. 48, no. 7, pp. 1804–1824, 2002.

[11] V. Tarokh, N. Seshadri, and A. R. Calderbank, "Space-time codes for high data rate wireless communication: performance criterion and code construction," *IEEE Transactions on Information Theory*, vol. 44, no. 2, pp. 744–765, 1998.

[12] G. J. Foschini, "Layered space-time architecture for wireless communication in a fading environment when using multi-element antennas," *Bell Labs Technical Journal*, vol. 1, no. 2, pp. 41–59, 1996.

[13] S. D. Blostein and H. Leib, "Multiple antenna systems: their role and impact in future wireless access," *IEEE Communications Magazine*, vol. 41, no. 7, pp. 94–101, 2003.

[14] F. Rashid-Farrokhi, L. Tassiulas, and K. J. R. Liu, "Joint optimal power control and beamforming in wireless networksusing antenna arrays," *IEEE Transactions on Communications*, vol. 46, no. 10, pp. 1313–1324, 1998.

[15] G. Song and K. Gong, "Performance comparison of optimum beamforming and spatially matched filter in power-controlled CDMA systems," in *Proceedings of IEEE International Conference on Communications (ICC '02)*, vol. 1, pp. 455–459, New York, NY, USA, April-May 2002.

[16] A. M. Wyglinski and S. D. Blostein, "On uplink CDMA cell capacity: mutual coupling and scattering effects on beamforming," *IEEE Transactions on Vehicular Technology*, vol. 52, no. 2, pp. 289–304, 2003.

[17] K. I. Pedersen and P. E. Mogensen, "Directional power-based admission control for WCDMA systems using beamforming antenna array systems," *IEEE Transactions on Vehicular Technology*, vol. 51, no. 6, pp. 1294–1303, 2002.

[18] J. Evans and D. N. C. Tse, "Large system performance of linear multiuser receivers in multipath fading channels," *IEEE Transactions on Information Theory*, vol. 46, no. 6, pp. 2059–2078, 2000.

[19] S. Brueck, E. Jugl, H.-J. Kettschau, M. Link, J. Mueckenheim, and A. Zaporozhets, "Radio resource management in HSDPA and HSUPA," *Bell Labs Technical Journal*, vol. 11, no. 4, pp. 151–167, 2007.

[20] Q. Liu, S. Zhou, and G. B. Giannakis, "Cross-layer combining of adaptive modulation and coding with truncated ARQ over wireless links," *IEEE Transactions on Wireless Communications*, vol. 3, no. 5, pp. 1746–1755, 2004.

[21] A. Sampath, P. S. Kumar, and J. M. Holtzman, "Power control and resource management for a multimedia CDMA wireless system," in *Proceedings of the 6th IEEE International Symposium on Personal, Indoor and Mobile Radio Communications (PIMRC '95)*, vol. 1, pp. 21–25, Toronto, Canada, September 1995.

[22] A. Wyglinski, "Performance of CDMA systems using digital beamforming with mutual coupling and scattering effects," M.S. thesis, Queen's University, Kingston, Ontario, Canada, 2000.

[23] H. C. Tijms, *Stochastic Modelling and Analysis: A Computational Approach*, John Wiley & Sons, Chichester, UK, 1986.